Herbicides and Plant Growth Regulators

W. W. Fletcher and R. C. Kirkwood

GRANADA
London Toronto Sydney New York

Granada Publishing Limited – Technical Books Division
Frogmore, St Albans, Herts AL2 2NF
and
36 Golden Square, London W1R 4AH
866 United Nations Plaza, New York, NY 10017, USA
117 York Street, Sydney, NSW 2000, Australia
100 Skyway Avenue, Rexdale, Ontario, Canada M9W 3A6
61 Beach Road, Auckland, New Zealand

British Library Cataloguing in Publication Data

Fletcher, W.W.
Herbicides and plant growth regulators.
1. Plant regulation
2. Herbicides
I. Title II. Kirkwood, R.C.
631.5'4 SB128

ISBN 0-246-11266-2

First published in Great Britain by Granada Publishing Ltd,
Technical Books Division

Printed in Great Britain by
Richard Clay (The Chaucer Press) Ltd, Bungay, Suffolk.

Contents

Acknowledgements

The authors express their thanks to John Wiley and Sons Inc. for permission to reproduce tables 7.1, 7.7, 7.8 and 7.9, from *The Use and Significance of Pesticides in the Environment* by F. L. McEwen and G. R. Stephenson and figs. 6.6 and 6.7 from *Progress in Pesticide Biochemistry*, Vol. I (Eds. D. H. Hutson and T. R. Roberts); to Blackwell Scientific Publications for permission to use fig. 7.1 from the *Weed Control Handbook*, 5th edn. (Eds. J. D. Fryer and S. A. Evans) and table 7.5 from the *Weed Control Handbook*, 6th edn. (Eds. J. D. Fryer and R. J. Makepeace); to Macmillan Journals Ltd for permission to reprint table 7.10 from *Nature*, **215**, 1406 (1967). Copyright (c) 1967 Macmillan Journals Limited; to the American Chemical Society for permission to use table 7.6 from *Advances in Chemistry Series*, No. 111, 143 (1972); to Marcel Dekker, Inc. for permission to use a quotation (p. 323) from *Pesticides in the Environment*, Vol. 1 (Ed. R. White-Stevens) and figs. 6.1 – 6.5 and 6.8 from *Herbicides, Chemistry, Degradation and Mode of Action*, Vols. I and II (Eds. P. C. Kearney and D. D. Kaufman); to Academic Press for permission to use figs, 5.3, 5.5, 6.10 and 6.11 from *Herbicides, Physiology, Biochemistry, Ecology*, Vol. 1 (Ed. L. J. Audus). Copyright by Academic Press Inc (London) Ltd; and to The Humana Press Inc for permission to use quotations (pp. 3 and 5) from *Pesticides, Contemporary Roles in Agriculture, Health and the Environment* (Eds. T. J. Sheets and D. Pimentel.)

Preface

This book has not been written without the help of a great many people. We are very grateful to the companies who responded so willingly to our request for technical data information on their products. They are Amchem Products Inc., B.A.S.F. A.G., Bayer A.G., The Boots Co. Ltd, Ciba Geigy A.G., The Dow Chemical Co., E.I.du Pont de Nemours & Co., Fisons Ltd, Hoechst, I.C.I. Ltd, Kumiai Chem. Industry, May & Baker Ltd, Monsanto Chemical Co., Philips Duphar, Sandoz Ltd, Schering A.G., Shell Chemicals, Union Carbide. In addition we would thank the Ministry of Agriculture, Fisheries and Food, the U.S. Dept. of Agriculture, and F.A.O. of the United Nations, the Library Staff of the University of Strathclyde, Mr J. D. Fryer, Director of the A.R.C. Weed Research Organisation and Mr John Hardcastle, Information Officer of the same oganisation who went to an immense amount of trouble to supply us with information. Our thanks are especially due to Professor B. J. Heywood, Consultant to May & Baker Ltd and Visiting Professor in the University of Strathclyde who read the manuscript and did a very great deal of work on chemical nomenclature and classification on our behalf. We would like to thank Mrs Valerie Crean for invaluable assistance with technical editing. We are grateful to Mrs Isobel Robertson and Mrs Effie McPhail for typing the manuscript and for cheerfully putting up with many changes and alterations. Theirs was no small task. Professor Leonard Broadbent, Pro-Vice Chancellor of the University of Bath and past Chairman of the British Crop Protection Council first suggested that we might write such a book. We trust that this confidence is not misplaced.

Introduction

Weeds have been part of the agricultural scene since Man first started cultivating crops more than 10 000 years ago and they are still a major problem today. They have been defined as 'plants growing in the wrong places' which means that every plant species is a potential weed. In other words the question of whether a plant is a weed or not is a subjective judgement. In addition, successful weeds are aggressive, competitive and adaptable. Their most important attributes are efficient reproduction combined with mechanisms that permit survival under temporarily unfavourable conditions. The vast majority of weeds reproduce by means of seeds and some of them are very efficient producers – e.g. rosebay willowherb (*Chamaenerion angustifolium*) 76 000 seeds per plant, ragwort (*Senecio jacobea*) 63 000 seeds per plant, rushes (*Juncus* spp.) 50 000 seeds per plant, field poppy (*Papaver rhoeas*) 17 000 seeds per plant. The significance of these figures is best appreciated when one considers that the multiplication rate for a cereal is commonly 25–30 times, grasses 30–80 times, white clover 40–70 times (Fletcher, 1974). One dense weed stand has been estimated to produce more than 12 000 million seeds per hectare and seed production by a dense stand of rushes (*Juncus* spp.) has been estimated at more than 9 million/m^2, and in a crop of winter wheat 62 million weed seeds/hectare were found at harvest (Roberts *et al.*, 1977).

Most perennial weeds, in addition to reproducing by seed, have vegetative methods of reproduction, e.g. wild onion (*Allium vineale*) produces offset bulbs, field horsetail (*Equisetum arvense*) produces tubers which readily regenerate when detached, creeping thistle (*Cirsium arvense*) has creeping roots capable of producing new plants, couch grass (*Agropyron repens*) has creeping stems and bracken (*Pteridium aquilinum*) has rhizomes.

Weeds are harmful to crops in many ways:

1. They compete for water.
2. They compete for light (especially those with large coarse leaves which shade the crop plants).

3. They compete for nutrients; thus a plant of yellow mustard needs twice as much nitrogen, twice as much phosphoric acid, four times as much potassium and four times as much water as a well-developed oat plant.
4. They compete for space above and below ground. Lucerne, for example, is slow to form a ground cover and may be overgrown by weeds before becoming established. Below ground wheat roots may be reduced by two-thirds when grown in the presence of wild oats. One small piece of rhizome of bracken can produce 300 fronds in only four years and below ground a strong stand of bracken may have 80 tons of rhizomes per hectare.
5. Some weeds, e.g. dodder (*Cuscuta* sp.) may be parasitic on crop plants.
6. They may reduce the value of produce, increase the difficulty of harvesting and entail seed cleaning.
7. Some, e.g. ragwort (*Senecio jacobea*) and water dropwort (*Oenanthe crocata*) are poisonous to stock.
8. They may harbour pests and diseases.
9. Even in areas not under crops they may be harmful, for example in railway tracks, canals and open grazing country (Fletcher, 1974).

Weeds have a very marked effect in reducing crop yield. Dr W. B. Ennis at an F.A.O. Symposium reported that in Peru it was shown that weeds reduced rice yields by 34–68% and in Latin America maize yields were reduced by 53% by allowing weeds to grow. In the U.S.A. yields of sugar from cane were reduced by 76%, 52% and 42% when weeds were permitted to compete with the crop for 3, 6 or 12 weeks respectively after planting. Pathak *et al.* (1976) have calculated that yield losses from weeds in rice range from 10–50% in transplanted rice and from 50–90% on upland fields. During the first 6 weeks of growth of cotton, weed control is absolutely essential and delay can reduce yields by as much as 450 kg/ha seed cotton (Matthews, 1976). McEwen and Stephenson (1979) have reported that corn losses in Illinois have been estimated at 81% (mean value) in unweeded plots (range 28–100%), that replicated studies in Minnesota over a 3 year period showed yield reduction in corn of 16–93% (average overall 51%) and that soybean yields in weedy plots in Nebraska and Minnesota were reduced by an average of 45%.

It has been calculated that we lose around 33% of agricultural production due to insects, diseases and weeds (the latter accounting for some 10%. To this must be added some 15% loss in store and transit. Obviously anything that we can do to reduce this heavy toll must be done for we face the problem of feeding not only our present population but a population that is rapidly increasing. Herbicides are technological tools which are primarily for use in helping to feed a hungry world. The rate of world population increase has been publicised to such an extent that every thinking person must now know the basic facts. They have been rather well summarised by Dr Fred Tschirley of Michigan State University (Tschirley, 1979)

'If you are given to commemorative celebrations you would have been delighted to celebrate in 1850 the first occasion of one billion people living on planet Earth. About 80 years later, in 1930, you would have been overjoyed to observe man's numbers pass the 2 billion mark. The third billion coming only 31 years later in 1961 may have suggested to you that the commemorative celebrations were occurring a bit too frequently. And only 15 years later in 1976 your fears would have been confirmed. On March 28th 1976 demographers told us that the population of the earth had passed the 4 billion mark. At the current rate of population growth there will be 5 billion souls on this planet in 1990, 6 billion in 2000, 7 billion in 2009 and 8 billion in 2017.'

These figures are frightening. There is some evidence that the rate of growth is slowing down. It must do or Man is set on a disaster course. The increase is in the main due to the use of pesticides and antibiotics in the control of insect-borne and other diseases. The remedy must lie in some form of control of population numbers, but meantime it is important that every scrap of food that can be produced should be produced and that we should not produce food to feed weeds, insects, fungi and rats. The immense task facing food producers may be gauged by the fact that the National Academy of Sciences has calculated that in 1977 as many as 450 million to a billion persons in the world did not receive enough food. Mechanical tillage and cultural practices are still used extensively throughout the world for weed control. According to Pimentel *et al.* (1979) weed control on about 80% of the crop land in the U.S.A. is still effected by mechanical and cultural practices. However, since the discovery of the selective weedkillers, 2,4-D and MCPA, in the early 1940s there has been a steady increase in the use of chemicals. It is becoming increasingly realised however that not only can bigger yields be attained by the use of herbicides but that costs are also lower. Eichers and Andrilenas (1979) report that many areas of the world are substituting chemicals for hand and mechanical weed control. Total farm use of pesticides more than doubled during the 10 year period from 1966 to 1976. Use of pesticides on crops increased from 328 million pounds of active ingredients to 665 million pounds, the most significant increase being for herbicides which in 1966 totalled 112 million pounds and by 1976, 400 million pounds; and herbicides have largely replaced mechanical cultivation in row crop production (Berry, 1979). The total crop acreage in the U.S.A. treated with herbicides increased from 99 million acres in 1966 to 159 million acres in 1971 (U.S.Dept.Agric., 1975.). In Korea and Taiwan a few years ago only 10% of rice fields were treated with herbicides. They are now used on 50% of the fields and a similar trend is developing in the tropics where herbicides are particularly useful on upland rice. Since it is mostly broadcast seeded and there are no rows, mechanical or manual weeders cannot be used (Pathak *et al.*, 1976).

The sales of herbicides in the U.K. were £50 million in 1975, £57.8 million in 1976, £72.9 million in 1977, £98.9 million in 1978 and £134 million in 1979 (Brit. Agrochem. Assoc. Ann. Rept., 1979/80). In 1976 the United States manufactured 557.9 million lb of herbicides, in 1977, 584.5 million lb and in 1978, 640 million lb worth respectively $1450.4 million, $1620.0 million and $1782.6 million (Fowler and Mahan, 1980).

In 1978 the world sale of herbicides was $3.7 billion, in 1980 it was $4.2 billion (estimated) and the estimate for 1984 is $4.8 billion (all at 1978 U.S. constant dollars). This represents a projected increase of 13.4% over the period 1980 to 1984. Herbicides continue to dominate, the corresponding figures for insecticides being $3 billion (1978), $3.34 billion (1980) and $3.8 billion (1984) and for fungicides $1.5 billion (1978), $1.68 billion (1980) and $1.96 billion (1984) (*Farm Chemicals*, 1979).

In 1955 the potato yield in the U.K. was under 20 tonnes per hectare; it is now more than 30 tonnes per hectare. Wheat yields in 1955 were less then 2.5 tonnes per hectare and are now more than 4.4 tonnes per hectare (Brit. Agrochem. Assoc. Ann. Rept., 1979/80). Many factors including machinery, plant breeding and fertilisers have all played their part but the major role has undoubtedly been that of pesticides and in particular herbicides. In the United States the yield of corn had remained fairly static at around 25 bushels per acre per year over the years from 1880 to 1940, then with the advent of pesticides there was a dramatic increase – 45 bushels by 1960, 70 bushels by 1970 and 90 bushels by 1975 (Dow Chem. Co., 1977). Some estimates indicate that pesticides were responsible for 20% of the gain in agricultural production in the U.S.A. since 1940 and that there would be a 25% overall reduction in crop output if pesticides could not be used (U.S. Dept. Agric., 1974). Jaske *et al.* (1977) estimate that U.S. agricultural production will increase by about 20% between 1975 and 1990. Between 1966 and 1976 crop output per man in the U.S.A. rose more than 40%. This coincided with a rapid increase in the use of chemical weed control. In the U.S.A. herbicide-treated corn acreage increased from 57% in 1966 to 90% in 1976, cotton acreage from 52% to 84%, soybean acreage from 27% to 88% and tobacco acreage from 2% to 55% (Berry, 1979). Some of the gains to be got from herbicide usage may be gauged from figures quoted by Ennis (1967) at an F.A.O. Symposium – the use of propanil on rice fields in the U.S.A. which were heavily infested by weeds increased average yields by 50% and in the Philippines, propanil increased rice yields by 43%. In the U.S.S.R. the yield of spring wheat in certain areas was increased by 65% by the use of 2,4-D. According to Berry (1979) about 40 million acres of wheat and other small grains were treated with herbicides in 1976 leading to an average yield increase of about 20% and since the introduction of 2,4-D wheat yield has doubled in North America (McEwen and Stephenson, 1979). Based on increased yields and total costs for pesticide purchase and application Pimentel *et al.* (1979) have calculated a four to one return to the farmer. The effect on retail prices of the withdrawal of pesticides would be an increase of about 12%. And the cost in energy terms of all this increased production? According to Berry (1979) U.S. food systems use about 16.5% of all the energy consumed in the nation. Pesticides although petroleum based and produced from crude oil distillates (e.g. benzene and naphtha) and from natural gas and its derivatives (e.g. ethane, propane, methanol, ethanol and amines) account for only about 5% of the total energy used in agricultural production. Herbicides have largely replaced mechanical cultivation in row crop production with a conservative estimate of energy saving of 20%. He believes that when mechanical control is used in place of herbicides the energy requirement may be about ten times as great or even greater. Dowling and Baker (1975) concluded from a study of costs of weeds in cotton in Georgia that the use of herbicides, or herbicides plus cultivation, reduced the total cost of mechanical energy for controlling weeds by $300 to $600 per hectare. Newsome (1979) believes that the increased cost of hand hoeing would eliminate cotton production from about two-thirds of its current acreage, and soybean production from about half of its current acreage, even if such labour were available – a total of 74 million hours of hoe labour would be required in Louisiana for the two crops mostly in May and June. The cost would be around $197 million as against $57 million for herbicides. He has written

'Agriculture cannot remain a viable enterprise in the United States without effective weed control. Those who would argue that effective weed control can be accomplished without overwhelming reliance upon use of herbicides under conditions existing in the United States, haven't done their arithmetic. It would be impossible to supply the labour required. If the labour were available it would be impossible for farmers to pay for it and remain in business.'

By the judicious use of pesticides, minimum tillage or no tillage at all can be practised on certain types of soils. The U.S. Department of Agriculture (1975) calculates that there will be approximately 75 million acres of no-tilled land in the U.S.A. by 1990 with an energy saving amounting to some 250 million gallons of diesel fuel equivalent. At the other end of the scale there have been outstanding examples of weed control by using biological agents, e.g. in Australia more than 60 million acres were cleared of the prickly pear weed (*Opuntia* spp.) by the introduction of the Argentinian moth borer (*Cactoblastis cactorum*) and the poisonous klamath weed (*Hypericum* sp.) was eliminated from millions of acres of range and pasture land in western U.S.A. by the introduction of the leaf beetle (*Chrysolina gemellata*) from Australia. The use of biological agents (including specific diseases) is a very important aspect of weed control. Biological agents whether introduced or, more commonly, occurring naturally, are constantly at work keeping all plant and animal life at 'natural' levels.

In the early days pesticides (including herbicides) were sprayed around rather indiscriminately. It was considered that they were the panacea for all agricultural ills (as antibiotics were once regarded as the cure-all for all bacterial illnesses). Wiser councils now prevail and although it is accepted that pesticides must be the cornerstone for most pest control programmes it is also recognised that they are but one factor (albeit the most important) in the totality of pest control and that cultural methods, biological control and plant breeding, and management have all important parts to play in what has been called 'integrated control' or 'pest control management'. Pilot studies on all major crops in the U.S.A. over the past 5 years have shown that the number of pesticide applications can often be cut in half or less when good management is introduced. It is recognised that the ideal system is one where many factors interact to produce the best results from the agro ecosystem and the recognition that this ecosystem is made up of a delicately balanced maze of ecological interactions. The disturbance of one aspect of the system can have profound effects upon other aspects. In the case of insect control, for example, care must be taken to ensure that insecticides do not adversely affect predators and parasites of the insect pest, and plant breeders producing higher yielding plants must ensure that they do not overlook resistance to insect and fungal attack. In the case of weeds it is important to integrate mechanical, chemical and biological control methods. These thoughts are in line withthe 7th Report *Agriculture and Pollution* of the Royal Commission on Environmental Pollution (1979). One of their conclusions was that research should point the way to the practical development of cost-effective and environmentally acceptable pest management.

In the 8th Report for 1978/79 of the Weed Research Organisation, Oxford (Fryer, 1980), the Director points out that the Institute's programme includes research towards understanding the population dynamics of important grass weeds such as wild oats (*Avena* spp.), blackgrass (*Alopecurus myosuroides*) and sterile brome (*Bromus sterilis*) in intensive cereal production, to enable strategies to be developed for their rational control by herbicides. He notes that the research is enormously detailed and time consuming but results have been of great assistance to farmers. This is another example of the multipronged approach that is necessary to this multidisciplinary problem of weed control. The problem is a dynamic one. It is fascinating to note the changes that have come about in weed species as a result of the use of herbicides. Some weeds formerly unimportant have become prominent and it is interesting to note the close taxonomic and morphological relationship of these 'new' weeds to the crop plants in which they occur. In the cereal crops in the U.K. for example the important weeds for centuries have been the broad-leaved species such as poppy (*Papaver rhoeas*), charlock (*Brassica sinapis*) and cornflower (*Centaurea cyanus*). With the advent of 2,4-D and MCPA in the 1940s these weeds were eliminated and their places taken in the ecosystem by the more resistant cleavers (*Galium aparine*), chickweed (*Stellaria media*) and *Polygonum* spp.

Many herbicides were developed in the 1950s to deal with these more difficult weeds but as they were attacked by herbicides such as mecoprop and dichlorprop increasing problems were posed as the 'ecological niches' left vacant due to the continuing control of the broad-leaved weeds were filled by the grass weeds, wild oats (*Avena fatua*), blackgrass (*Alopecurus myosuroides*) and sterile brome (*Bromus sterilis*).

One of the great triumphs of weed technology has been the development of herbicides which are able to kill these weeds selectively in cereal crops. There is even one herbicide that is able to distinguish between two varieties of the same species. We are getting close to the limits of selectivity. It is here perhaps timely to sound a warning. Induced resistance to herbicides is not yet a serious problem, although *Senecio vulgaris*, *Amaranthus retroflexus* and *Chenopodium album* have developed substantial levels of resistance to atrazine (Holliday *et al.*, 1976) and our biological experience tells us that more and more cases will be reported as time goes on. It is said that one of the main causes for fungi quickly developing resistance to some of the systemic fungicides (apart from the obvious plasticity of the fungi) is the fact that since both host and parasite have a number of metabolic pathways in common then selectivity has to depend upon the inactivation of only a few enzyme systems that are different (rather than the 'shot-gun' effect of a number of fungicides which are not so selective). As a result the fungi are able to 'find their way around' the affected systems by means of alternative biochemical pathways, and thus become resistant to the fungicide. If this is so then there is the obvious danger that a parallel situation will develop where a herbicide distinguishes between say wild oats and wheat. They cannot have many enzyme systems that are different and one should be on the alert for the possibility of the development of resistance. One of the major developments has been the use of mixtures of herbicides (and other pesticides). Not only does this widen the spectrum of activity and reduce the possibility of resistance but it also means that fewer sprayings have to be carried out by the farmer.

Plant growth regulators

The history of agriculture is the story of Man's attempts to modify plants and animals in order to produce more and better food. There is nothing more 'artificial' than the farm and hopefuls looking for a 'natural' life in the country had best look elsewhere. Plants (and animals) are capable of being extensively modified and Man has made use of this variability in his farming practice. In doing so he has made use of both 'nature' and 'nurture' – selecting and breeding from plants which have the most desirable characteristics and by cultural practices such as weeding, tilling, manuring, irrigating, pruning and disbudding he has provided the best conditions for the hereditary material of the desirable plants to realise their potential. Always of course he has been limited by the genetic makeup of the plant in question and breeding for desirable characteristics can often be a long, time-consuming, process and it may be years before the desired result is obtained. Now, however, with the introduction of chemical plant growth regulators a new dimension has been added to the possibilities of modifying plant growth and they offer the possibility of compensating for genotypic deficiencies that might take many years of breeding to alter by genetical methods.

Plant growth regulators are organic compounds, other than nutrients, that, in low concentrations, affect the morphological structure and/or physiological processes of plants. Plant hormones or phytohormones, are naturally occurring growth regulators that in low concentrations control physiological processes in plants. The synthetic growth regulators are used by Man to control such processes as fruit development, fruit thinning, defoliation, growth stimulation and retardation, rooting of cuttings and many other processes. Over the past 30 years the investigation and development of plant growth regulators has been one of the most active areas of fundamental and applied botanical research. Kemp (1979) in the *PANS Plant Growth Regulator Index* under the List of Common and Trade Names and Code Numbers has 492 entries (excluding herbicides except where these are used specifically for some growth regulatory purpose other than weedkilling).

The first international conference on plant growth substances was held in 1949 and the proceedings were subsequently published (Skoog, 1951). It is of interest to note that in his paper on 'Plant hormones in practice' delivered at the conference, Zimmerman (1951) dealt with the only naturally occurring plant growth substance that was recognised at that time, namely β-indole-acetic acid (IAA), and the only synthetic ones in practical use were those chemically related to IAA such as α-naphthylacetic acid (NAA) and its derivatives; β-naphthyloxyacetic and β-naphthyloxypropionic acid and their derivatives; unsaturated hydrocarbons, such as ethylene and acetylene; and substituted phenoxyalkanoic acids such as 2,4-D, MCPA, 2,4,5-T and 2,4,5-TP which had been in use since the mid-1940s as herbicides. He did note however that 'there are many natural hormones which have not been isolated and definitely identified but which are known to exist', and that 'they (plant hormones) are rapidly becoming a part of our everyday life, and as with the telephone and the X-ray we shall soon wonder how we ever got along without them'. His words were prophetic and it is for example certainly difficult to see how modern agriculture could operate without 'hormone' (and other) herbicides. Plant regulators have not yet made such a major impact but they are now the most rapidly expanding sector of the agricultural chemical industry and it has been predicted that their use will eventually outstrip that of herbicides. Although they currently account for only a small share (5–10%) of the total worldwide agrochemical market they are still in their infancy (Nickell, 1978). Great changes have occurred during the past 30 years in the commercial plant growth regulator field, but Morgan (1979) believes that the current questions and opportunities may be more critical to the future of this aspect of agricultural chemistry than any single event that has gone before.

Although ethylene has been used since the 1920s to ripen fruit (Denny, 1924), and auxins have been used to promote the rooting of cuttings (Thimann and Went, 1934), the development of plant growth regulators has been overshadowed by the development of herbicides. However, the immense amount of fundamental work that has been done on the naturally occurring auxins, the unfolding of the importance of ethylene as a plant hormone, the development of the gibberellins (Brian *et al.*, 1955), the discovery of the cytokinins (Jablonski and Skoog, 1954), the isolation of abscisic acid (Robinson *et al.*, 1963; Okhuma *et al.*, 1963), the synthesis of morphactins (Schneider, 1970) and the development of growth retardants such as CCC (Cyclocel) (Tolbert, 1960) have led to the recognition of the fundamental roles of the natural and practical applications of the synthetic plant growth regulators. Nickell (1978) lists twenty-seven regulators that are in use in the United States at the present time. They affect a great variety of plant growth processes, including the following (some of the growth regulators in common use are in brackets):

rooting of cuttings (indole-butyric acid); promotion of flowering in pineapples (1-naphthaleneacetic acid; β-hydroxyethylhydrazine; ethephon); prevention of pre-harvest drop of apples (NAA; daminozide); inhibition of turf growth (maleic hydrazide; mefluidide-diethanolamine); prevention of sprouting of potatoes (maleic hydrazide); floral induction in apple, pear, peach (succinic acid-2,2-dimethylhydrazine; 2,3,5-tri-iodobenzoic acid); early flowering of 'long day' plants, e.g. lettuce, radish, mustard, dill (gibberellins); flowering of many biennials which normally require low temperatures to flower (gibberellins); improvement of yield of sugar-cane by prevention of flowering (diuron; diquat); delay in flowering in almond and peach to avoid adverse weather conditions (daminozide); induction of abscission of mature citrus fruits (cyclohexim; 5-chloro-3-methyl-4-nitro-1*H*-pyrazole); defoliation of cotton leaves to aid harvesting of bolls (ethephon); thinning of fruit, e.g. grapes, peaches (gibberellic acid; ethephon; 3-chlorophenoxy-α-propionamide); prevention of pre-harvest drop of citrus (2,4-dichlorophenoxyacetic acid); induction of fruit set, e.g. in tomato, squash, eggplant, fig (4-chlorophenoxyacetic acid; 2-naphthyloxyacetic acid); increase in size and quality of grapes (gibberellins); induction of amylase in barley for malting (gibberellins); stimulation of growth of sugar-cane (gibberellins); reduction of stem length in cereals (2-chloroethyl trimethylammonium chloride); development of female flowers, e.g. in pumpkins (NAA; ethephon; daminozide); promotion of male flowers, e.g. in hops (gibberellins); bioregulation of plant composition, e.g. colour in citrus, sugar in sugar-cane, vitamin content in vegetables, increase in dry weight, timing of crop development, increased latex from rubber trees (various growth regulators).

Morgan (1979) has analysed the present status and future potential of plant growth substances and he has outlined some opportunities for their further development in agriculture. These include treatments for seed or seedlings for transplanting which will promote early growth and root development; substances to improve quality (usually protein levels and amino acid balance) of grain crops; substances to improve yield and quality of forages; opportunities in forestry, such as seedling survival and growth, early seed production and accelerated growth rates; systems to reduce energy costs by maximising response to cultivation, fertilisers (i.e. uptake, mobilisation, etc.) and irrigation water; compounds to inhibit ethylene action or production and thus reduce young fruit abscission in indeterminately fruiting crops; new gibberellins with species- or function-specific effects; new applications of known substances based on understanding hormone interactions and storage/inactivation systems ('slow release' compounds) and substances to manipulate natural conjugation reactions; substances to alleviate or minimise effects of plant diseases and insects or to facilitate systems of integrated pest management; substances to modify productivity by reducing photorespiration, dark respiration, or by promoting nitrogen metabolism/fixation, photosynthesis, translocation; substances that intensify synthesis of specific highly desired end-products (oil, protein, cellulose); substances to increase productivity by shifting developmental patterns, such as extending the period of inflorescence differentiation or seed development. He remarks that this list should serve to illustrate the wide range of opportunities available to creative agricultural chemists. He notes however that only the gibberellins, cytokinins and the auxins, of all the commercial growth regulators, are true growth promoters and many of their current applications involve a modification of development rather than a simple effect on growth rate or final size. Several uses of auxins involve growth inhibition rather than growth promotion.

Plant tissue culture pioneered by White, Steward, Skoog and others, beginning almost as a botanical curiosity, has with the help of growth-regulatory chemicals become a powerful tool in the hands of the plant breeder. It is now possible to tissue culture almost any plant and to develop uniform plantlets from such cultures. Even pollen grains can be used and the subsequent haploid plants made polyploid by the use of suitable chemical agents. Together with apical meristem culture there is an unending supply of material.

Hudson (1976) in discussing future roles for growth regulators considers that one of their main applications will be in the provision of vegetative material for planting. The raising of plants by the rootings of cuttings is a common practice with many species but Hudson points out that there is at least one important exception. Some of the palms, including species of great economic value such as oilpalm and coconut, never branch and have always to be grown from seed, with the obvious drawback that this leads to variability in the resulting plants. He considers that the best hope for the future lies in single cell culture using a sequence of growth regulators. Looking further ahead he considers it possible that plantlets raised in this way may be encapsulated with a supply of nutrients and a kit of growth regulators that make them virtually analogous to seeds in their ability to be space sown in open ground. It would be a real breakthrough, says Hudson, if future forests could be set out with uniform plantlets all with the growth potential of the very best seedlings now available. It is interesting to note that Unilever have announced (April 1980) the development of palm tree plants from tissue culture. They believe that this development will increase productivity by about 20% per acre.

In chapter 2 the various groups of plant growth regulators are discussed and some examples are given. This is a very rapidly expanding field and it is not possible to do full justice to growth regulators in one short chapter, but it should be a useful introduction. It is hoped that this book on herbicides and plant growth regulators will be of interest and value to University and College staff and students of various disciplines, and to scientists in research institutes, advisory services and industry. The difficulty in writing a book on weed control is that so much information is available that it is difficult to know what to include and what to leave out for this multidisciplinary subject has something to offer at all levels of intellectual effort. Since the early 1940s it has been recognised that herbicides have a major role to play in maintaining and increasing food production. What has been a little longer in being recognised (despite their parentage in fundamental studies of tropic responses in plants) is that it is 'academically respectable' to study herbicides. There have, however, been a number of outstanding academics in the field for many years including Professor A. Crafts, Professor T. S. Sheets, Professor G. Blackman, Professor C. L. Wain and Professor L. J. Audus and others who have never accepted this and who have brought the study of herbicides to the academic forefront.

This is not a book of recommendations and readers should note that the word 'introduced' as used frequently in chapters 1 and 2 does not imply that the herbicide or growth regulator is on general sale, nor that it has necessarily been approved by appropriate official bodies in the various countries. Many of them have been approved and are on sale but some may still be at the experimental stage. Anyone reading this book who is proposing to use any of these chemicals should always check the status of the herbicide and use according to label instructions. In the U.K. under the Agricultural Chemicals Approval

Scheme a list of 'Approved Products for Farmers and Growers' is published each year by the Ministry of Agriculture, Fisheries and Food. Similar booklets are available in other countries. Because it is not a book of recommendations we have not mentioned dosages. These are readily obtained when required. Essentially this is a teaching book which we hope will be valuable to lecturer and student alike and we hope that the referencing of many statements will induce further study as required. There is much material that we could have included but did not as there must be a limit to the size of all books. Thus there is little on the husbandry side of weed control, nor on biological control, though the importance of these subjects is recognised and discussed, however briefly, in this introduction. In writing this book we have consulted many sources which are acknowledged in the list of references at the end of each chapter but we would mention some sources that have been invaluable to us. These are *The Pesticide Manual*, 6th edn. (1979) published by the British Crop Protection Council; *Weed Abstracts* published by the Commonwealth Agricultural Bureau; *Plant Growth Regulator Abstracts* published by the Commonwealth Agricultural Bureau; *Residue Reviews* published by Springer-Verlag; *Weed Control Handbook*, Vols. 1 and 2 published by Blackwell Scientific Publications; and Technical Data sheets supplied by many firms.

References

BERRY, J. H. (1979). 'Pesticides and energy utilization' in *Pesticides, Contemporary Roles in Agriculture, Health and the Environment* (Eds. T. J. and D. Pimental), Humana Press.

BRIAN, P. W. *et al.* (1955). *J. Sci. Fd. Agric.*, **5**, 602.

Brit, Agrochem. Assoc. Ann. Rept. (1979/1980).

DENNY, F. E. (1924). *Bot. Gaz.*, **77**, 322.

Dow Chemical Co. (1977). *Silent Autumn.*

DOWLING, C. C. and BAKER, S. H. (1975). *Proc. S. Weed Sci. Soc.*, **28**, 133.

EICHERS, T. R. and ANDRILENAS, P. A. (1979). Evaluation of Pesticide Supply and Demand for 1979, *U.S. Dept. Agric. Economics Rept.*, No. 422.

ENNIS, W. B. (1967). *F.A.O. Symposium on Crop Losses*, 2nd–6th Oct.

Farm Chemicals (1979). **142** (9), 61.

FLETCHER, W. W. (1974). *The Pest War*, Blackwell and Mott.

FOWLER, D. L. and MAHAN, J. N. (1980). *The Pest Review 1978*, U.S. Dept. Agric.

FRYER, J. D. (1980). In *Weed Research Organisation, 8th Rept., Oxford, 1978/79*,

HOLLIDAY, R.J., PUTWAIN P. D. and DAFNI, A. (1976). *Proc. Brit. Crop Prot. Conf. – Weeds*, **3**, 937.

HUDSON, J. P. (1976). *Outlook on Agric.*, **9**(2), 95.

JABLONSKI, J. R. and SKOOG, F. (1954). *Physiol. Plant*, **7**, 16; **3**, 754.

JASKE, M. R. *et al.* (1977). *Econ. Res. Serv. U.S. Dept. Agric.*

KEMP, P. J. (1979). *PANS Plant Growth Regulator Index*, **25** (2), 211 and 213.

McEWEN, F. L. and STEPHENSON, G. R. (1979). *The Use and Significance of Pesticides in the Environment*, John Wiley & Sons.

MATTHEWS, G. A. (1976). In *Pesticides and Human Welfare* (Eds. G. L. Gunn and J. G. R. Stephens), Oxford Univ. Press.

MORGAN, P. W. (1979). *Pl. Growth Reg. Bull.*, **7**(3), 20.

NEWSOME, L. D. (1979). 'Role of pesticides in pest management' in *Pesticides, Contemporary Roles in Agriculture, Health and the Environment* (Eds. T. J. Sheets and D. Pimentel), Humana Press.

NICKELL, L. G. (1978). *Chem. Eng. News*, **56**(41), 18.

OKHUMA, K. *et al.* (1963). *Science*, **142**, 1592.

PATHAK, M. D., OU, S. H. and DE DATTA, S. K. (1976). In *Pesticides and Human Welfare* (Eds. D. L. Gunn and J. G. R. Stephens), Oxford Univ. Press.

PIMENTEL, D. *et al.* (1979). 'Cost-benefit analysis of pesticide use in U.S. food production' in *Pesticides, Contemporary Roles in Agriculture, Health and the Environment* (Eds. T. J. Sheets and D. Pimentel), Humana Press.

ROBERTS, H. A. *et al.* (1977). In *Weed Control Handbook* (Eds. J. D. Fryer and R. J. Makepeace), Vol. I, Blackwell.

ROBINSON, M. *et al.* (1963). *Nature (London)*, **199**, 874.

Royal Commission on Environmental Pollution (1979). *7th Rept. Agriculture and Pollution,* H.M.S.O.

SCHNEIDER, G. (1970). *Ann. Rev. Pl. Physiol.*, **21**, 499.

SKOOG, F. (1951). *Plant Growth Substances*, Univ. Wisc. Press.

THIMANN, K. W. and WENT, F. W. (1934). *Proc. Kon. Akad. Wetensch, Amsterdam,* **37**, 456.

TOLBERT, N. E. (1960). *Pl. Physiol., Lancaster*, **35**, 380.

TSCHIRLEY, F. H. (1979). 'Role of pesticides in increasing agricultural production' in *Pesticides, Contemporary Roles in Agriculture, Health and the Environment* (Eds. T. J. Sheets and D. Pimentel), Humana Press.

U.S. Dept. of Agric. (1974). *Impact of Loss of Pesticides on Agricultural Production and Food Supply*, Ag. Res. Serv., 22 pp.

U.S. Dept. of Agric. (1975). *Minimum Tillage*. Senate Comm. on Agric. and Forestry Print, 57–398.

Weed Research Organisation, Oxford (1980). *8th Report for 1978/79*.

ZIMMERMAN, P. W. (1951). 'Plant hormones in practice' in *Plant Growth Substances* (Ed. F. Skoog), Univ. Wisc. Press.

CHAPTER ONE

Classification, Discovery and Uses of Herbicides

No classification is perfect and the classification of herbicides is a particularly difficult task. Many options are available on which to base it – method of use, crop to which applied, mode of action, chemical structure. We have chosen to open this chapter with a short description of the different terms that are used in connection with herbicide application and then to classify the herbicides into major groups according to their chemical structure. In doing this we have in large part, with minor alterations, used the system proposed by Silk *et al.* (1977). Some herbicide molecules contain more than one classifying group and these have been put into the group most likely to be contributing activity, e.g. in methabenzthiazuron the herbicide activity resides in the urea structure $>$N—CO—N$<$ and not the nitrogen heterocyclic ring (unclassified). Within each group we have, however, listed each herbicide, with a few minor exceptions where it would have appeared artificial to have done so, according to the date of its introduction. Information relating to discovery, commercial introduction and usage have been derived from a number of sources mentioned in the Introduction. Readers requiring further information on the synthesis of herbicides are referred to *The Pesticide Manual* edited by C. R. Worthing (1979) and *Chemicals for Crop Protection and Pest Control* by Green, Hartley and West (1979).

The uses of herbicides

Herbicides may be used in a variety of ways and in a variety of situations. Some may be more effective in an aquatic environment, others when applied to the soil, and still others when applied to the foliage of land plants. Within these various environments some herbicides (e.g. sodium chlorate) may be '*non-selective*', i.e. they kill all of the vegetation; others may be '*selective*' in that certain plants are killed off while others are unaffected. Some of these selective herbicides will kill only those parts of the plant on which they are sprayed – the '*contact*' herbicides (e.g. DNOC); others (e.g. 2,4-D) are '*translocated*', i.e. they move within the plant and may kill off parts which have not been covered by the herbicide.

PRE-SOWING OR PRE-PLANTING APPLICATION

The time of application depends upon the type of herbicide being used and the particular weed problem to be dealt with. Thus some herbicides may be applied to the soil in order to kill off any weeds that may be present and thus clear the ground before sowing or planting. Both contact and translocated herbicides, e.g. MCPA, may be used for this purpose. It is important that the herbicide should either be quickly inactivated in the soil or should be of such a nature as not to cause injury to the subsequent crops. In the latter case it is possible to use a relatively persistent *residual* herbicide which will remain in the soil and control weed seeds which may subsequently germinate, e.g. tri-allate for the control of wild oats in certain crops.

PRE-EMERGENCE APPLICATION

Related to the pre-sowing techniques is that known as pre-emergence which means the application of a herbicide to kill weeds prior to the emergence of the crop. Again contact, translocated and residual herbicides may be used for this purpose. In the case of *contact herbicides* the essential requirements are first that the majority of weed seedlings should have emerged before the crop. It is a good system for use with crop seeds which take some time to germinate such as onions, leeks, sugar-beet. Paraquat is a good example of a herbicide that may be used in this way. There are drawbacks in that there is little latitude in timing the spray and a spell of bad weather could mean that the operation has to be abandoned. Further, all of the weed seedlings may not appear at the same time but germinate over an extended period.

Occasionally *translocated* foliage herbicides may be used such as the application of dalapon to control grass weeds in a potato field. *Residual* herbicides may also be employed. They must have a low solubility in water so that they remain in the top few inches of the soil where most of the weed seeds germinate. The crop seeds must be sown deeply so that their roots are not in range of the herbicide or they must have a high resistance to the herbicide. The effectiveness of these residual herbicides is also very dependent on rainfall and on soil type. Outstanding results have been obtained by using simazine or atrazine in this way for the control of weeds in maize which shows remarkable resistance to these two herbicides.

POST-EMERGENCE APPLICATION

This is the commonest method of application and contact, translocated and residual herbicides may be used. *Contact foliage* consists of an overall spray when both the crop and the weeds are above ground. Two good examples of herbicides used in this way are ioxynil and bromoxynil which are used for the control of many weeds in cereals. Among the important factors governing success are the volume rate, the drop size, the uniformity of application, the stage of growth of the weeds and the crop and the prevailing weather conditions. In some cases, e.g. in fruit orchards, selectivity may be obtained by directing the spray towards the weeds between the fruit trees. In the case of *translocated foliage* herbicides the basis for selectivity rests on the resistance of the crop and the susceptibility of the weed rather than volume rate, drop size, etc. The whole of the weed plant does not need to be covered since the herbicide moves in either the phloem or the xylem to all parts, including any new growth which has taken place subsequent to spraying. These herbicides are generally fairly slow acting. A good example is the use of MCPA for weed control in cereals.

Residual post-emergence treatment is closely allied to residual pre-emergence and the same types of herbicides are used. Most of them affect seedling weeds and have little effect on the mature plants. Directed spray applications are often used and granular formulations which bounce off the crop foliage are very useful.

Formulation of herbicides

Herbicides may be applied as liquid sprays or as solid particles. The formulated material supplied by the manufacturer or agent is generally in a concentrated form. It has to be diluted before use, generally with water and many chemicals are formulated in such a way as to be readily dispersible in water. Herbicides which are insoluble in water are usually formulated as emulsifiable concentrates, wettable powders or suspension concentrates such as pastes and slurries. *Emulsifiable concentrates* contain the herbicide dissolved in an organic solvent such as oil and when diluted with water they form emulsions. *Wettable powders* consist of the finely divided herbicide, a 'filler' such as clay and a dispersing agent which provides a suspension of the solid particles when the powders are diluted with water.

Flowables (pastes and slurries) contain finely divided herbicide mixed with surface-active agents and dispersed in a small amount of water. They are diluted with water in the same way as emulsifiable concentrates. Most formulated herbicides contain, in addition to the active ingredient, surface-active agents – wetters, stickers or humectants, which increase the activity of the herbicide.

Solid particles such as *dusts* and *granules* are generally for soil application though granules may on occasions be used for foliage application. Dusts and granules are prepared by spraying a solution of the herbicide on to pre-formed granules, or by agglomeration starting from a powdered mixture of herbicide and carrier. Dusts are prepared in a similar manner, the distinction between the two being generally based on an arbitrary size (dust particles, less than 80μm in diameter).

Most herbicides are applied as sprays and many factors influence their activity. These include the volume of liquid applied, the mean diameter of the spray drops, the surface tension of the leaf surface, the orientation of the foliage, the leaf shape and size, and the nature of the leaf surface. Many different types of spraying equipment have been devised including portable hand-operated sprayers, portable powered sprayers, tractor-powered sprayers and aircraft equipment. Nozzles of various size and design are also available. For a full and authoritative account of these various matters the reader is referred to Evans *et al.* (1977).

Classification of herbicides

INORGANIC COMPOUNDS

Among the earliest of inorganic salts to be used as a selective herbicide in cereal crops was **copper sulphate** ($CuSO_4$). It was introduced almost simultaneously in 1896 by Bonnet in France, Schultz in Germany and Bolley in the United States.

It is probable that its use as a fungicide led to the observation that it had a greater adverse effect upon dicotyledonous plants than it had upon cereals.

Soon afterwards **sulphuric acid** (H_2SO_4) was used in cereals after being recommended as a herbicide by Woods and Bartlett (1909). Its selective killing action depends upon the fact that it runs off the leaves of the cereals but clings to, and scorches, the relatively rough surfaces of certain weeds. It is still occasionally used for pre-harvest desiccation of certain crops but not as a selective herbicide, not only because of the difficulty in handling it but also because, being a contact herbicide, it killed where it touched. If a susceptible plant was not completely covered then unsprayed parts could effect a recovery and possibly set seed. Perennial weeds were not killed because even if the above-ground parts were scorched, the underground perennating organs were protected and were thus able to grow.

A few inorganic salts are still used for 'total' weed control on non-agricultural land.

The most widely used is **sodium chlorate** ($NaClO_3$). According to Brian (1976) the early interest in chlorates as toxic agents stemmed from the occasional injury that plants suffered from the application of sodium nitrate as a fertiliser which was known to have sodium perchlorate ($NaClO_4$), as an impurity. Among the earliest recorded experiments on the toxicity of sodium chlorate are those by de Caluwe (1900) who looked at the effects of perchlorate and chlorate salts on rye (*Secale cereale*). Guthrie and Helms (1904) found that as little as 0.001% sodium chlorate mixed into soil was toxic to wheat (*Triticum vulgare*) and maize (*Zea mays*). Later it was used under the name 'mort-herbes' in France for the killing of plants on roadside verges. Aslander (1926) showed that it was capable of killing deep-rooted perennial weeds. It is readily translocated and is most effective when applied to plants which are actively growing, causing chlorosis and scorch followed by a dying-back of the leaves. Because of its high solubility in water (79 g/100 ml) sodium chlorate moves readily in soils being rapidly leached and it may move from treated onto non-treated areas. It is a strong oxidising agent and in strong sunlight it may react with organic materials thus constituting a considerable fire hazard. Fire depressants are now normally included for formulation purposes and it may be used in conjunction with borates such as **borax** (disodium tetraborate decahydrate), **sodium metaborate** (the sodium salt of boracic acid – HBO_2) and **disodium octaborate** which are potent 'total' herbicides in their own right, the latter two being marketed by the U.S. Borax and Chemical Corp. **Sodium tetraborate pentahydrate** ($Na_2B_4O_7.5H_2O$) is a crude product introduced as a general contact weedkiller and soil sterilant as 'Tronabor' by Trona Chemicals.

Another non-selective herbicide, **ammonium sulphamate** ($NH_4SO_3NH_2$) for use particularly for the control of woody plants, was introduced by E. I. du Pont de Nemours & Co. in 1945 as 'Ammate'. It may be applied as a foliar spray, as crystals to cut bark or stumps, or as a contact herbicide and temporary soil sterilant.

Inorganic compounds – formulae

$CuSO_4$	H_2SO_4	$NaClO_3$
Copper sulphate	Sulphuric acid	Sodium chlorate

$Na_2B_4O_7.10H_2O$	$NH_4SO_3NH_2$
Borax	Ammonium sulphamate

HALOALKANOIC ACIDS

The haloalkanoic acids are very active against grasses inhibiting growth and causing chlorosis and necrosis of the leaves. Probably the first herbicide to deal specifically with grass weeds was **TCA** (sodium trichloroacetate; CCl_3COONa) marketed by E. I. du Pont de Nemours and the Dow Chemical Co. in 1947. It may be used pre-emergence, together with cultivation techniques for the control of couch grass (*Agropyron repens*). It has been produced by May & Baker Ltd in the form of granular pellets as 'Varitox' for the control of grass weeds in sugar-cane plantations.

In 1953 the Dow Chemical Co. produced **dalapon** (2,2,-dichloropropionic acid; CH_3CCl_2COOH) also for the control of couch (*Agropyron repens*) and other grasses. It may be used in arable stubble where it should be applied not later than November and then ploughed in some 14 days later. For spring treatment of arable land ploughing should be 1 month after treatment and crops such as beet, carrots, potatoes and kale (but not cereals) sown 7 weeks after spraying. The herbicide may be used on carrots (at the 'pencil' stage), and on sugar-beet at the 2–4 leaf stage. It may be used in established orchards, on certain soft fruits, and for the control of grasses, reeds and sedges in or near water.

An important herbicide, **chlorfenprop-methyl** (methyl 2-chloro-3-[4-chlorophenyl]propionate), to deal specifically, post-emergence, with wild oats (*Avena fatua*) in barley, spring oats and wheat, was introduced by Bayer A.G. in 1968 as 'Bisidin'. It is contact acting and to be really effective it must reach the meristematic tissue of the wild oat plant either by the direct route of the spray run down or indirectly by diffusion from the proximal end of the foliage. Time of application in relation to the stage of development of *A. fatua* is critical, good results being achieved when the weed is at the 1–4 leaf stage, provided that tillering has not commenced, in both spring barley and spring wheat.

Haloalkanoic acids – formulae

CCl_3COONa — TCA

CH_3CCl_2COOH — Dalapon

$Cl-C_6H_4-CH_2CHClCOOCH_3$

Chlorfenprop-methyl

PHENOXYALKANOIC ACIDS

It was during the 1939–45 war that the foundations were laid for the massive development of organic herbicides that was to take place with the coming of peace and which has continued almost unabated through to the present time.

Phenoxyacetics

The development of herbicides in their present form is due to the discovery of the 'hormone' herbicides. It is worth recalling the circumstances of their discovery. In the 1920s it was shown that plants produce a hormone – indole-acetic acid (IAA) which plays a major part in controlling their growth (see chapter 2). It was not however until the early days of the 1939–45 war that this knowledge was applied to the killing of weeds. Slade, Templeman and Sexton (1945) at the Jeallots Hill Research Station of Imperial Chemical Industries while investigating the role of hormones in plant growth, sprayed mixed stands of oats and charlock (*Brassica sinapis*) with α-naphthylacetic acid (NAA) and found that it killed off the latter but left the cereal unaffected. They tried out NAA against a great variety of dicotyledonous weeds, finding that many of them were killed; and against a variety of cereals finding that all of them were unharmed. They turned their attention to chemicals related in structure to NAA, and with appropriate substitutions, in an attempt to find some that might be even more potent than NAA. One of these **MCPA** ([4-chloro-*o*-tolyloxy] acetic acid) was a very active compound and could kill many weeds selectively at low concentrations.

Over the same period, Nutman, Thornton and Quastel (1945) at Rothamsted Experimental Station had reached similar conclusions by a different route. While investigating the role of indole-acetic acid in the formation of root nodules by rhizobia they found that, at low concentrations, IAA killed off red clover (*Trifolium pratense*) whilst cereals were unharmed. Like the I.C.I. workers they realised that IAA could be used as a selective weedkiller and looking at related chemicals they tried out **2,4-D** (2,4-dichlorophenoxy acetic acid) which had been first mentioned by Pokorny (1941) in the U.S.A. though not as a weedkiller. In 1942 the results of this preliminary work by these British teams was communicated to the Agricultural Research Council who asked G. E. Blackman (1945) of Oxford University to carry out a series of field trials with these two chemicals. Both MCPA and 2,4-D were shown to be potent selective herbicides in the field. Publication of the Rothamsted, I.C.I. and Oxford findings was held up for security reasons and they did not appear until 1945.

Two Americans, Zimmerman and Hitchcock (1942) described the use of 2,4-D as a plant growth regulator but not as a herbicide. Marth and Mitchell (1944) in the U.S.A. reported that it killed dandelions (*Taraxacum officinale*) and plantains (*Plantago* spp.) in lawns and in the same year Hamner and Tukey (1944), also in the U.S.A., described field trials using 2,4-D as a herbicide. It is therefore difficult to establish precedence for the precise discovery of 2,4-D as a herbicide (Fletcher, 1974).

Both MCPA and 2,4-D were quickly adopted, the former being almost exclusively preferred in the United Kingdom whereas in the U.S.A. 2,4-D was the chemical of choice. They are widely used for selective weed control in turf and cereals. Both are cheap to produce; they are active at very low concentrations, they are readily translocated so that the whole of the plant need not be covered by the spray and they are virtually non-poisonous to Man and his animals. The characteristic symptoms on affected plants are epinasty, tumour formation and the development of secondary roots from the stems. There is a swelling of the parenchyma cells producing callus tissue. Root elongation is halted and there is a swelling of the root tips (Ashton and Crafts, 1973).

Among the weeds which are controlled by 2,4-D and MCPA are charlock (*Brassica sinapis*), annual nettle (*Urtica* sp.), corn buttercup (*Ranunculus* sp.), docks (*Rumex* spp.), fat hen (*Chenopodium album*), runch (*Brassica alba*), shepherds purse (*Capsella bursa-pastoris*), poppy (*Papaver rhoeas*), cornflower (*Centaurea cyanus*), fumitory (*Fumaria officinalis*), daisy (*Bellis perennis*). Both herbicides are often used in conjunction with other herbicides in order to extend the range of weeds controlled, for example, MCPA (as the iso-octyl ester) may be formulated with flurecol-butyl (9-hydroxy-fluorene-9-carboxylic acid butyl ester) (see under Aromatic Acids) and it was introduced in this combination as 'Aniten' by E. Merck in 1964 for use post-emergence in cereals. The two ingredients act synergistically, the latter affecting the growth of meristematic tissues and the former, the elongation growth. Among the 'difficult' weeds controlled are speedwell (*Veronica* sp.), hemp-nettle (*Galeopsis* sp.), cleavers (*Galium aparine*), dead nettle (*Lamium* sp.), chickweed (*Stellaria media*), *Matricaria* sp. (4–6 leaf stage) and *Polygonum* sp.

Sodium 2-(2,4-dichlorophenoxy) ethyl sulphate (2,4-DES sodium) which is not in itself herbicidal, but which is converted in the soil to 2,4-D, was introduced in 1951 by the Union Carbide Corp. as 'Crag Herbicide I' and 'Sesone' for the control of germinating weeds in horticultural crops. It may be used as a mixture with simazine to control annual weeds pre-emergence in maize, roses, soft fruits and certain established perennials.

MCPA and 2,4-D are not active against all dicotyledonous weeds and in 1944 Amchem Ltd introduced **2,4,5-T** (2,4,5-trichlorophenoxy-acetic acid) as 'Weedone'. It resembles 2,4-D in its herbicidal properties but is much more active against many woody species. Combined with mecoprop as 'Ban-Dock' (Shell) it is used for the control of established docks (*Rumex* spp.) in grassland. The compounds 2,4-D, MCPA and 2,4,5-T are formulated as salts with the alkali metals and amines which are water soluble. They are also formulated as esters which are water insoluble but are soluble in oils.

Phenoxybutyrics

Wain and Wightman (1955) investigating the herbicidal properties of a series of ω-phenoxyalkanecarboxylic acids found that herbicide activity alternated in the series. Those members which had an even number of methylene groups in the side chain were inactive whereas those with an odd number were active, due to being converted, inside the plant by the enzyme β-oxidase, to the corresponding substituted phenoxyacetic acid. This suggested to Wain (1955a, 1955b) that here was a new and fundamental principle upon which selective weed control might operate, namely, that the susceptibility of a plant to a phenoxybutyric compound might depend on its ability to convert the phenoxybutyric acid to the corresponding phenoxyacetic acid. It was found that some species (e.g. *Pisum sativum*, *Trifolium pratense*) lacked the ability to effect this conversion and could therefore tolerate being sprayed with a chlorine-substituted butyric acid, whereas others, e.g. charlock (*Sinapis arvensis*), fat hen (*Chenopodium album*), creeping buttercup (*Ranunculus repens*) and others possessing the

β-oxidase enzyme system and being susceptible to the corresponding substituted phenoxyacetic acid, were killed.

The first substituted phenoxybutyric herbicide – **MCPB** (4-[4-chloro-*o*-tolyloxy]butyric acid) was introduced by May & Baker Ltd in 1954 as 'Tropotox'. It may be used on undersown wheat, barley and oats, on certain varieties of peas, on soft fruits when bushes have ceased to make new growth, apples, pears and hops as a directed spray. **2,4-DB** (4-[2,4-dichlorophenoxy]butyric acid) was introduced as 'Embutox' by the same company in 1957. Both are formulated as alkali metal and amine salts, and as esters. Red and white clovers (*Trifolium pratense* and *T. repens*) can be sprayed at the seedling stages (both direct and undersown in cereals) with MCPB and 2,4-DB. Established crops of red and white clovers can be sprayed with MCPB but 2,4-DB should not be used in established crops of red clover although it is perfectly safe in established white clover. Many varieties of lucerne (*Medicago sativa*) can be treated, especially at the seedling stages, with 2,4-DB (but not by MCPB). MCPB, on the other hand, can be safely used on several important varieties of peas (*Pisum sativum*) but 2,4-DB is not recommended for use on this crop. Both can be used on groundnuts (*Arachis hypogea*). Both are commonly used in conjunction with MCPA or 2,4-D.

Phenoxypropionics

These are all used post-emergence. The growth regulatory activities of the phenoxypropionic compounds were first described by Zimmerman and Hitchcock (1944) and Synerholm and Zimmerman (1945). The first substituted phenoxypropionic acid, **fenoprop** (±2-[2,4,5,-tri-chlorophenoxy]propionic acid), was introduced as a herbicide by the Dow Chemical Co. as 'Kuron' in 1953 for the control of woody and shrubby plants, for annual weeds in cereals and for aquatic weed control. It was followed by **mecoprop** (±-2-[4-chloro-2-methyl-phenoxy]propionic acid) introduced in 1957 by The Boots Co. Ltd. as 'Iso-Cornox', a foliar-absorbed translocated herbicide, to deal with a number of weeds such as chickweed (*Stellaria media*) and cleavers (*Galium aparine*) that are resistant to the phenoxyacetic acids. Its properties were described by Fawcett *et al.* (1953). It may be used in combination with a number of other herbicides, e.g. with chlortoluron (for the control of *Galium, Veronica* and *Papaver* spp.), with 2,4-D, dicamba, 3,6-dichloropicolinic acid and others. In 1961 the same Company introduced **dichlorprop** (±-2-[2,4-dichlorophenoxy]propionic acid) as 'Cornox RK', a post-emergence, translocated herbicide, for use in cereals for the control of a number of *Polygonum* spp. (*P. persicaria,*

P. lapathifolium and *P. convolvulvus*), chickweed (*Stellaria media*), cleavers (*Galium aparine*) and other broad-leaved weeds. It may also be used in combination with a number of other herbicides.

In 1971 the Stauffer Chemical Co. marketed **napropamide** (*N,N*-diethyl-2-[1-naphthyloxy propionamide]) as 'Devrinol', a pre-emergence or soil-incorporated herbicide, active against both annual and perennial grasses and a range of broad-leaved weeds in a wide variety of horticultural and agronomic crops, tree fruits and vines. Its properties have been described by Van den Brink *et al.* (1969).

An important recent discovery is that some 4-substituted phenoxypropionic acids applied post-emergence are capable of controlling grasses selectively in dicotyledonous crops. In 1975 Hoescht A.G. marketed **diclofop-methyl**, the methyl ester of diclofop (±-2-[4-(2,4-dichlorophenoxy)phenoxy]propionic acid), as 'Illoxan', 'Hoe-Grass' and 'Hoelon', a systemic post-emergence herbicide for the control of wild oats (*Avena* spp.), wild millets (*Panicum* spp.) and other grassy weeds in wheat, rye, barley and some fodder grasses. It is tolerated by most dicotyledonous crops except cotton. In the same year they produced **clofop-isobutyl** (isobutyl ±-2-[4-(chlorophenoxy)phenoxy] propionate) as 'Hoe 22870', a post-emergence herbicide for the control of annual grasses in cereals and nearly all dicotyledonous crops. In 1977 the same Company introduced **Hoe 29125** (methyl 2-[4-(chloromethylphenoxy)phenoxy]propionate) which shows good post- and pre-emergence control of *Avena* spp., *Alopecurus myosuroides*, weed millets and volunteer cereals, and post-emergence control of *Agropyron repens*, *Cynodon dactylon* and *Sorghum halepense*. Most annual and all perennial dicotyledonous crops are tolerant. **Fluazifop-butyl** (butyl 2-[4-(4-trifluoromethyl-2-pyridyloxy)phenoxy]propionate) was introduced by Ishihara Sangyo Kaisha Ltd in 1980 and is being jointly developed by I.C.I. Ltd as a post-emergence herbicide for the selective control of annual and perennial grass weeds, primarily for use in dicotyledonous crops. Extensive field trials have shown that it can be used in over sixty different crops to control the major grass weeds.

Phenoxyalkanoic acids – structural formulae

Phenoxyacetics

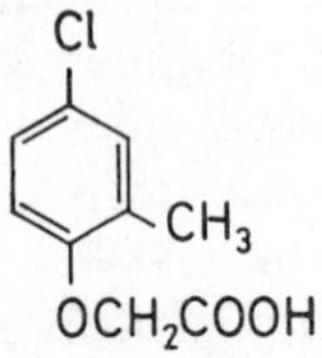

Cl
CH_3
OCH_2COOH
MCPA

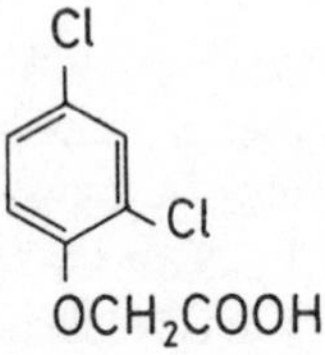

Cl
Cl
OCH_2COOH
2,4-D

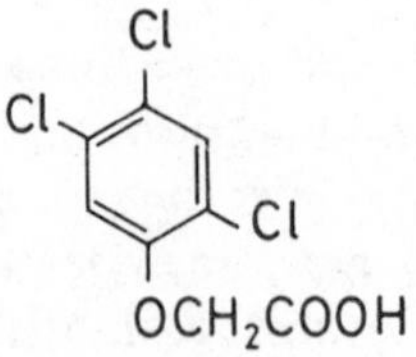

Cl
Cl
Cl
OCH_2COOH
2,4,5-T

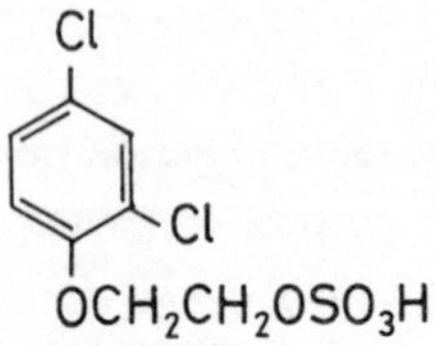

Cl
Cl
$OCH_2CH_2OSO_3H$
2,4-DES

Phenoxybutyrics

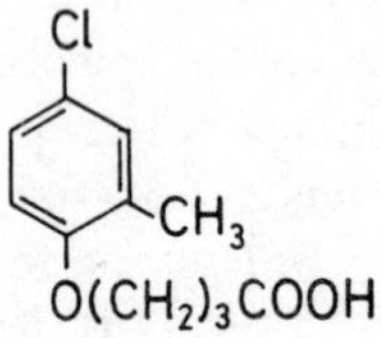

Cl
CH_3
$O(CH_2)_3COOH$
MCPB

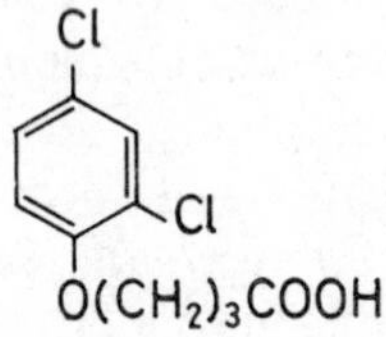

Cl
Cl
$O(CH_2)_3COOH$
2,4-DB

Phenoxypropionics

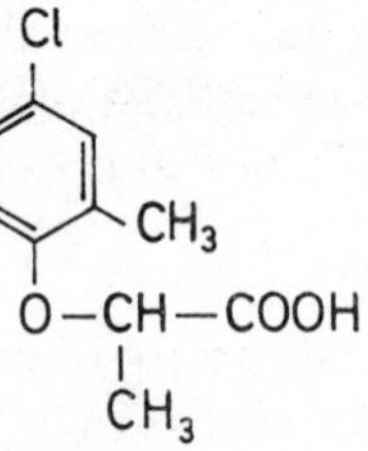

Cl
CH_3
$O-CH(CH_3)-COOH$
Mecoprop

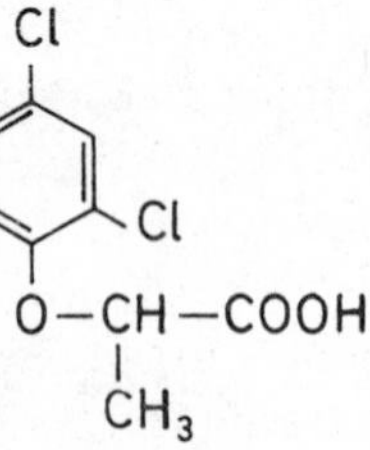

Cl
Cl
$O-CH(CH_3)-COOH$
Dichlorprop

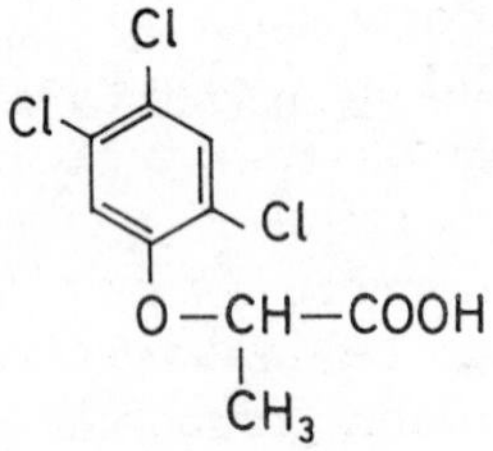

Cl
Cl
Cl
$O-CH(CH_3)-COOH$
Fenoprop

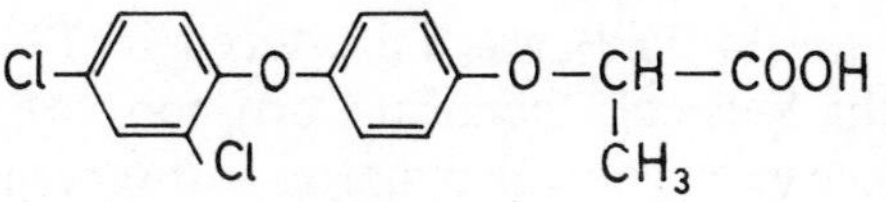

Diclofop-methyl

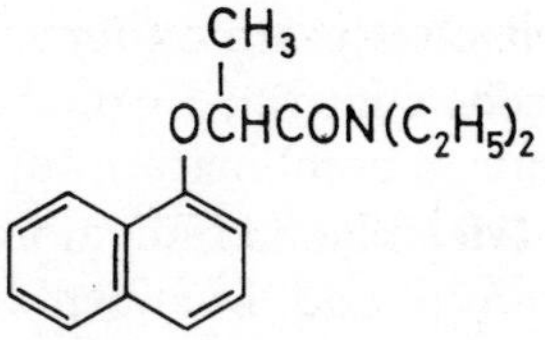

Napropamide

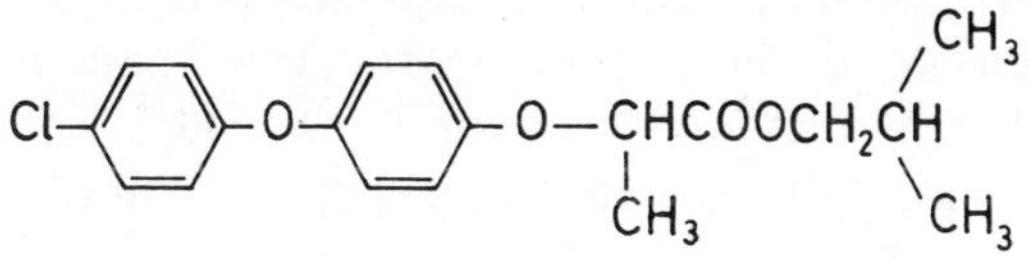

Clofop-isobutyl

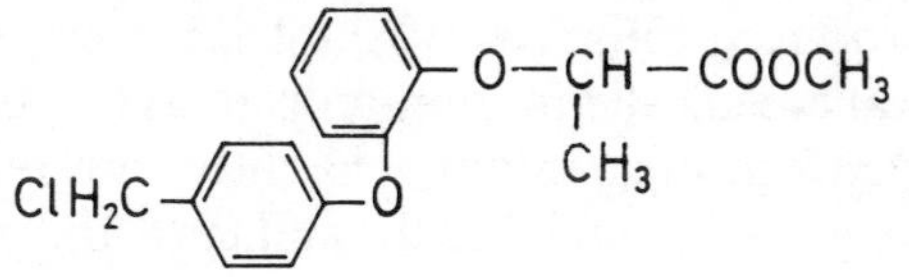

Hoe 29125

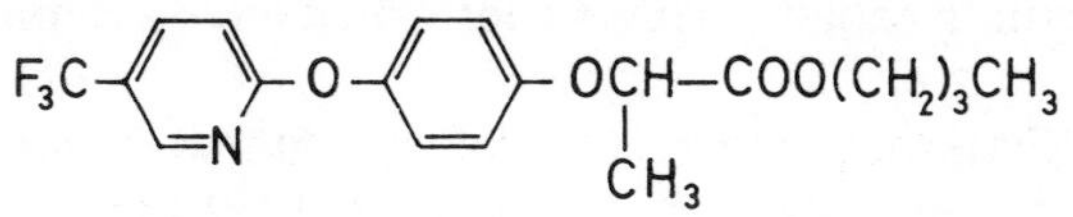

Fluazifop-butyl

AROMATIC ACIDS

The first herbicide based on the aromatic acids was **2,3,6-TBA** (2,3,6-trichlorobenzoic acid) introduced by the Heyden Chemical Corp. as HC-1281 and by Du Pont in 1954 as 'Trysben'. It is used mixed with other herbicides, e.g. MCPA, for the control of annual and perennial weeds in cereals.

Chloramben (3-amino-2,5-dichlorobenzoic acid) was introduced by Amchem Ltd in 1958 as 'Vegiben' for pre-emergence control of weeds in soybeans, groundnuts, maize, carrots and other crops. In 1970 a chloramben/atrazine combination was marketed by the same company for use pre-emergence in maize, and a chloramben/dinoseb combination as 'Dynoram' for weed control in soybeans. A similar formulation is marketed by the Dow Chemical Co. as 'Premerge'. They may be used where chloramben alone has been erratic because of soil moisture requirements. There may also be a wider spectrum of weeds controlled. Amchem Ltd have also introduced a chloramben/linuron combination for use in soybeans.

Dicamba (3,6-dichloro-*o*-anisic acid) was introduced in 1965 as 'Banvel' and 'Mediben' by the Velsicol Chemical Corp. for both pre-emergence and post-emergence weed control in maize, post-emergence weed control in small grains, post-emergence weed control in perennial grasses grown for seed, and on golf courses. It kills many annual grass and broad-leaved weeds as they start to germinate. It may be combined with other herbicides, e.g. with 2,4-D and 2,4,5-T, as 'Banvel-D' (Shell Chemicals), for the control of perennial nettles and broad-leaved weeds in pasture.

The above benzoic acid derivatives affect dicotyledonous plants in a similar way to the phenoxyalkanoic acids.

In 1958 Amchem Ltd introduced **chlorfenac** (2,3,6-trichlorophenylacetic acid) as 'Fenac', a 'total' weedkiller for non-crop areas (though it may also be used pre-emergence in sugar-cane). It has little or no effect on weeds and grasses already growing but it has the ability to 'fix' in the soil when weed seeds sprout so that when developing roots come into contact with the herbicide in the soil these weeds are killed. The Company produced in 1968 a combined mixture with 2,4-D ('Fenatrol Plus') which kills seedling broad-leaved weeds which have germinated prior to treatment.

The Diamond Shamrock Corp. marketed **chlorthal-dimethyl** (dimethyltetrachloroterephthalate) as 'Dacthal' in 1959 for the pre-emergence control of many annual weeds in a variety of crops.

In 1964 E. Merck introduced **flurecol-butyl** (the butyl ester of 9-hydroxy-fluorene-9-carboxylic acid) under the code number 'IT 3233', as an inhibitor of plant growth which potentiates the phenoxyalkanoic acid herbicides. The following year the same company introduced **chlorflurecol-methyl** (methyl 2-chloro-9-hydroxyfluorene-9-carboxylate) as a general growth retardant and for soil application for weed suppression (see also Morphactins – chapter 2).

Aromatic acids – structural formulae

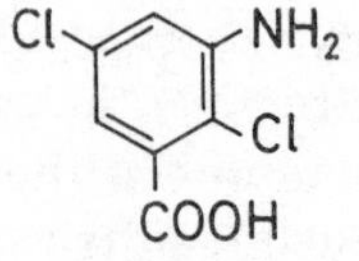

Cl, NH_2, Cl, COOH — Chloramben

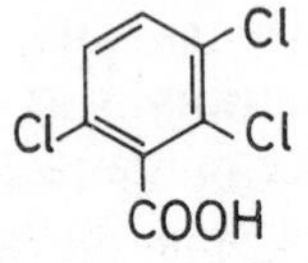

Cl, Cl, Cl, COOH — 2,3,6-TBA

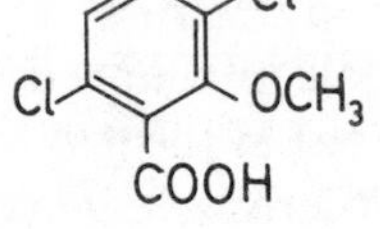

Cl, Cl, OCH_3, COOH — Dicamba

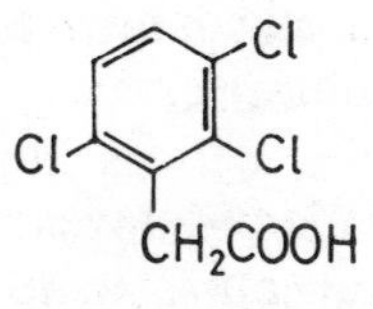

Chlorfenac 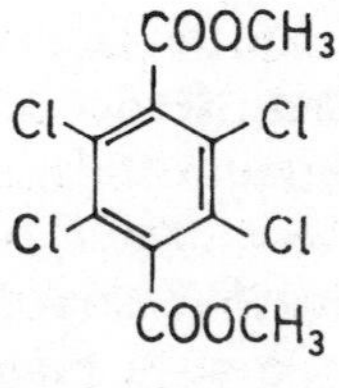 Chlorthal-dimethyl

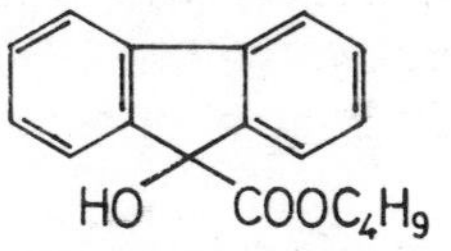

Flurecol-butyl 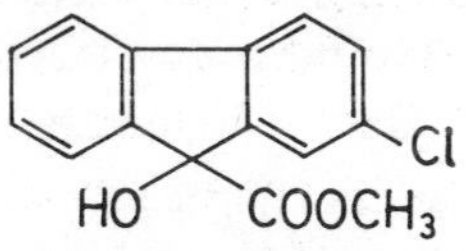 Chlorflurecol-methyl

AMIDES

The members of this group are primarily soil acting especially against annual grasses. Many are important seed-germination inhibitors.

Susceptible species show a severe stunting of shoots with grasses often failing to emerge from the coleoptile. At low doses the foliage tends to be very dark green (Silk *et al.*, 1977). The properties of a number of *N*-arylphthalamic acids were described by Hoffman and Smith (1949) and **naptalam** (*N*-1-naphthylphthalamic acid) was the first of the group to be introduced by Uniroyal Inc. in 1950 as 'Alanap' for use pre-emergence on a number of crops including potatoes and groundnuts controlling annual weeds and grasses.

In 1960 **diphenamid** (*N,N*-dimethyl-2,2-diphenylacetamide) was introduced as 'Dymid' by Eli Lilly and Co. and as 'Enide' by The Upjohn Co. for the control of germinating annual grasses and some dicotyledons after sowing or planting a range of crops including beans, ornamental trees and shrubs, potatoes, fruit trees, cotton peanuts, soybeans, tobacco and many other agronomic and horticultural crops. Diphenamid may be used in combination with a number of other herbicides, e.g. prometryne, paraquat, linuron and others.

Propyzamide (3,5-dichloro-*N*-[1,1-dimethyl-2-propynyl]benzamide), introduced by the Rohm and Haas Co. as 'Kerb' in 1965, is unusual in the group in that not only can it be used pre-emergence in many crops, e.g. cotton, soybean, peanut, squash, sunflower and others, controlling many annual and perennial grasses, chickweed (*Stellaria media*), dodder (*Cuscuta* sp.),etc., but it can also be applied post-emergence for the control of many grasses in lucerne (*Medicago sativa*), other small-seeded legumes, sugar-beet, apples, strawberries and perennial ornamentals. In granular form as 'Clanex' (Shell) it is used for weed control in forestry.

Butam (*N*-benzyl-*N*-isopropylpivalamide) was introduced by Gulf Oil Chemicals Co. as a pre-emergence herbicide for the control of grasses and some broad-leaved weeds, e.g. *Amaranthus retroflexus* and fat hen (*Chenopodium album*), in many crops, e.g. cotton, soybeans and potatoes. Affected weeds emerge but soon die. Its properties have been described by Schwartzbeck (1976).

Amides – structural formulae

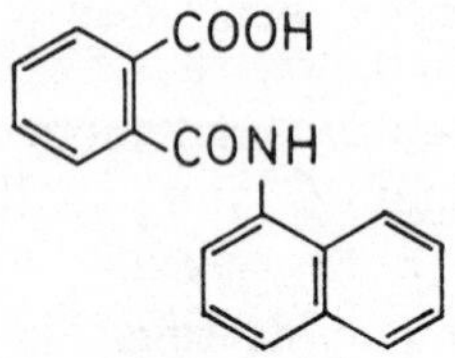

COOH
CONH

Naptalam

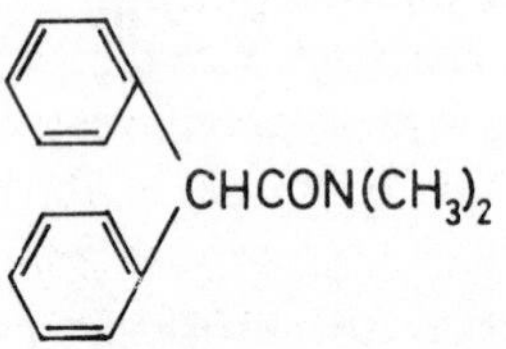

$CHCON(CH_3)_2$

Diphenamid

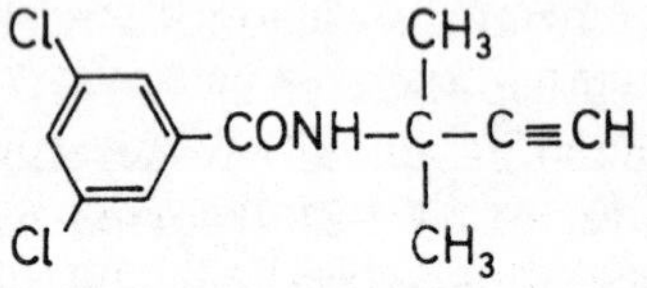

Cl
Cl
CH_3
CONH–C–C≡CH
CH_3

Propyzamide

CH_2N
$COC(CH_3)_3$
$CH(CH_3)_2$

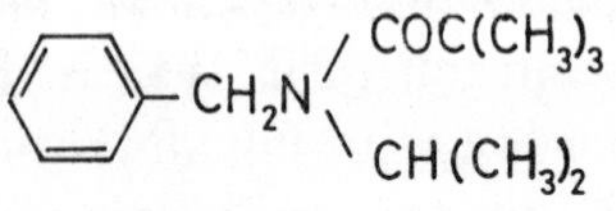

Butam

NITRILES

Dichlobenil (2,6-dichlorobenzonitrile) was the first member of this group to be introduced as 'Casoron' by Philips Duphar in 1960. It can be used both pre- and post-emergence for the control of weeds in many crops but it is mainly soil acting where it controls germinating annual weeds and buds of perennial weeds such as *Pteridium aquilinum*, *Agropyron*, *Artemisia* and *Cynodon* spp. It may also be used for aquatic weed control It inhibits actively dividing meristems and causes a swelling of affected parts followed by browning and death. Affected leaves may become dark green.

Chlorthiamid (2,6-dichloro[thiobenzamide]) is classified here because it is converted in the soil to dichlobenil. It was introduced by Shell in 1963. It is toxic to germinating seeds and it is highly effective against a number of hard-to-kill weeds such as *Tussilago*, *Aegopodium*, *Rumex* and *Equisetum* spp. It is used for 'total' weed control and for selective control in apples, blackcurrants and gooseberries. It is also used for 'spot' treatment of docks (*Rumex* spp.) and thistles (*Cirsium* spp.) and for aquatic weed control in water courses and dry ditches.

Two interesting nitriles were introduced by May & Baker Ltd and by Amchem Ltd in 1963.

Ioxynil (4-hydroxy-3,5-di-iodobenzonitrile), 'Actril', is a contact herbicide with some systemic activity for the control of dicotyledonous weeds in cereals being particularly active against young seedlings of the Polygonaceae, Compositae and certain Boraginaceae.

Bromoxynil (3,5-dibromo-4-hydroxybenzonitrile), 'Buctril', is used in similar situations. Both show a high degree of selectivity to graminaceous crops. Both are active within 24 hours of spraying showing necrotic spots on affected leaves. These spread until the plant dies. Both have also been marketed as octanoates. **Ioxynil octanoate** (4-cyano-2,6-di-iodophenyl octanoate) was introduced as 'Totril' by May & Baker Ltd in 1963 as a post-emergence contact herbicide for weed control in onions and related crops, controlling among others, such 'difficult' weeds as *Amaranthus* spp., *Chenopodium album*, *Stellaria media*, *Senecio* spp., *Polygonum convolvulus*, *Veronica* spp. and *Solanum* spp. **Bromoxynil octanoate** (2,6-dibromo-4-cyanophenyl octanoate) was introduced in 1963 by May & Baker Ltd as 'Buctril' and by Amchem Ltd as 'Brominil' (which has been changed to 'Brominal') for the control of annual broad-leaved weeds in cereals, flax and newly seeded grasses. It is used in conjunction with other herbicides. Formulations containing both ioxynil and bromoxynil, 'Oxitril', are also marketed, and mixtures in varying proportions with phenoxyacetic acids such as 2,4-D, MCPA, dichlorprop or mecoprop have been formulated for particular weed problems. **Bromofenoxim** (3,5-dibromo-4-hydroxybenzaldehyde 2,4-dinitrophenyl oxime) was introduced by Ciba Geigy in 1969 as 'Fareron', a post-emergence contact herbicide active against many dicotyledonous weeds in cereals. It is active particularly against certain members of the Compositae family which are 'difficult' weeds – *Matricaria* spp., *Tripleurospermum maritimum* and *Chrysanthemum segetum*. It is not active against grasses but Ciba Geigy have marketed it in combination with terbuthylazine which is taken up through the roots and acts as a check on the growth of certain grasses especially silken bent grasses (*Agrostis* sp.) and annual meadow grass (*Poa annua*).

Nitriles – structural formulae

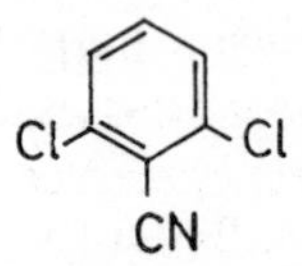

Dichlobenil 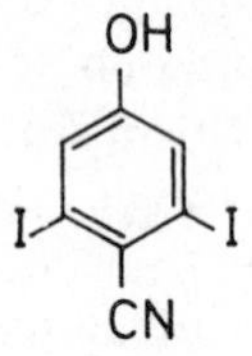 Ioxynil 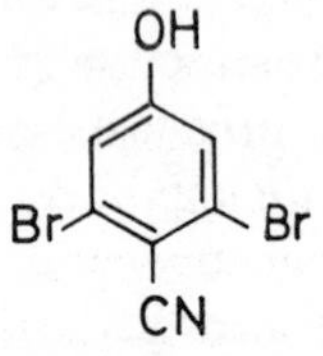 Bromoxynil

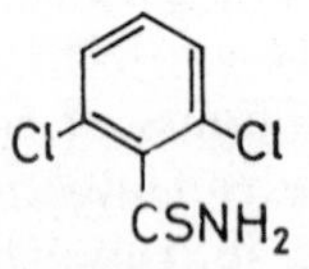

Chlorthiamid 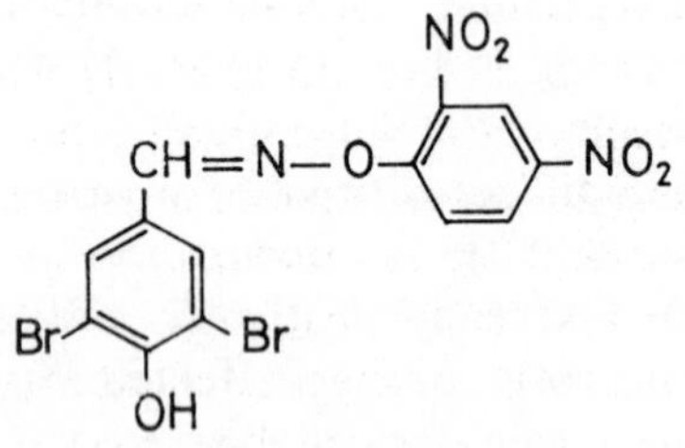 Bromofenoxim

ANILIDES

The type of activity and the range of weed control varies greatly within this group some being used post-emergence while others are active through the soil.

They may be conveniently put together into groups on the basis of chemical structure and activity.

Group 1

Pentanochlor (3-chloro-2-methylvaler-*p*-toluidide) was the first anilide to be introduced in 1958 by F.M.C. for use as a post-emergence herbicide on a variety of horticultural crops. It is also used pre-emergence with chlorpropham. It was followed by **propanil** (3,4-dichloropropionanilide) introduced by Rohm and Haas Co. in 1960 as 'Stam F-34' and subsequently by Bayer A.G. as 'Surcopur'. It was formerly manufactured by Monsanto Chemical Co. as 'Rogue'. It is a contact herbicide for use post-emergence for the control of annual grasses and broad-leaved weeds in paddy rice and potatoes. In 1963 Schering A.G. marketed **monalide** (4-chloro-2-dimethylvaleronilide) as 'Potablan' for post-emergence weed control in umbelliferous crops.

Group 2

In 1965, 1966 and 1969 Monsanto introduced three anilides for pre-emergence control of annual weeds. These were **propachlor** (2-chloro-*N*-isopropylacetanilide) as 'Ramrod' which shows a high degree of specificity for annual grass weeds and certain broad-leaved weeds in maize, soybeans, sugar-cane, peanuts and certain vegetables; **alachlor** (2-chloro-2′,6′-diethyl-*N*-[methoxymethyl]acetanilide) as 'Lasso' for use in maize, cotton, soybeans, sugar-cane, peanuts and certain vegetable crops where it shows very good activity against annual grasses, particularly *Echinochloa crus-galli, Setaria* spp. and *Digitaria* spp. and some broad-leaved weeds (especially when mixed with atrazine); and **butachlor** (*N*-[butoxymethyl]-2-chloro-2′,6′-diethylacetanilide) as 'Machete' for the control of most annual grasses, certain broad-leaved species such as *Monochoria vaginalis, Rotala indica, Dopatrium junceum* and *Alisma canaliculatum* and several important annual and perennial sedge species such as *Cyperus serotinus, Cyperus difformis, Eleocharis acicularis* and *Scirpus hotarui* in transplanted rice.

Diethatyl-ethyl (the ethyl ester of *N*-chloroacetyl-*N*-[2,6-diethylphenyl]glycine) was introduced by Hercules Inc. in 1974 as 'Antor' for the control of grasses, e.g. *Echinochloa, Digitaria, Panicum, Alopecurus* and *Setaria* spp., in beet, winter wheat, soybeans and spinach. It is used pre-emergence.

MON 097 (α-chloro-*N*-[ethoxymethyl]-6-ethyl-*o*-acetotoluidide) was introduced by Monsanto. It shows good pre-emergence activity against annual broad-leaved weeds and certain perennials – *Cyperus* sp. and *Agropyron repens.* Crop tolerance is shown by maize, groundnuts, soybeans and sugar-cane.

In 1974 Ciba Geigy introduced **metolachlor** (2-chloro-6-ethyl *N*-[2-methoxy-1-methylethyl]acet-*o*-toluidide) as 'Dual', a pre-emergence germination inhibitor active mainly on grasses, for use in maize, soybeans, groundnuts. Two combination products 'Primextra' and 'Primagram' consisting of metolachlor and atrazine have been marketed by Ciba Geigy for use in maize, metolachlor being active against annual grasses and atrazine active against annual dicotyledons. In 1977 they also introduced **dimethachlor** (2-chloro-*N*-[2-methoxyethyl]acet-2,6-xylidide) as 'Teridox', a selective herbicide for use on rape against annual grass and dicotyledonous weeds.

AC 206 784 (α-chloro-*N*-isopropyl-2,6-acetoxylidide) was introduced by the American Cyanamid Co. in 1979 as an experimental selective herbicide for the control of annual grasses in maize, soybean, cotton, cereals, rice and others. The herbicide is incorporated before sowing or used pre-emergence and it acts by inhibiting seed germination and seedling development. It may be used in mixtures with compounds active against broad-leaved weeds for broad spectrum control.

In 1979 B.A.S.F. A.G. introduced two more anilides: **BAS 47900H** (α-chloro-*N*-(1-pyrazolylmethyl)-2,6-acetoxylidide) for the control of a number of grasses and dicotyledonous weeds in a range of crops including cotton, strawberries, groundnuts, potatoes and soybeans; and **LAB 114253** (α-chloro-*N*-[(3,5-dimethylpyrazol-1-yl)methyl]-2,6-acetoxylidide) for the control of the same spectrum of weeds in cabbages, green beans, onions, maize, sorghum and soybeans.

Group 3

In 1969, 1972 and 1974 Shell introduced three very important herbicides for the post-emergence control of wild oats (*Avena fatua*, *A. ludoviciana, A. sterilis, A. barbata*) namely **benzoylprop-ethyl** (ethyl *N*-benzoyl-*N*-[3,4-dichlorophenyl]-DL-alaninate) as 'Suffix' for use in wheat, field and broad beans and rye grass grown for seed; **flamprop-isopropyl** (isopropyl *N*-benzoyl-*N*-[3-chloro-4-fluorophenyl]-DL-alaninate) as 'Barnon' for use in spring-sown barley; the enantiomorph, the 'R' form, is known as 'Suffix BW'; and **flamprop-methyl** (methyl *N*-benzoyl-*N*-[3-chloro-4-fluorophenyl]-DL-alaninate) as 'Mataven' for use in wheat and sugar-beet.

Anilides – structural formulae

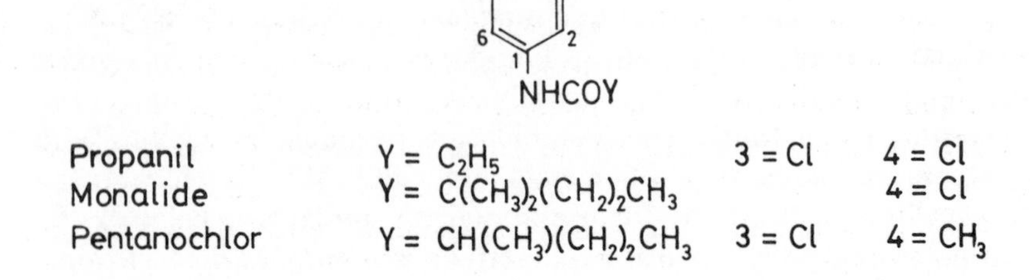

Group 2

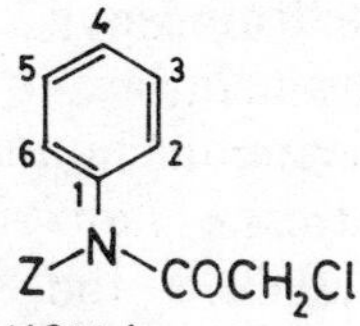

(Ring positions 1–6; at position 1: Z–N–$COCH_2Cl$)

	Z	2	6
Propachlor	Z = $CH(CH_3)_2$		
Alachlor	Z = CH_2OCH_3	2 = C_2H_5	6 = C_2H_5
Butachlor	Z = $CH_2O(CH_2)_3CH_3$	2 = C_2H_5	6 = C_2H_5
Diethatyl-ethyl	Z = $CH_2COOC_2H_5$	2 = C_2H_5	6 = C_2H_5
MON 097	Z = $CH_2OC_2H_5$	2 = CH_3	6 = C_2H_5
Metolachlor	Z = $CH(CH_3)CH_2OCH_3$	2 = CH_3	6 = C_2H_5
Dimethachlor	Z = $CH_2CH_2OCH_3$	2 = CH_3	6 = CH_3
AC 206 784	Z = $CH(CH_3)_2$	2 = CH_3	6 = CH_3
BAS 47900H	Z = CH_2–N (pyrazol-1-yl) 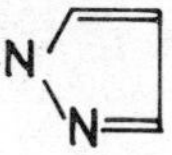	2 = CH_3	6 = CH_3
LAB 114 253	Z = CH_2–N (3,5-dimethylpyrazol-1-yl; CH_3, CH_3) 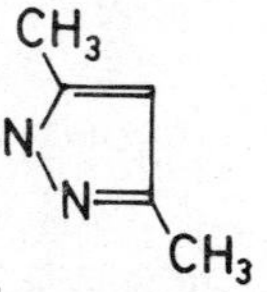	2 = CH_3	6 = CH_3

Group 3

(Ring positions 1–6; at position 1: N(CO–C_6H_5)–$CH(CH_3)COOY$)

Benzoylprop-ethyl	Y = C_2H_5	3 = Cl	4 = Cl
Flamprop-isopropyl	Y = $CH(CH_3)_2$	3 = Cl	4 = F
Flamprop-methyl	Y = CH_3	3 = Cl	4 = F

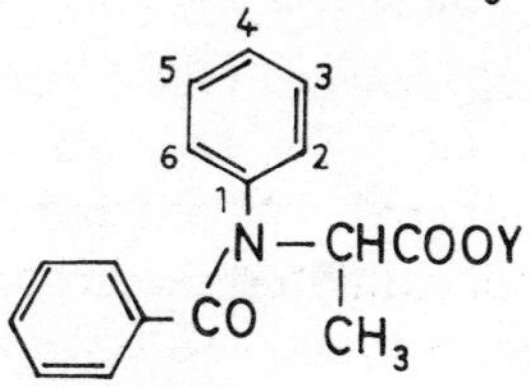

NITROPHENOLS

The first organic selective herbicide to be widely used was a nitrophenol **DNOC** (4,6-dinitro-*o*-cresol) first used as an insecticide in 1892 and introduced as a herbicide by Truffaut et Cie as 'Sinox' in 1932. It was much more efficient as a contact post-emergence selective herbicide in cereals than was sulphuric acid which had been widely used up till then, and it was quickly adopted. It had its drawbacks in that the dry salt was rather inflammable and the compound was toxic and easily absorbed through the skin so that protective clothing had to be worn. It has been largely supplanted but may be used for pre-harvest desiccation of potatoes and legume seed crops.

In 1945 Crafts described **dinoseb** (2-s-butyl-4,6-dinitrophenol), a post-emergence herbicide, and it was introduced by Dow as 'Premerge' for the control of dicotyledonous annual weeds in peas, lucerne and cereals and for pre-emergence control in beans, peas, clover and undersown cereals. In 1958, Hoechst marketed **dinoseb acetate** (2-s-butyl-4,6-dinitrophenyl acetate) as 'Aretit' and 'Ivosit' for use post-emergence against dicotyledonous annual weeds in similar crops to those above. It is also used pre-emergence with linuron for weed control in potatoes and dwarf beans. **Dinoterb** (2-t-butyl-4,6-dinitrophenol) was described by Poignant and Crisinel (1967) and was marketed by Pepro (now a subsidiary of Rhône-Poulenc) for post-emergence control of annual weeds in cereals.

Nitrophenols – structural formulae

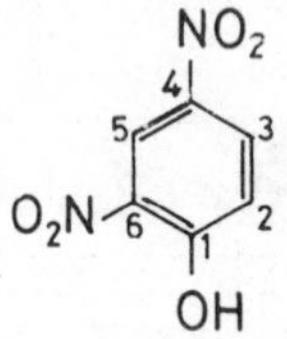

DNOC $2 = CH_3$

Dinoseb $2 = CH(CH_3)CH_2CH_3$

Dinoterb $2 = C(CH_3)_3$

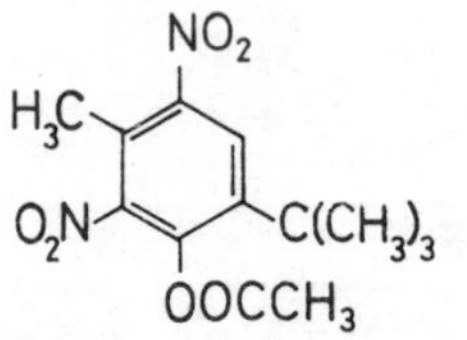

Medinoterb acetate

NITROPHENYL ETHERS

Members of this group have got two benzene rings joined together through oxygen, and they are used pre-emergence being applied to the soil surface to control germinating seeds. They cause shoot inhibition.

Nitrofen (2′,4′-dichlorophenyl 4-nitrophenyl ether) was introduced by Rohm and Haas Co. in 1964 as 'Tok E-25' and 'Tokkorn' for use in cereals. It is applied pre-emergence to the crop and pre-emergence or early post-emergence to the weed population. It is most effective when applied as a thin layer on top of the soil, as activity is quickly lost when it is incorporated into soil. It may also be used on a number of vegetable crops.

Fluorodifen (4-nitrophenyl α,α,α-trifluoro-2′-nitro-*p*-tolyl ether) was introduced as 'Preforan' by Ciba Geigy in 1968 for use in rice, cotton and beans and **bifenox** (methyl 5-[2,4-dichlorophenoxy]-2-nitrobenzoate) was introduced by the Mobil Chemical Co. in 1970 as 'Mowdown' for use pre- and post-emergence against important grass and broad-leaved weeds in soybeans, maize and rice.

Chlornitrofen (2′,4′,6′-trichlorophenyl 4-nitrophenyl ether) was introduced by the Diamond Shamrock Corp. in 1978. It forms a thin layer on the soil surface and is absorbed by germinating seedlings coming into contact with it. It is effective against most annual broad-leaved weeds for selective pre-emergence control in wheat (especially with linuron) and in paddy rice before or immediately after transplanting.

Rohm and Haas Co. (1) have described a potent diphenyl ether herbicide (2′-chloro-4′-trifluoromethylphenyl 4-nitrophenyl ether) with great potential for use in soybeans, cotton, groundnuts, sunflower, wheat and rice and for total weed control. In 1980 the above Company introduced **oxyfluorfen** (2′-chloro-4′-trifluoromethylphenyl 3-ethoxy-4-nitrophenyl ether) having mainly pre-emergence activity both on grasses and dicotyledons but it can also be used post-emergence. It is safe on cotton and soybeans in dry areas. Transplanted onions are tolerant and wheat, corn and tomatoes are unaffected. It may be used post-emergence in vineyards, orchards and coffee plantations in combination with low rates of paraquat or dalapon.

Nitrophenyl ethers – structural formulae

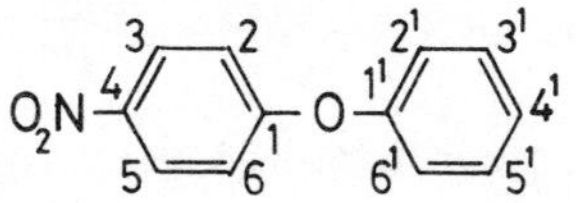

Nitrofen	$2^1 = Cl$	$4^1 = Cl$	
Fluorodifen	$2^1 = NO_2$	$4^1 = CF_3$	
Bifenox	$2^1 = Cl$	$4^1 = Cl$	$3 = COOCH_3$
Chlornitrofen	$2^1 = Cl$	$4^1 = Cl$	$6^1 = Cl$
Rohm and Haas (1)	$2^1 = Cl$	$4^1 = CF_3$	
Oxyfluorfen	$2^1 = Cl$	$4^1 = CF_3$	$3 = OC_2H_5$

NITROANILINES

All of these herbicides are active through the soil and they are applied before germination of the weed seeds. They are particularly active against grass seeds although they may affect the buds of underground stems. They inhibit cell division so that there is stunting of the shoots and roots due to swelling of the meristematic regions. The first herbicide of this group was **trifluralin**(*α,α,α*-trifluoro-2,6-dinitro-*N,N*-dipropyl-*p*-toluidine) introduced as 'Treflan' by Eli Lilly in 1960 for the control of annual weeds in cotton, groundnuts and brassicas. It is a soil-incorporated pre-emergence herbicide which prevents the germination of susceptible weed seeds and prevents weed growth by inhibition of root development. Fields must be clean cultivated before application as it is not recommended for control of established weeds. It is also available in a formulation for use in rice and with linuron as 'Chandor' or 'Trinulan' for use in winter cereals. It was followed by **benfluralin** (*N*-butyl-*N*-ethyl-*α,α,α*-trifluoro-2,6-dinitro-*p*-toluidine) as 'Balan' and 'Bonalan' from the same company in 1965 for use in lettuce, tobacco and lucerne.

In 1966 Shell marketed **nitralin** (4-methylsulphonyl-2,6-dinitro *N,N*-dipropylaniline) as 'Planavin' for use in a wide variety of crops including cotton, soybeans, groundnuts, water melons, sunflower, tobacco and others. In 1969 Eli Lilly introduced **isopropalin** (4-isopropyl-2,6-dinitro-*N,N*-dipropylaniline) as 'Paarlan' for the pre-emergence control of many grass and broad-leaved weeds in a number of crops including tobacco, tomatoes and peppers. In the same year the same Company marketed **oryzalin** (3,5-dinitro-*N, N*-dipropylsulphanilamide) as 'Dirimal' and 'Surflan' as a selective pre-emergence herbicide for the control of both annual grasses and certain broad-leaved weeds in a wide range of crops such as soybeans, cotton and groundnuts. It may be used combined with linuron and other herbicides.

Ciba Geigy introduced **profluralin** (*N*-[cyclopropylmethyl]-*α,α,α*-trifluoro-2,6-dinitro-*N*-propyl-*p*-toluidine) as 'Tolban' in 1970. It is used as a pre-emergence spray, and maize, sorghum, rice, squash, tomatoes, show excellent tolerance. It controls a wide range of grasses and some broad-leaved species. In 1971 Amchem Ltd marketed **butralin** (*N*-s-butyl-4-t-butyl-2,6-dinitroaniline) as 'Amex' and 'Tamex' which shows good selectivity for many agronomic and horticultural crops including cotton, soybeans, water melons, peanuts, cantaloupes, tobacco, snap beans and others, controlling a wide range of annual grasses including crabgrass (*Digitaria sanguinalis*), barnyard grass (*Echinochloa crus-galli*) and others; also a number of broad-leaved weeds such as pigweed (*Amaranthus retroflexus*), fat hen (*Chenopodium album*) and purslane (*Portulacea oleracea*). In the same year Borax Corp. introduced **dinitramine** (*N′,N′*-diethyl-2,6-dinitro-4-trifluoromethyl-*m*-phenylenediamine) as 'Cobex', a selective pre-plant herbicide incorporated into the soil for the control of annual grass and broad-leaved weeds at the time of germination. It has little effect on established weeds. It is used for a number of crops including cotton, soybeans, sunflowers, groundnuts and transplanted rice, cabbage, tomatoes, peppers, tobacco, swedes and turnips.

In 1974 Eli Lilly marketed **ethalfluralin** (*N*-ethyl-α,α,α-trifluoro-*N*-[2-methylallyl]-2,6-dinitro-*p*-toluidine) as 'Sonalan' a selective pre-emergence herbicide for the control of annual grasses and broad-leaved weeds in cotton and soybeans. It is soil incorporated and it controls susceptible species as they germinate. It is particularly recommended for use where *Solanum nigrum* is a problem. In the same year the American Cyanamid Co. introduced **pendimethalin** (*N*-ethyl-2,6-dinitro-*N*-propyl-3,4-xylidine) as "Prowl', 'Stomp' and 'Herbadox' for the selective control of annual broad-leaved weeds including some traditionally difficult species such as *Viola* spp., *Veronica* spp. and *Galium aparine* and annual grasses in many agronomic and horticultural crops.

Nitroanilines – structural formulae

Trifluralin	1 = $N(CH_2CH_2CH_3)_2$	4 = CF_3
Nitralin	1 = $N(CH_2CH_2CH_3)_2$	4 = CH_3SO_2
Oryzalin	1 = $N(CH_2CH_2CH_3)_2$	4 = H_2NSO_2
Isopropalin	1 = $N(CH_2CH_2CH_3)_2$	4 = $(CH_3)_2CH$
Benfluralin	1 = $N(C_2H_5)(CH_2)_3CH_3$	4 = CF_3
Profluralin	1 = N<($CH_2CH_2CH_3$)(cyclopropylmethyl, CH_2–◁)	4 = CF_3
Butralin	1 = $NHCH(CH_3)CH_2CH_3$	4 = $(CH_3)_3C$
Ethalfluralin	1 = $N(C_2H_5)CH_2{-}C(CH_3){=}CH_2$	4 = CF_3
Dinitramine	1 = $N(C_2H_5)_2$ 3 = NH_2	4 = CF_3
Pendimethalin	1 = $N(C_2H_5)CH_2CH_2CH_3$	3,4 = CH_3

CARBAMATES

Templeman and Sexton (1945, 1946) investigating a series of arylurethanes as plant growth substances found that **propham** (isopropyl *N*-phenylcarbamate) was active against a number of grasses whereas it was relatively harmless to many dicotyledonous crop plants. In 1946 it was marketed by I.C.I. Ltd for the control of annual grasses in peas and beet.

A characteristic feature of the effect of the carbamate herbicides on affected plants is mitotic disturbance which interferes with meristematic activity frequently causing deformation of roots, shoots and buds. There may also be interference with photosynthetic activity.

In 1951, Columbia Southern introduced **chlorpropham** (isopropyl-[*N*-3-chlorophenyl]carbamate) as a pre-emergence herbicide for the control of many weeds in bulb crops, soft fruits and some vegetables. It is often used in combination with other herbicides. It may also be used as a sprout inhibitor in potatoes.

One major problem facing farmers had long been the control of annual grasses such as wild oats (*Avena fatua*) and blackgrass (*Alopecurus myosuroides*) in cereals and the introduction for use post-emergence of **barban** (4-chlorobut-2-ynyl-3-chlorocarbanilate) as 'Carbyne' by the Spencer Chemical Co. in 1958, was a major step. Applied when the majority of the wild oats are in the 2-leaf stage it gives effective control without harming seedling wheat, barley and a number of other crops including lucerne, beans, sugar-beet. It may also be formulated with other herbicides. In the same year B.A.S.F. introduced **chlorbufam** (1-methylprop-2-ynyl-3-chlorophenyl carbamate) with cycluron as 'Alipur' for pre-emergence use in beet and a number of vegetable crops.

Asulam (methyl 4-aminophenylsulphonylcarbamate) was introduced by May & Baker Ltd as 'Asulox' in 1968. Its herbicidal properties have been described by Cottrell and Heywood (1965). It is a systemic herbicide active both pre- and post-emergence against a range of broad-leaved and grass weeds in a number of crops including sugar-cane, bananas, grassland, linseed, plantation crops and oil seed poppy. A special formulation 'Asulox F' has been developed for the control of wild oats in flax. Asulam is particularly active against established perennials such as docks (*Rumex* spp.) and Johnson grass (*Sorghum halepense*). It is very active against bracken (*Pteridium aquilinum*) being readily translocated from the fronds to the underground rhizomes and in the year following treatment it will give a 90% or better control.

Within the next 10 years Schering A.G. introduced many new

carbamates including (1) **phenmedipham** (methyl *N*-[3-m-tolylcarbamoyloxy]phenylcarbamate) as 'Betanal' in 1968, the first post-emergence weedkiller available for the sugar-beet crop. It is used after emergence of both crop and weeds and shows a high degree of crop safety on a wide range of sugar-beet weeds including fat hen (*Chenopodium album*), chickweed (*Stellaria media*), charlock (*Sinapis arvensis*) and speedwell (*Veronica* spp.). (2) **Desmedipham** (ethyl *N*-[3-phenylcarbamoyloxy]phenylcarbamate) as 'Betanal A.M.' also for use post-emergence for the control of broad-leaved weeds in beet crops (desmedipham and phenmedipham are often used in combination). (3) **phenisopham** (isopropyl 3-[*N*-ethyl-*N*-phenylcarbamoyloxy]phenylcarbamate) as 'Verdinal', a post-emergence herbicide controlling a great variety of dicotyledonous weeds in cotton. It may be used in conjunction with a number of other herbicides such as dinitramine, trifluralin, alachlor and pendimethalin for the extension of the weed spectrum. (4) **SN 40 624** (*N,N*-dimethyl-*N*-[3-(*N*-methyl-*N*-phenylcarbamoyloxy)phenyl]urea) controls a large number of mono- and dicotyledonous weeds and is particularly suitable for pre-emergence use in cereals. Good pre-emergence selectivity is reported in groundnuts, peas, field beans and soybeans. (5) **SN 40 454** (3-methoxycarbonylaminophenyl-*N*-methyl-*N*-phenylcarbamate) controls a wide spectrum of dicotyledonous weeds but is less effective against grasses. It may be used pre-emergence in groundnuts, cotton and soybeans and as a directed post-emergence spray in cotton and soybeans. (6) **SN 45 311** (3-methoxycarbonylaminophenyl-*N*-[3-methyl-2-butylphenyl]carbamate) controls a wide range of weeds and promising results have been obtained in groundnuts, maize, beans and peas when the herbicide is used pre-emergence and as a post-emergence directed spray in maize.

Carbamates – structural formulae

NHCOOR (benzene ring, positions 1–6)

Propham	R = $CH(CH_3)_2$	
Chlorpropham	R = $CH(CH_3)_2$	3 = Cl
Barban	R = $CH_2C{\equiv}CCH_2Cl$	3 = Cl
Chlorbufam	R = $CH(CH_3)(C{\equiv}CH)$	3 = Cl

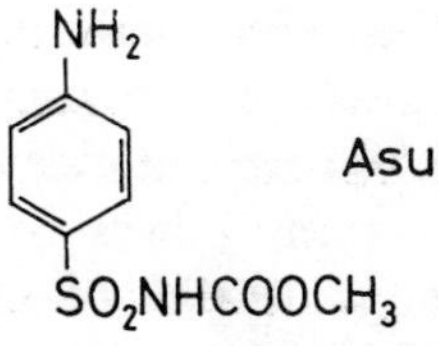

NH_2 (benzene ring) $SO_2NHCOOCH_3$ — Asulam

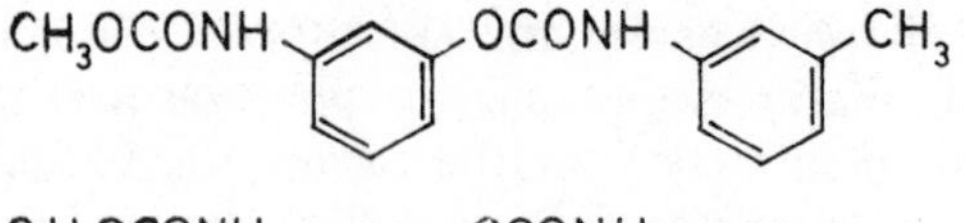

CH_3OCONH $OCONH$ CH_3 Phenmedipham

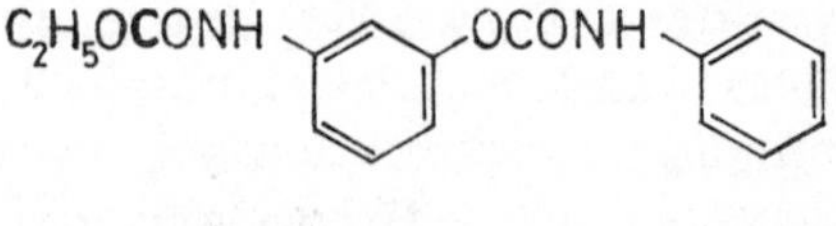

C_2H_5OCONH $OCONH$ Desmedipham

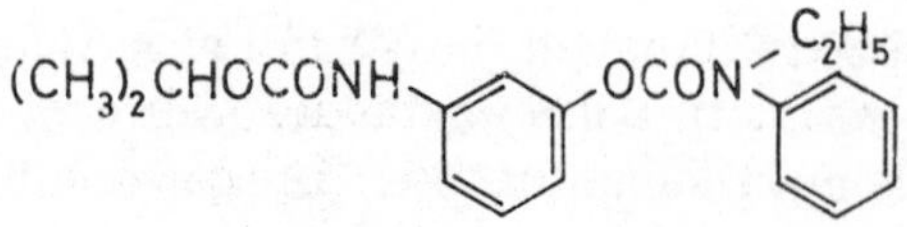

$(CH_3)_2CHOCONH$ $OCON$ C_2H_5 Phenisopham

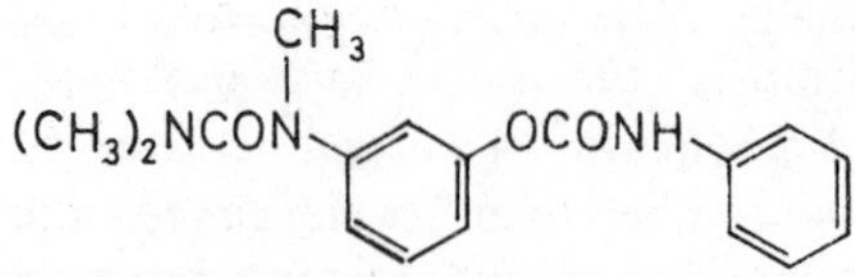

CH_3 $(CH_3)_2NCON$ $OCONH$ SN 40 624

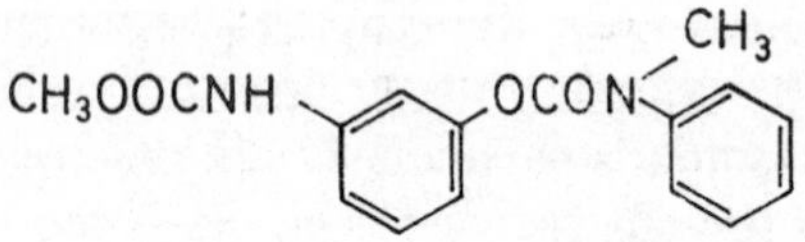

CH_3OOCNH $OCON$ CH_3 SN 40 454

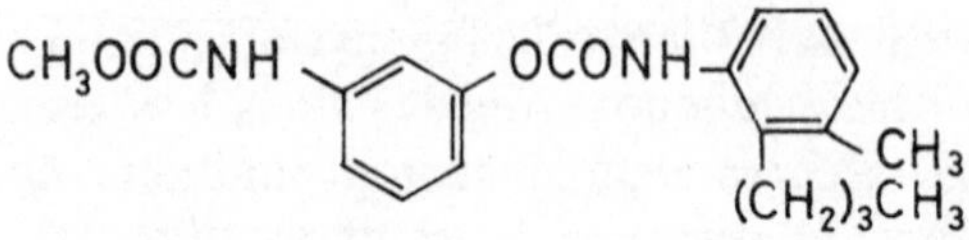

CH_3OOCNH $OCONH$ CH_3 $(CH_2)_3CH_3$ SN 45 311

THIOCARBAMATES

The first thiocarbamates were marketed by Stauffer A.G. in 1954 and these were followed by a long series of herbicides from this company for use mainly as pre-plant incorporation into the soil. Except where otherwise stated the compounds mentioned in this section were introduced by the Stauffer Chemical Co.

All thiocarbamates are active through the soil. Affected plants show an inhibition of shoots, buds and roots. Foliage is generally darker green and shiny due to lack of wax. Grasses may fail to emerge from the coleoptile forming a loop with the leaf tip fixed within the coleoptile (Silk *et al.*, 1977). **EPTC** (*S*-ethyl-*N,N*-dipropylthiocarbamate), 'Eptam', was introduced in 1954. It kills germinating seeds, a number of annuals and inhibits bud development in the underground organs of perennial weeds such as couch grass (*Agropyron repens*) and sedges (*Cyperus* spp.). It may be used soil incorporated 3 weeks before planting potatoes, field beans, sugar-beet and others. Stauffer have also marketed a combination of 'Eptam' and **R-25 788** (*N,N*-diallyl-

2,2-dichloroacetamide) as 'Eradicane'. R-25 788 increases the tolerance of maize to 'Eptam'. **Vernolate** (*S*-propyl-*N*,*N*-dipropylthiocarbamate) was marketed as 'Vernam' in the same year for pre-emergence weed control in soybeans, peanuts, sweet potatoes and tobacco. It controls annual grasses and many annual broad-leaved weeds as their seeds germinate by interfering with normal germination and seedling development. These were folowed by **butylate** (*S*-ethyl-*N*,*N*-di-isobutylthiocarbamate) as 'Sutan' for the control of germinating grass and broad-leaved weeds in maize; **pebulate** (*S*-propyl-*N*-butyl-*N*-ethylthiocarbamate) as 'Tillam' for soil incorporation in crops of sugar-beet and tobacco where it is very effective in controlling grass weeds and certain broad-leaved weeds such as red-root pigweed (*Amaranthus retroflexus*), purslane (*Portulacea oleracea*) and fat hen (*Chenopodium album*); **cycloate** (*S*-ethyl-*N*-cyclohexyl-*N*-ethylthiocarbamate) as 'Ro-Neet' for the control of annual grasses especially *Avena fatua* in beet crops; and **molinate** (*S*-ethyl-*N*,*N*-hexamethylenethiocarbamate) as 'Ordram' for pre- and post-emergence weed control in rice. It is particularly active against barnyard grass (*Echinochloa* spp.)

Monsanto introduced three important thiocarbamates beginning with **sulfallate** (*S*-2-chloroallyl-*N*,*N*-diethyldithiocarbamate) as 'Vegadex' in 1954 for the control of certain annual grass and broad-leaved weeds in a wide range of vegetable crops by application during or immediately after planting; **di-allate** (*S*-2,3-dichloroallyl-*N*,*N*-di-isopropylthiocarbamate) as 'Avadex', a volatile herbicide for pre-plant control of *Avena fatua* and *Alopecurus myosuroides* in brassica and beet crops in 1960; and **tri-allate** (*S*-[2,3,3-trichloroallyl]-*N*,*N*-di-isopropylthiocarbamate) in 1961 as 'Avadex BW' and 'Fargo' for the control of these grasses and others in cereals and peas. The latter may also be effective when applied post-emergence (as granules) to wild oats from the 1–2 leaf up to the tillering stage.

In 1970 **thiobencarb** (also known as benthiocarb) (*S*-4-chlorobenzyl-*N*,*N*-diethylthiocarbamate) was introduced by the Kumiai Chemical Industry Co. Ltd as 'Saturn' and 'Bolero' and by Chevron Chemical Co. as 'Bolero'. It is an important herbicide for the control of weeds in rice showing very high selectivity between rice and barnyard grass (*Echinochloa crus-galli*). In addition it controls many other grass, cyperaceous and broad-leaved weeds. It can be used from pre-emergence until the 2-leaf stage of the weeds and may be applied to the surface of the water 3–5 days before or 5–10 days after sowing (direct seeded rice) or to the water 3–7 days after transplanting. It may be used in conjunction with propanil as 'Satunil'. The mixture is synergistic acting rapidly and strongly and inhibiting the emergence of weeds over a long period. In 1972 Montedison marketed **Tiocarbazil** (*S*-benzyl-*N*,*N*-di-s-butylthiocarbamate) as 'Drepamon' for use pre- and post-emergence in rice particularly for the control of *Echinochloa* spp., *Cyperus* spp., *Lolium perenne* and *Leptochloa fascicularis*.

Thiocarbamates – structural formulae

EPTC	$(CH_3CH_2CH_2)_2NCOSC_2H_5$
Vernolate	$(CH_3CH_2CH_2)_2NCOSCH_2CH_2CH_3$
Butylate	$[(CH_3)_2CHCH_2]NCOSC_2H_5$
Pebulate	$(C_2H_5)(C_4H_9)N-COSCH_2CH_2CH_3$
Cycloate	$C_6H_{11}(H_5C_2)N-COSC_2H_5$
Molinate	$(CH)_6NCOSC_2H_5$
Sulfallate	$(C_2H_5)_2N-CSSCH_2CCl=CH_2$
Di-allate	$[(CH_3)_2CH]_2NCO-S-CH_2CCl=CHCl$
Tri-allate	$[(CH_3)_2CH]_2NCOSCH_2CCl=CCl_2$
Thiobencarb	$(C_2H_5)_2NCOSCH_2-C_6H_4-Cl$
Tiocarbazil	$[C_2H_5(CH_3)CH]_2NCOSCH_2-C_6H_5$

UREAS

In 1951 Bucha and Todd reported that a substituted urea, monuron, was very active in the control of many weeds especially grasses. This led to the production of a series of substituted urea herbicides for use in a variety of situations ranging from 'total' weed control in non-crop areas to selective control in a variety of crops, their persistence in the soil depending on their solubility and the extent to which they are bound in the soil. They are absorbed by both roots and by foliage. The general effect is a leaf-tip die-back followed by progressive chlorosis and inhibition of growth with 'waterlogging' effects (Bucha and Todd, 1951).

Although **dichloralurea** (1,3-bis[2,2,2-trichloro-1-hydroxyethyl]-urea), introduced by the Union Carbide Corp. in 1950 for pre-emergence control in annual grasses for a number of crops, was the first urea herbicide, it has now been superseded and is little used.

In 1952 Du Pont introduced **monuron** (3-[4-chlorophenyl]-1,1-dimethylurea) as 'Telvar'. It was originally used mainly as a 'total' herbicide on non-crop land where it will control most annual and perennial weeds (and is still widely used for this purpose) but because of its low solubility in water (230 p.p.m.) it remains largely in the top layer of soil and can therefore be used as a selective pre-emergence weedkiller in deep-rooted crops such as citrus trees and cotton. It was followed by **diuron** (3-[3,4-dichlorophenyl]-1,1-dimethylurea) as 'Karmex' produced by the same company in 1954. Having an even lower solubility in water (42 p.p.m.) than monuron it can, besides being a very useful total weedkiller, also be used as a selective pre-emergence herbicide on a wide range of crops. **Fenuron** (1,1-dimethyl-3-phenylurea) as 'Dybar' and **neburon** (1-butyl-3-[3,4-dichlorophenyl]-1-methylurea) as 'Kloben' were introduced by Du Pont in 1957. The former is used for the control of woody plants by basal application and also with chlorpropham for the pre-emergence control of many weeds in vegetable crops and the latter for pre-emergence use in a variety of crops including wheat and lucerne. It has a low solubility in water (5 p.p.m.) and this with a high degree of adsorption means that little neburon moves beyond the first inch of soil.

In 1958 two new substituted ureas were introduced, **monolinuron** (3-[4-chlorophenyl]-1-methoxy-1-methylurea) by Hoechst as 'Aresin' and 'Afesin' and **cycluron** (3-cyclo-octyl-1,1-dimethylurea) by B.A.S.F. Monolinuron is used as a pre-emergence herbicide for the selective control of broad-leaved weeds and annual grasses on asparagus, beans, potatoes, wheat, ornamentals and ornamental shrubs and in orchards and vineyards. Cycluron is used in mixture with chlorbufam as 'Alipur', a pre-emergence herbicide for the control of weeds, including *Polygonum* spp., *Atriplex* spp., *Matricaria* spp. and *Poa* spp., in sugar-beet and other vegetables. **Fluometuron** (1,1-dimethyl-3-[α,α,α-trifluoro-*m*-tolyl]urea) as 'Cotoran', **chloroxuron** (3-[4-(4-chlorophenoxy)phenyl]-1,1-dimethylurea) as 'Tenoran' and **linuron** (3-[3,4-dichlorophenyl]-1-methoxy-1-methylurea) as 'Afalon' and 'Lorox' were all introduced in 1960, the first two by Ciba Geigy, and the last by Hoechst and Du Pont. Fluometuron, absorbed by the roots, is used for the control of weeds in cotton; chloroxuron absorbed by roots and leaves controls chiefly dicotyledonous weed seedlings in strawberries, soybeans, vegetables and ornamentals; and linuron is used for pre-and post-emergence in a number of crops including carrots, winter wheat, and for grain, maize and soybeans (pre-emergence only). It is absorbed mainly by the roots but also by the foliage.

Ciba Geigy produced **chlorbromuron** (3-[4-bromo-3-chlorophenyl]-1-methoxy-1-methylurea) as 'Maloran' in 1961 as a selective herbicide with contact and residual action for use in potatoes, carrots and parsnips. It can be used pre- or post-emergence and controls many annual broad-leaved weeds and grasses. **Metobromuron** (3-[4-bromophenyl]-1-methoxy-1-methylurea) as 'Patoran' was produced by the same Company in 1963 for pre-emergence weed control of potatoes, soybeans and transplanted tomatoes. It controls many annual broad-leaved weeds and some annual grasses. The year 1964 saw the beginning of a number of very important substituted ureas for selective use in cereals against grass weeds. **Siduron** (1-[2-methylcyclohexyl]-3-phenylurea) marketed as 'Tupersan' was produced by Du Pont in this year. It is toxic to many annual grasses and crabgrass (*Digitaria sanguinalis*) and it is used for pre-emergence control in turf, in bluegrass (*Poa pratensis*), fescue, redtop, smooth brome, perennial ryegrass and others which tolerate it well. It may also be used in cereals and many dicotyledonous crops. **Buturon** (3-[4-chlorophenyl]-1-methyl-1-[1-methylprop-2-ynyl]urea) produced by B.A.S.F. in 1966 as 'Eptapur' is used for the control of shallow germinating grasses and other weeds in cereals and maize. In the same year **benzthiazuron** (1-[benzothiazol-2-yl]-3-methylurea) was introduced by Bayer A.G. as 'Gatnon', a pre-emergence herbicide for use on sugar-beet.

In 1968 **metoxuron** (3-[3-chloro-4-methoxyphenyl]-1,1-dimethylurea) was marketed as 'Purivel' by Sandoz A.G. for use post-emergence in cereals (and carrots) against a variety of important grass weeds including blackgrass (*Alopecurus myosuroides*), wild oats (*Avena* spp.), ryegrass (*Lolium perenne*) and canary grass (*Phalaris canariensis*) and most annual dicotyledons. In the same year Bayer A.G. marketed **methabenzthiazuron** (1-[benzothiazol-2-yl]-1,3-dimethylurea) as 'Tribunil' for pre- and early post-emergence use in cereals. Applied in autumn it will control *Alopecurus myosuroides*, *Poa annua* and *P. trivialis*, *Sinapis arvensis*, *Stellaria media*, *Papaver rhoeas*, *Veronica* spp. and others.

In 1969 Ciba Geigy marketed **chlortoluron** (3-[3-chloro-*p*-tolyl]-1,1-dimethylurea) as 'Dicurane', a residual and foliar acting herbicide for the control of annual grass weeds, in particular, blackgrass (*Alopecurus myosuroides*) and wild oats (*Avena* spp.) and certain broad-leaved annual weeds in certain varieties of winter wheat and winter barley. In 1970 Ciba Geigy produced **difenoxuron** (3-[4-(4-methoxyphenoxy)phenyl]-1,1-dimethylurea) as 'Lironion' for selective use in onions. Another very important herbicide for the control of annual grasses including *Alopecurus myosuroides*, *Avena fatua* and *Poa annua* and many annual broad-leaved weeds in cereals was marketed by three companies – Hoechst (as 'Arelon'), Ciba Geigy (as 'Graminon') and Rhône-Poulenc (as 'Tolkan') in 1972. This was **isoproturon** (3-[4-isopropylphenyl]-1,1-dimethylurea). It may be used alone or in association with dinoterb or the hydroxybenzonitriles (ioxynil and bromoxynil). In the same year Ciba Geigy introduced **thiazafluron** (1,3-dimethyl-1-[5-trifluoromethyl-1,3,4-thiadiazol-2-yl] urea) as 'Erbotan', a total weedkiller for industrial sites controlling annual and perennial herbs as well as woody plants. In 1973 Bayer A.G. introduced **isocarbamid** (*N*-isobutyl-2-imidazolidone-1-carboxamide) as 'Merpelan AZ' for use pre-emergence in beet crops.

In 1974 the Veliscol Chemical Co. introduced **buthidazole** (3-[5-t-butyl-1,3,4-thiadiazole-2-yl]-4-hydroxy-1-methyl-2-imidazolidone) as 'Ravage', a residual herbicide for use in non-crop areas and also as a selective herbicide in maize, sugar-cane and pineapple. In the same year Eli Lilly marketed **tebuthiuron** (1-[5-t-butyl-1,3,4-thiadiazol-2-yl]-1,3-dimethylurea) as 'Spike' and 'Perflan' also for total weed control, though it can also be used for the control of woody plants in pastures and for weed control in sugar-cane. **Ethidimuron** (1-[5-ethylsulphonyl-1,3,4-thiadiazol-2-yl]-1,3-dimethylurea) introduced in 1975 as 'Ustilan' by Bayer A. G. is used in similar situations.

In 1980 the Sumitomo Chemical Co. Ltd introduced **S-3552** (*N*′-4-[4-methylphenethyloxy]phenyl-*N*-methoxy-*N*-methylurea) as a selective post-emergence herbicide for use in soybeans. It is active against such troublesome dicotyledonous weeds as *Xanthium sturamium*, *Ipomoea purpurea*, *Abutilon theophrasti*, *Datura stramonium*, *Ambrosia* spp., *Euphorbia helioscopia* and many others. It is weaker against grass weeds.

Ureas – structural formulae

(Benzene ring, positions 1–6, with position 1 bearing) $-NHCON(CH_3)_2$

Fenuron	unsubstituted	
Monuron		4 = Cl
Diuron	3 = Cl	4 = Cl
Fluometuron	3 = CF_3	
Chlortoluron	3 = Cl	4 = CH_3
Metoxuron	3 = Cl	4 = CH_3O
Isoproturon		4 = $(CH_3)_2CH$

Chloroxuron 4 = Cl–C_6H_4–O

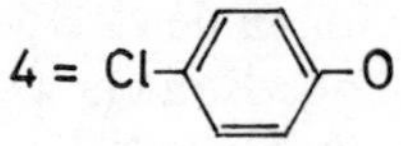

Difenoxuron 4 = CH_3O–C_6H_4–O

(General structure: 3,2,4,1,5,6-substituted phenyl–NHCON(X)(Y))

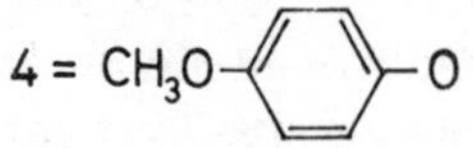

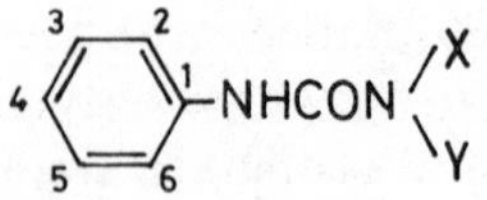

Neburon	3 = Cl	4 = Cl	X = CH_3	Y = $CH_3(CH_2)_3$
Monolinuron		4 = Cl	X = CH_3	Y = CH_3O
Linuron	3 = Cl	4 = Cl	X = CH_3	Y = CH_3O
Chlorbromuron	3 = Cl	4 = Br	X = CH_3	Y = CH_3O
Metobromuron		4 = Br	X = OCH_3	Y = CH_3
Buturon		4 = Cl	X = CH_3	Y = CH≡C–CH(CH_3)
SN 40 624	3 = C_6H_5–N(CH_3)COO			
Siduron			X = 2-methylcyclohexyl (CH_3)	
S-3552		4 = H_3C–C_6H_4–$(CH_2)_2O$	X = CH_3	Y = CH_3O

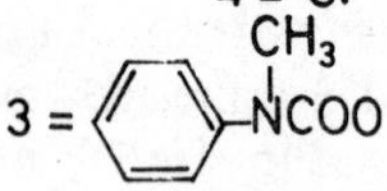

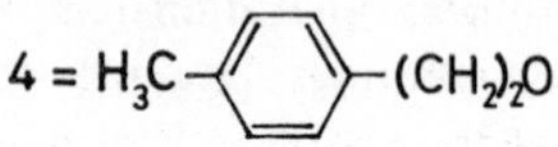

$(CCl_3CH(OH)NH)_2CO$

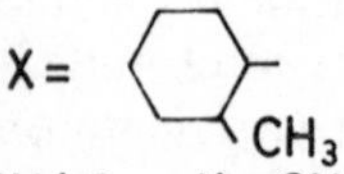

Dichloralurea

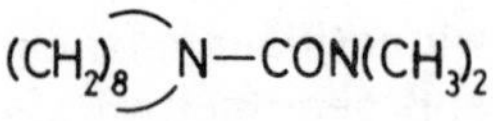

$(CH_2)_8$ N–CON$(CH_3)_2$

Cycluron

(Benzothiazol-2-yl)–N(R)CONHCH$_3$

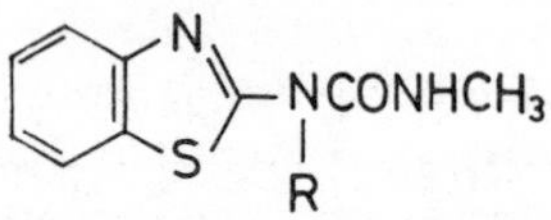

Benzthiazuron R = H

Methabenzthiazuron R = CH_3

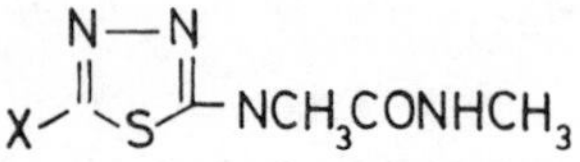

X–(1,3,4-thiadiazol-2-yl)–NCH_3CONHCH$_3$

Thiazafluron X = CF_3

Tebuthiuron X = $(CH_3)_3C$

Ethidimuron X = $C_2H_5SO_2$

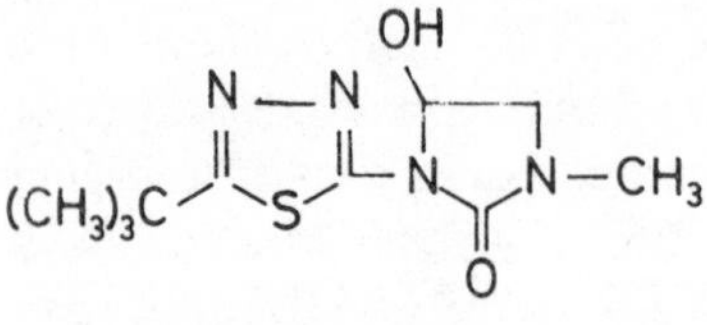

$(CH_3)_3C$–(thiadiazolyl: N—N, S)–N(imidazolidinone, OH, C=O)N–CH_3

Buthidazole

HN(imidazolidinone, C=O)NCONHCH$_2$CH$(CH_3)_2$

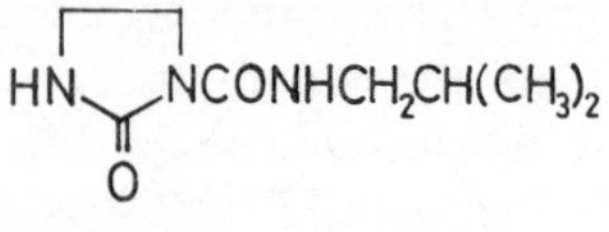

Isocarbamid

HETEROCYCLIC NITROGEN COMPOUNDS – TRIAZINES

According to Brian (1976) the preparation of the compound now known as simazine was made by Hoffman (1885). Further triazine compounds were synthesised by Pearlman and Banks (1948) but it was in the 1950s that J. R. Geigy S.A. (now Ciba Geigy A.G. and hereafter referred to as such) introduced triazine compounds as herbicides. This was followed by the synthesis of a long series of triazines by Gysin and Knüsli (1957, 1958, 1960) in Geigy's laboratories. It was to prove a most fruitful seam and many were shown to be potent herbicides. In general they have little effect on germination. They are taken up by the roots or leaves and due to an inhibition of the Hill reaction of photosynthesis (see chapter 5) affected plants turn yellow and necrotic symptoms develop. With a few exceptions the symmetrical triazines have got substituted amino groups at two of the carbon atoms while the third carbon has a chloro, a thioether or a methoxy function.

The chloro compounds (Cl) end in *azine*, the thioethers (–S–) end in *tryne(e)* and the methoxy ones (CH_3O) in *ton*.

In the list of herbicides which follows all were introduced by Ciba Geigy A.G. unless stated otherwise.

The first commercial triazine **simazine** (2-chloro-4,6-bis[ethyl-amino]-1,3,5-triazine) was introduced as 'Gesatop' in 1956 for the selective residual pre-emergence control of a great many annual grass and broad-leaved weeds in a variety of deep-rooted crops (including citrus fruits, coffee, tea and cocoa), due in part to its low solubility in water (3.5 mg/l at 20° C). It is also used for the control of most annual and perennial weeds in non-crop areas. It is remarkably selective for use on maize because of the ability of this crop to degrade it non-enzymically to the non-active hydroxy derivative (Castelfranco *et al.*, 1961).

The second introduction **atrazine** (2-chloro-4-ethylamino-6-isopropylamino-1,3,5-triazine), introduced as 'Gesaprim' and 'Primatol' in 1958, is both foliar and soil acting being taken up both by leaves of emerged weeds and by the roots of weed seedlings emerging after spraying. In maize, where it is also degraded in a manner similar to simazine, it is preferred to the latter especially in dry years. Being more water soluble (30 mg/l at 20° C) it is more suitable for the dry soils on which this crop is grown where it will effectively control couch grass (*Agropyron repens*) and other perennial grasses. It is also used in raspberries and roses, for selective use in coniferous forests and for non-selective use on non-crop land and industrial sites.

Prometon (2,4-bis[isopropylamino]-6-methoxy-1,3,5-triazine), the first methoxy triazine, was introduced as 'Primatol' and 'Pramitol' in 1959 as a non-selective herbicide for the control of annual and perennial weeds in non-agricultural land.

In 1960 **propazine** (2-chloro-4,6-bis[isopropylamino]-1,3,5-triazine) was introduced as 'Gesamil' and 'Milogard' as a pre-emergence herbicide for the control of weeds in millet and umbelliferous crops.

The first of the important thioether group, **prometryne** (2,4-bis[isopropylamino]-6-methylthio-1,3,5-triazine), was marketed as 'Gesagard' and 'Caparol' in 1962 for the selective pre-or post-emergence control of most annual weeds in cotton, sunflowers, peas, early potatoes, carrots and certain other vegetable crops. It has both a contact and residual action being absorbed by leaves of emerged weeds and by the roots of weed seedlings.

It was followed in 1964 by **desmetryne** (2-isopropylamino-4-methylamino-6-methylthio-1,3,5-triazine) as 'Semeron' for the selective post-emergence control of many annual weeds including fat hen (*Chenopodium album*) in many brassica crops. Perennial weeds are only temporarily affected since persistence of the herbicide in the soil is short.

In the same year **ametryn** (2-ethylamino-4-isopropylamino-6-methylthio-1,3,5-triazine) was marketed as 'Gesapax', a pre- and post-emergence selective herbicide for the control of most annual monocotyledonous and dicotyledonous weeds, as well as some perennial species, in a number of established crops including banana, citrus fruits, coffee, tea, pineapple and sugar-cane. It is also used with atrazine as 'Gesapax combi'.

In the following year **methoproptryne** (2-isopropylamino-4-[3-methoxypropylamino]-6-methylthio-1,3,5-triazine) was introduced as 'Gesaran' for use post-emergence with simazine for the control of grasses in winter-sown cereals.

Four herbicides were marketed in 1966: (1) **terbuthylazine** (2-t-butylamino-4-chloro-6-ethylamino-1,3,5-triazine) as 'Gardoprim' for pre-emergence use in vineyards, citrus and apple orchards and sorghum; (2) **terbutryne** (2-t-butylamino-4-ethylamino-6-methylthio-1,3,5-triazine) as 'Igran', 'Clarosan' and 'Prebane' for the selective pre-emergence control of most annual dicotyledonous weeds (including many difficult weeds such as chickweed (*Stellaria media*), mayweeds (*Matricaria* spp.), speedwells (*Veronica* spp.), annual meadow grass (*Poa annua*) and rough stalked meadow grass (*Poa trivialis*)) in winter wheat and winter barley; (3) **secbumeton** (2-s-butylamino-4-ethylamino-6-methoxy-1,3,5-triazine) as 'Etazine' is taken up by leaves and roots and it controls annual and perennial monocotyledonous and dicotyledonous weeds. It is used in lucerne (alfalfa); (4) **terbumeton** (2-t-butylamino-4-ethylamino-6-methoxy-1,3,5-triazine) as 'Caragard'. Mixed with terbuthylazine it is known as 'Caragard combi' and is used for post-emergence control of annual weeds and a number of perennials in orchards, vineyards and in forestry.

In 1967 **aziprotryn** (2-azido-4-isopropylamino-6-methylthio-1,3,5-triazine) was introduced as 'Mesoranil' and 'Brasoran'. It is both foliar and soil acting and was introduced for weed control in Brussels sprouts, cabbage, peas, onions and leeks.

Dimethametryn (2-[1,2-dimethylpropylamino]-4-ethylamino-6-methylthio-1,3,5-triazine) was introduced in 1969. In combination with piperophos it is marketed as 'Avirosan' for the control of the most important monocotyledonous and dicotyledonous weeds in rice.

In 1971 Shell Chemicals introduced **cyanazine** (2-[4-chloro-6-ethylamino-1,3,5-triazine-2-ylamino]-2-methylpropionitrile) as 'Bladex' and 'Fortrol', a residual herbicide for the control of some of the more 'difficult' annual grass and dicotyledonous weeds in winter wheat and winter barley, for pre-emergence use on maize, for post-emergence use in onions and for pre- and post-emergence control of weeds in peas. It is combined with atrazine as 'Holtox' for the control of grasses and dicotyledonous weeds in young forest and woodland plantations of certain species and for use pre-emergence in maize and for weed control in raspberries.

In the same year Ciba Geigy continued their prolific output with **dipropetryn** (2-ethylthio-4,6-bis[isopropylamino]-1,3,5-triazine) as 'Cotofor' for use pre-emergence in the control of numerous dicotyledonous and several monocotyledonous weeds in cotton and melons. In 1972 Fisons Ltd marketed **trietazine** (2-chloro-4-diethylamino-6-ethylamino-1,3,5-triazine) in combination with other herbicides for weed control in potatoes and peas. Trietazine had first been discovered by Ciba Geigy A.G.

In the same year Ciba Geigy continued their prolific output with (2,4-bis[ethylamino]-6-methylthio-1,3,5-triazine) which is now marketed as 'Gybon' for use as a mixture with thiobencarb for the control of dicotyledonous weeds in rice and (2) **procyazine** (2-chloro-6-[2-cyano-2-n-propylamino]-4-cyclopropylamino-s-triazine) for the control of annual grasses in maize.

In 1972 Nitrkemia Ipartelepek introduced two new triazines – **eglinazine-ethyl**, the ethyl ester of (*N*-[4-chloro-6-ethylamino-1,3,5-triazine-2-yl]glycine) for use pre-emergence in cereals and **proglinazine-ethyl**, the ethyl ester of (*N*-[4-chloro-6-isopropylamino-1,3,5-triazine-2-yl]glycine) for use pre-emergence in maize.

In 1971 an unsymmetrical triazine, **metribuzin** (4-amino-6-t-butyl-4,5-dihydro-3-methylthio-1,2,4-triazine-5-one), was introduced by Bayer as 'Sencor' and by Du Pont as 'Lexone' for use pre- and post-emergence in the control of grass and broad-leaved weeds in potatoes, tomatoes, peppers, lucerne, raspberry and others. It is less effective against deep-rooted perennials. Affected leaves are pale due to an interference with photosynthesis. It gives good overall weed control, particularly from before crop emergence when the majority of weeds are in the cotyledon to two true leaf stage. It deals with a number of 'problem' weeds including fat hen (*Chenopodium album*), fumitory (*Fumaria officinalis*), black bindweed (*Polygonum convolvulus*) and knotgrass (*Polygonum aviculare*) at this stage. The herbicide undergoes fast degradation in the soil.

Bayer also produced a related compound **metamitron** (4-amino-4,5-dihydro-3-methyl-6-phenyl-1,2,4-triazin-5-one) as 'Goltix' in 1975. It has a high selectivity in beet crops where it controls many annual weeds both pre- and post-emergence.

In 1980 Du Pont introduced **DPX 4189** (2-chloro-*N*-[4-methoxy-6-methyl-1,3,5-triazin-2-yl-aminocarbonyl] benzenesulphonamide), a very active herbicide with excellent crop selectivity in wheat, barley, oats and rye. It is active against most dictotyledonous weeds and gives suppression of a few grass weeds. Several species tolerant to phenoxy and benzoic herbicides are sensitive. It is taken up through both foliage and root systems.

Triazines – structural formulae

(1,3,5-triazine ring, positions 2, 4, 6)

	2	4	6
Simazine	Cl	C_2H_5NH	C_2H_5NH
Atrazine	Cl	C_2H_5NH	$(CH_3)_2CHNH$
Prometon	CH_3O	$(CH_3)_2CHNH$	$(CH_3)_2CHNH$
Propazine	Cl	$(CH_3)_2CHNH$	$(CH_3)_2CHNH$
Prometryne	CH_3S	$(CH_3)_2CHNH$	$(CH_3)_2CHNH$
Desmetryne	CH_3S	CH_3NH	$(CH_3)_2CHNH$
Ametryn	CH_3S	C_2H_5NH	$(CH_3)_2CHNH$
Methoproptryne	CH_3S	$(CH_3)_2CHNH$	$CH_3O(CH_2)_3NH$
Terbuthylazine	Cl	C_2H_5NH	$(CH_3)_3CNH$
Terbutryne	CH_3S	C_2H_5NH	$(CH_3)_3CNH$

Secbumeton	CH_3O	C_2H_5NH	$C_2H_5CH(CH_3)NH$
Terbumeton	CH_3O	C_2H_5NH	$(CH_3)_3CNH$
Aziprotryn	CH_3S	$(CH_3)_2CHNH$	N_3
Dimethametryn	CH_3S	C_2H_5NH	$(CH_3)_2CHCH(CH_3)NH$
Cyanazine	Cl	C_2H_5NH	$NCC(CH_3)_2NH$
Dipropetryn	C_2H_5S	$(CH_3)_2CHNH$	$(CH_3)_2CHNH$
Trietazine	Cl	C_2H_5NH	$(C_2H_5)_2N$
Simetryn	CH_3S	C_2H_5NH	C_2H_5NH
Eglinazine-ethyl	Cl	C_2H_5NH	$C_2H_5OOCCH_2NH$
Proglinazine-ethyl	Cl	$(CH_3)_2CHNH$	$C_2H_5OOCCH_2NH$
DPX 4189	CH_3O	CH_3	Cl-C_6H_4-$SO_2NHCONH$
Procyazine	Cl	(cyclopropyl)NH	$NCC(CH_3)_2NH$

Unsymmetrical triazines

Metribuzin

Metamitron

HETEROCYCLIC NITROGEN COMPOUNDS – PYRIDINES

In 1957 and 1958 I.C.I. introduced two very important bipyridilium quaternary herbicides. Both are broad spectrum, rapidly acting causing wilt and desiccation, and are translocated to some degree, although kill usually stops at ground level.

Diquat (1,1′-ethylene-2,2′-bipyridilium ion formulated as the dibromide) under various trade names – 'Reglone', 'Weedol', 'Pathclear', is used for potato haulm desiccation, for seed crop desiccation and for aquatic weed control. The Chapman Chemical Co. market it as 'Aquacide'. **Paraquat** (1,1′-dimethyl-4,4′-bipyridilium ion normally formulated as the dichloride) as 'Gramoxone', 'Dextrone X', 'Esgram' and 'Weedol', destroys photosynthetic tissues and is used for a variety of purposes including stubble cleaning, inter-row weed control, desiccation of various crops, and killing out of old pastures which can then be resown without ploughing. It is very fast acting, the first effects being noticeable after a few hours and kill is usually completed in 3–4 days. It is quickly adsorbed on to soil (particularly clay) particles so that sowing can follow soon after application (Calderbank, 1968).

The Dow Chemical Co. has introduced three interesting foliar-applied, selective, growth-regulatory herbicides which produce symptoms on susceptible plants very similar to those produced by the auxin-type herbicides, namely tissue proliferation, epinasty, leaf curling and production of adventitious roots. They may be used in combination with other herbicides. **Picloram** (4-amino-3,5,6-trichloropyridine-2-carboxylic acid) as 'Tordon' for the control of annual weeds and deep-rooted perennials was marketed in 1963. Most grasses are resistant and it has little or no effect on cruciferous plants. Although it may be used alone it is often used in combination with 2,4-D or 2,4,5-T for the control of problem weeds in cereals, maize and sugar-cane. **Triclopyr** (3,5,6-trichloro-2-pyridinyloxyacetic acid) marketed as 'Garlon' in 1970 may be used for the control of certain 2,4-D resistant weeds in cereals, in conifer re-afforestation programmes and for the control of unwanted brush and perennial weeds in industrial areas. It is highly active against woody plants including ash (*Fraxinus*) which is normally difficult to control. **3,6-Dichloropicolinic acid** (3,6-dichloropyridine-2-carboxylic acid) marketed as 'Lontrel' is used for the control of many difficult weeds of the Leguminosae and Polygonaceae families in graminaceous and brassica crops, sugar-beet and flax. It was introduced in 1975. It has been combined with a number of other herbicides such as phenmedipham for use in sugar-beet and benazolin and propyzamide for use in oil seed rape. I.C.I. have mixed it with mecoprop as 'Seloxone', a broad spectrum herbicide for use in all winter cereals. In the same year I.C.I. released **diethamquat** (1,1′-bis[diethylcarbamoyl-methyl]-4,4′-bipyridilium cation) as a post-emergence contact selective herbicide for the control of dicotyledonous weeds in cereals and grassland. In 1976 Eli Lilly marketed **fluridone** (1-methyl-3-phenyl-5-[α,α,α-trifluoro-*m*-tolyl]-4-pyridone) as a pre-emergence herbicide for use principally on cotton where it controls a wide range of weeds. **Uniroyal S734** (2-[1-(2,5-dimethylphenyl)ethyl]sulphonyl pyridine-1-oxide) was introduced by Uniroyal Inc. in 1979 as a pre-sowing incorporated herbicide for the control of grasses and *Cyperus* spp. in cotton, soybeans, sugar-beet and potatoes.

Pyridines – structural formulae

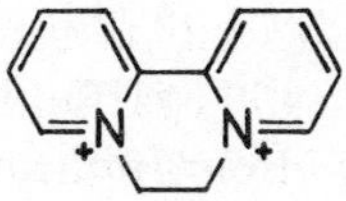

Diquat 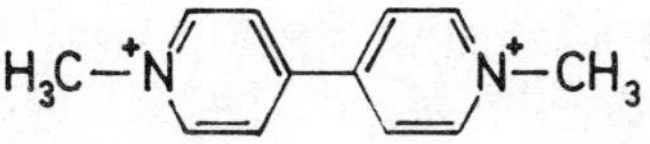 Paraquat

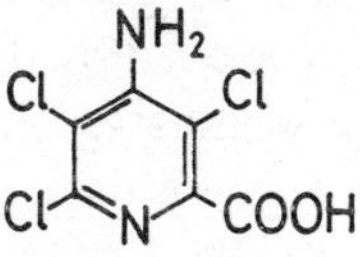

Picloram 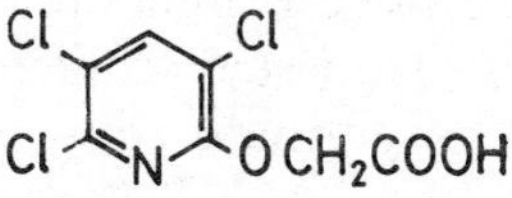 Triclopyr 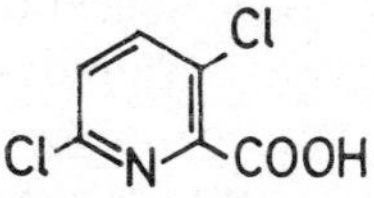 3,6-Dichloropicolinic acid

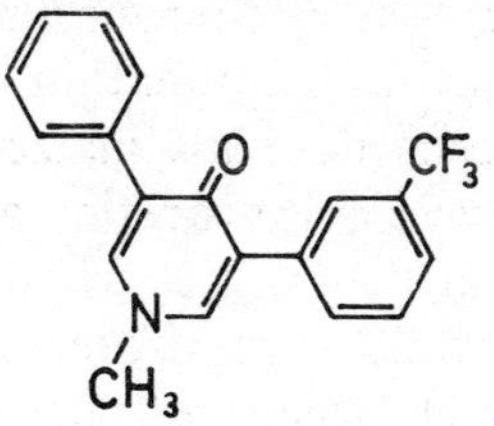

Fluridone 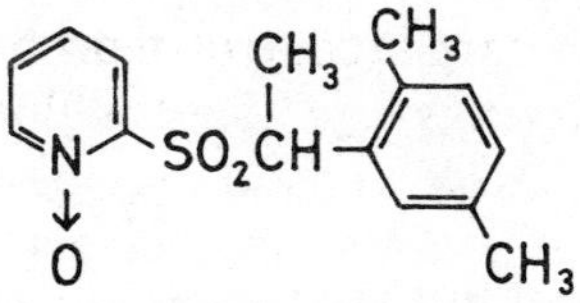 Uniroyal S734

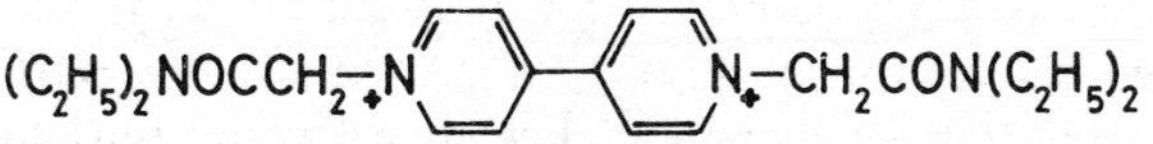

Diethamquat

HETEROCYCLIC NITROGEN COMPOUNDS – PYRIDAZINES

The first herbicide of this group to be introduced was **chloridazon** (pyrazon) (5-amino-4-chloro-2-phenylpyridazin-3-one) as 'Pyramin', by B.A.S.F. in 1962, for use pre-emergence or after the late cotyledon stage on beet crops for the control of dicotyledonous weeds and surface-germinating grasses. Combined with chlorbufam as 'Alicep' it is used for the control of weeds in onions, leeks and bulbs. Another soil-applied member of this group **3-*o*-tolyloxypyridazine** was introduced by the Sankyo Chemical Co. in 1970 as 'Dusakira' for the control of annual grasses and some other weeds in tomatoes and strawberries. **Norflurazon** (4-chloro-5-methylamino-2-[α,α,α-trifluoro-*m*-tolyl]pyridazin-3-one) was marketed as 'Zorial', 'Evital' and 'Solicam' by Sandoz A.G. in 1971 for weed control in cotton, stone fruits and cranberries being particularly active against grasses and sedges. The properties of **pyridate** (6-chloro-3-phenylpyridazin-4-yl-*S*-octyl thiocarbonate) were described by Diskus *et al.* (1976) and it was introduced by Chemie Linz for use in cereals and certain other crops being very active against *Galium aparine* and *Amaranthus retroflexus*.

Pyridazines – structural formulae

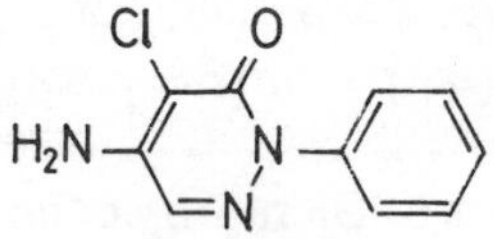

Cl O H_2N N N

Chloridazon

H_3C O N—N

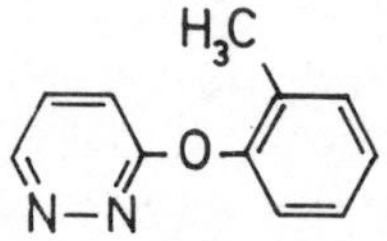

3-o-Tolyloxypyridazine

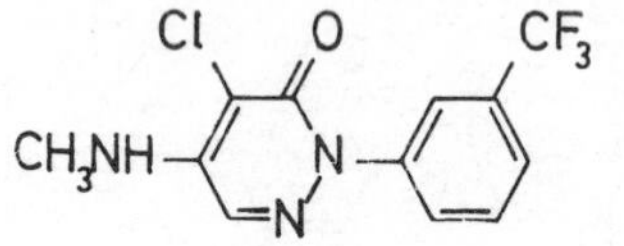

Cl O CF_3 CH_3NH N N

Norflurazon

$OCOSC_8H_{17}$ Cl N—N

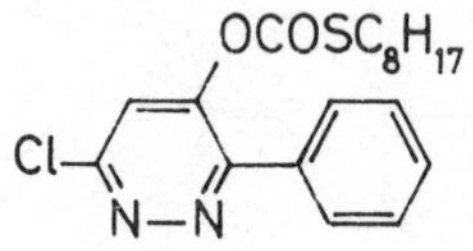

Pyridate

HETEROCYCLIC NITROGEN COMPOUNDS – PYRIMIDINES (URACILS)

These herbicides are derived from uracil. They are applied to the soil and are absorbed via the roots but eventually inhibit the Hill reaction of photosynthesis causing chlorosis and death. Thier properties have been described by Bucha *et al.* (1962). All three have been introduced by Du Pont. **Bromacil** (5-bromo-3-s-butyl-6-methyluracil) was introduced in 1963 as 'Hyvar X', a foliar and root-acting residual 'total' herbicide for use on non-agricultural land; and as a selective herbicide in raspberries and other cane fruit and in citrus and pineapple plantations. It is active against a wide range of annual and perennial weeds including established couch (*Agropyron repens*) and bent grasses (*Agrostis* spp.). It was followed in 1966 by **terbacil** (3-t-butyl-5-chloro-6-methyluracil) as 'Sinbar' for weed control in asparagus, hops, roses, sugar-cane, apple and citrus orchards. It gives good control of established couch and bent grasses, Bermuda grass (*Cynodon dactylon*) and Johnson grass (*Sorghum halepense*). **Lenacil** (3-cyclohexyl-6,7-dihydro-1*H*-cyclopentapyrimidine-2,4-dione) as 'Venzar' was marketed in the same year for the control of a wide range of weeds in beet crops, spinach and strawberries.

Pyrimidines – structural formulae

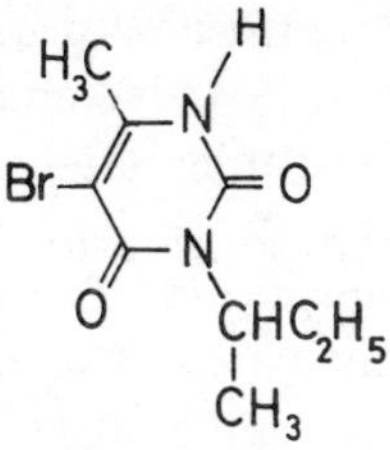

Bromacil

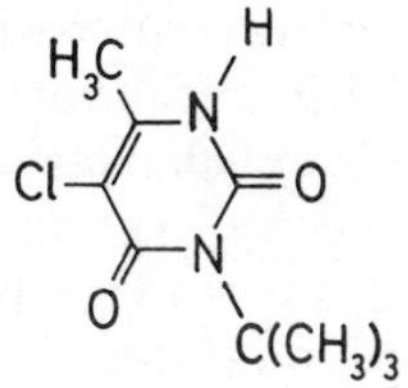

Terbacil

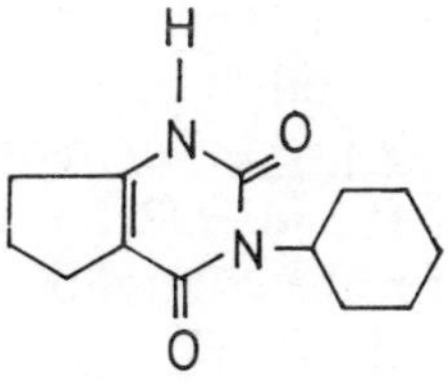

Lenacil

HETEROCYCLIC NITROGEN COMPOUNDS – UNCLASSIFIED

This group of compounds has a variety of chemical structures and types of herbicidal activity. **Aminotriazole** (1,2,4-triazol-3-ylamine) as 'Weedazol' was introduced by Amchem Ltd in 1955. It is a non-selective herbicide being absorbed by both roots and foliage; it is readily translocated being active against annual and perennial weeds where it inhibits regrowth from buds. Commercial preparations usually contain ammonium thiocyanate ('Weedazol TL') which enhances its activity. It is very effective for the control of *Agropyron repens*, *Rumex* spp. and others, and is recommended for use in fallows, autumn stubbles, as a pre-planting spring treatment and in established orchards. It may be combined with simazine as 'Amizine' for 'total' weed control in non-agricultural areas. **Benazolin** (4-chloro-2-oxobenzothiazolin-3-ylacetic acid) introduced by Boots in 1965 as 'Cornox CWK' is a post-emergence, translocated, growth regulator herbicide very active against chickweed (*Stellaria media*) and cleavers (*Gallium aparine*) and others. It synergises dicamba, and a number of products are available comprising benazolin and dicamba together with various phenoxy acid herbicides which provide broad spectrum weed control in non-undersown cereals.

In 1968 B.A.S.F. introduced **bentazone** (3-isopropyl-1*H*-benzo-2-yl-3-thiodiazin-4-one-2,2-dioxide) as 'Basagran', a post-emergence contact herbicide for the control of a number of difficult weeds, e.g. mayweeds (*Matricaria* spp.), chamomile (*Anthemis cotula*), cleavers

(*Galium aparine*), chickweed (*Stellaria media*), corn marigold (*Chrysanthemum segetum*) and nipplewort (*Lapsana communis*), in cereals. It may be used in combination with a number of other herbicides including MCPB, dichlorprop, mecoprop, MCPA and 2,4,5-T. In the same year **methazole** (2-[3,4-dichlorophenyl]-4-methyl-1,2,4-oxadiazolidine-3,5-dione) was marketed as 'Probe' by Velsicol and as 'Paxilone' by Fisons Ltd for use pre-emergence in cotton and potatoes. In 1969 **oxadiazon** (5-t-butyl-3-[2,4-dichloro-5-isopropyloxyphenyl]-1,3,4-oxidazolin-2-one) as 'Ronstar' was introduced by Rhône-Poulenc. It is active both pre- and post-emergence against grasses and broad-leaved weeds, being selective in rice, cotton, soybean, sunflower, sugar-cane, peanuts, various bulbs and transplant plants and on turf. **TO 2** (5-chloro-4-methyl-2-propionamidothiazole) was introduced by Mitsu Toatsu Chemicals Inc. in 1972. It may be used for post-emergence application in wheat for the control of many grasses including *Avena* spp., *Alopecurus aequalis, Poa annua, Digitaria ciliaris*, and many dicotyledons including *Stellaria* spp., *Chenopodium* spp., *Polygonum* spp., *Rorippa islandica* and others. Barley and oats are sensitive. **Difenzoquat** (1,2-dimethyl-3,5-diphenyl-pyrazolium ion) was introduced by Cyanamid in 1973 as 'Avenge' and 'Finaven' specifically to deal post-emergence with wild oats (*Avena* spp.) in wheat and barley. It is absorbed through the leaves and axils and translocated to meristematic regions where it disrupts cell division so that shoot growth is inhibited. **RP 23 465** (2-t-butyl-4-[2-chloro-4-(3,3-dimethylureido)phenyl]-1,3,4-oxadiazolin-5-one) was introduced by Rhône-Poulenc in 1974 for the control of a wide range of grasses and broad-leaved weeds in sugar-cane, cotton, sunflower, cereals, lucerne, groundnuts, field beans, peas and others. It shows both pre- and post-emergence activity, the latter varying with the rate applied. It may also be used for total weed control. **BTS 30 843** (1-[*N*-ethyl-*N*-propylcarbamyl]-3-propylsulphonyl-1*H*-1,2,4-triazole) was introduced by The Boots Co. in 1974. Applied pre-emergence it is active against all major annual weeds in soybeans, groundnuts, cotton, maize and small grains. It inhibits germination of weed seeds and causes twisting and stunting of seedlings. Early post-emergence applications halt the growth of seedling weeds followed by their death. **TH 2946** (2-[3-chlorobenzylthio]-5-methyl-1,3,4-oxadiazole) was introduced by Takeda Chemicals in 1974. It is a most interesting chemical in that it shows intra-specific toxicity to two rice cultivars and to two *Echinochloa* spp. It could prove to be a useful tool for studying inter-genetic, inter-specific and intra-specific selectivity between rice cultivars and *Echinochloa* spp. **HOE 19 070** (6-difluoro-methylsulphonyl-2-trifluoromethylbenzimidazole) was introduced by

Hoechst A.G. in 1975. It shows good post-emergence activity against annual broad-leaved weeds in cereals. It may be used in a mixture with dichlorprop. In 1975 Bayer A.G. marketed **isomethiozin** (6-t-butyl-4-isobutylideneamino-3-methylthio-1,2,4-triazine-5-one) as 'Tantizon', a post-emergence herbicide giving good control of numerous grass species including wild oat (*Avena fatua*, *A. ludoviciana*) and blackgrass (*Alopecurus myosuroides*) and many broad-leaved weeds in winter barley when applied in the Spring. A number of weeds such as *Galium* spp. and *Matricaria* spp. are however resistant. For them a formulation containing dichlorprop is very effective and is especially suitable for weed control on both spring barley and spring wheat.

Heterocyclic nitrogen compounds (unclassified) – structural formulae

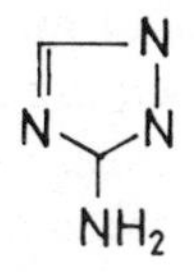

Aminotriazole

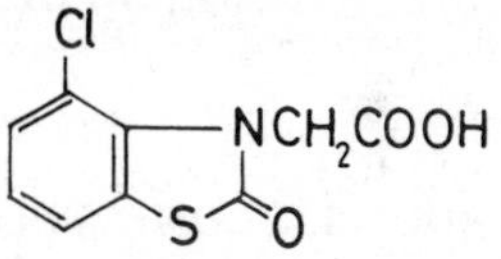

Benazolin

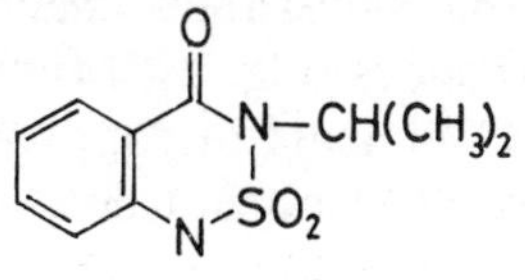

Bentazone

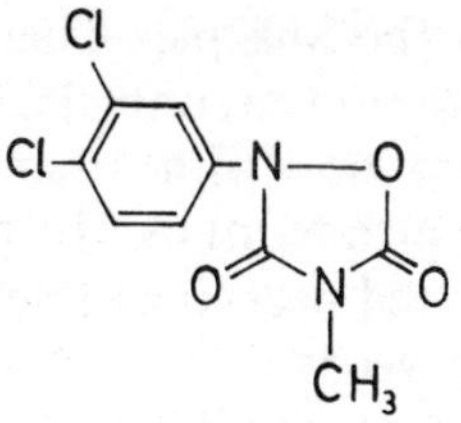

Methazole

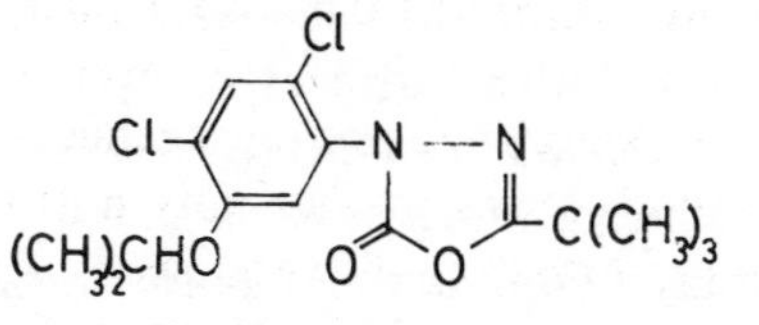

Oxadiazon

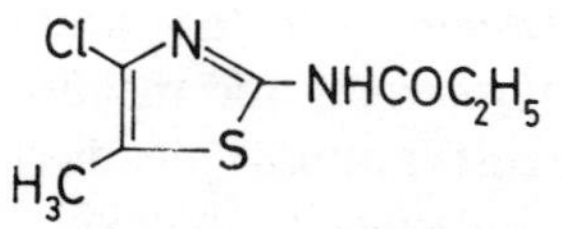

TO 2

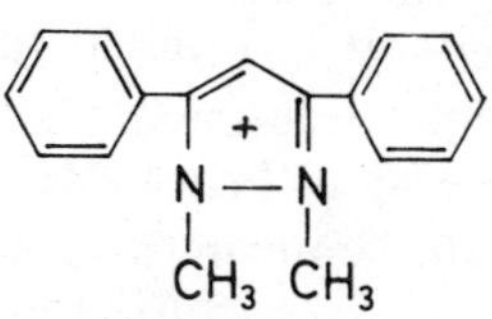

Difenzoquat

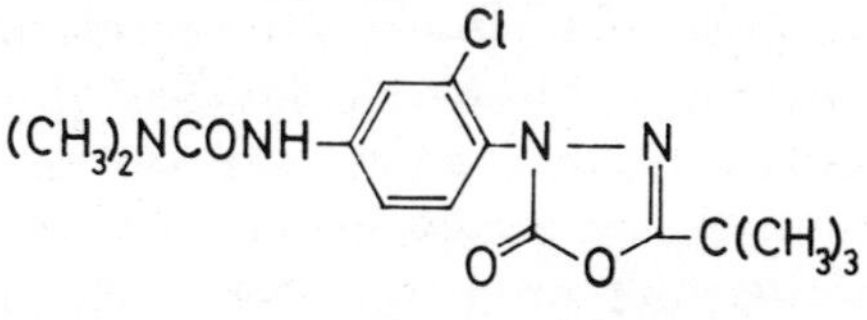

RP 23 465

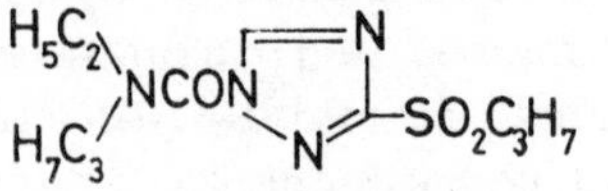

BTS 30 843

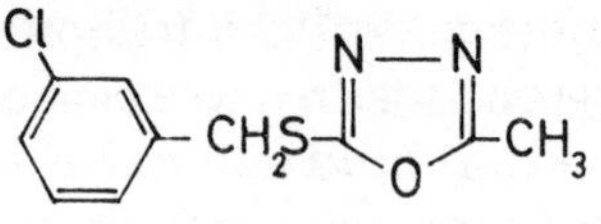

TH 2946

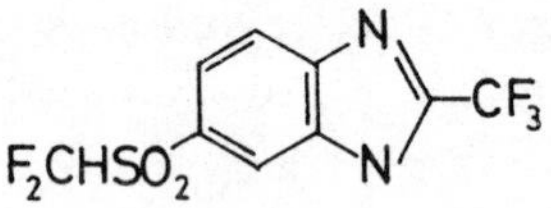

HOE 19 070

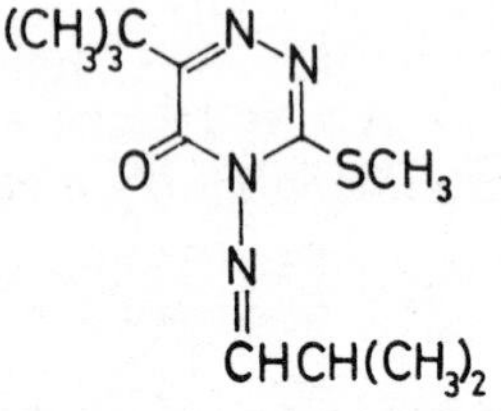

Isomethiozin

HETEROCYCLIC COMPOUNDS – OTHER HETERO ATOMS

This is a very varied group both in chemical structure and in herbicidal activity.

Endothal (7-oxabicyclo(2,2,1)-heptone-2,3-dicarboxylic acid) was introduced by the Sharples Chemical Corp. in 1954 for use pre- and post-emergence for the control of weeds in beet and spinach. It may be used in combination with other herbicides, e.g. as 'Murbetex Plus' (endothal and propham) by Murphy Chemical Ltd, for pre-emergence in sugar, beet, red beet, mangolds and fodder beet where it controls a wide range of weeds.

In 1974 **ethofumesate** (±-2-ethoxy-2,3-dihydro-3,3-dimethylbenzo-furan-5-yl methanesulphonate) was introduced by Fisons Ltd as 'Nortron'. This has the most interesting ability to selectively control weed grasses in desirable sown species. It does not affect the important pasture species such as timothy (*Phleum pratense*), cocksfoot (*Dactylis glomerata*) and meadow fescue (*Festuca pratensis*), but it is active against annual meadow grass (*Poa annua*), barley grass (*Hordeum murinum*), blackgrass (*Alopecurus myosuroides*), soft brome (*Bromus mollis*), wild oats (*Avena fatua*) and Yorkshire fog (*Holcus lanatus*). It may also be used pre-emergence for the control of weeds in beet. The herbicide is best applied October to mid-December and pastures should be cut or grazed before treatment. Fisons have also marketed a new pre-emergence herbicide 'Morlex' containing ethofumesate, propham, chlorpropham and fenuron for use on beet, sugar-beet and mangolds.

In 1980 Fisons Ltd introduced **NC 20484** (2,3-dihydro-3,3-dimethylbenzofuran-5-yl ethanesulphonate) a pre-plant incorporated or pre-emergence selective herbicide for the control of *Cyperus esculentus* and *C. rotundus* and a wide range of annual grasses and dictotyledonous weeds in cotton and many other crops including sugar-cane, tobacco and rice.

Heterocyclic compounds (other hetero atoms) – structural formulae

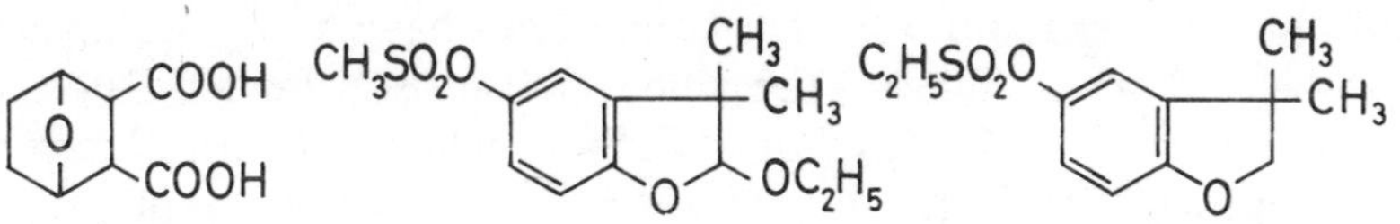

Endothal Ethofumesate NC 20 484

ORGANOARSENIC COMPOUNDS

For many years **sodium arsenite** was used as a contact herbicide but because of the dangers involved its use was abandoned. Some organic arsenic compounds whose mammalian toxicity is low are however still used. They have been developed in the U.S.A. but they are not approved for use in the U.K. They cause chlorosis, cessation of growth and progressive browning followed by dehydration and death. Several salts of **methylarsonic acid (MAA)** have been used since 1956 as herbicides. They include **MSMA** (sodium hydrogen methylarsonate), **DSMA** (disodium methylarsonate) and **CMA** (calcium methyl-arsonate). They were introduced by Ansul Chemical Co. (a company which no longer exists). These herbicides are now produced by the Diamond Shamrock Corp., MSMA as 'Daconate' and 'Bueno' and CMA as 'Calcar'. They are used as selective pre-emergence herbicides and may also be used for weed control on uncropped land, both DSMA and MSMA being well known for their control of Johnson grass (*Sorghum halepense*) as well as a number of other important weeds.

In 1958 the same company introduced **cacodylic acid** (dimethyl-arsinic acid) as 'Phytar' for non-selective post-emergence use on non-crop areas, for pasture renovation, as a desiccant and defoliant for cotton and for killing trees by injection. A special formulation 'Broadside' contains MSMA, sodium cacodylate and cacodylic acid as a general broad spectrum, post-emergence herbicide providing a quick burn down of vegetation without sacrificing the systemic effect necessary to control certain deep-rooted perennial grasses and weeds. Some synergism has been noted.

Organoarsenic compounds – formulae

$NaAsO_2$	$CH_3AsO(OH)_2$	$(CH_3)_2AsOOH$
Sodium arsenite	Methylarsonic acid	Cacodylic acid

ORGANOPHOSPHORUS COMPOUNDS

A number of organophosphorus compounds have been developed as herbicides. They include: (1) **bensulide** (*O,O*-di-isopropyl-*S*-2-phenylsulphonylaminoethyl phosphorodithioate) which was introduced by the Stauffer Chemical Co. in 1964 as 'Prefair' for pre-plant pre-emergence use on cucurbits, brassicas, lettuce and cotton, and as 'Betasan' for pre-emergence control of annual grasses and broad-leaved weeds in lawns. (2) **Piperophos** (*S*-2-methylpiperidinocarbonylmethyl-*O,O*-dipropyl phosphorodithioate) was introduced by Ciba Geigy in 1969. It can be used pre-emergence in rice, maize, cotton, soybeans and groundnuts for the control of many monocotyledonous weeds including *Cyperus* spp., *Echinochloa* spp., *Dopatrium junceum, Trianthema portulacastrum, Eleocharis acicularis* and *Monochoria vaginalis*. It may be used in a mixture with dimethametryn, as 'Avirosan', which will control most dicotyledonous weeds and give improved control of monocotyledonous weeds. Piperophos may also be used in mixtures with 2,4-D isopropyl ester and combinations based on these two at different rates control the most important weeds in transplanted and direct seeded rice. The mixture may be spread broadcast into standing water some 5–14 days after transplanting. (3) **Butamifos** (*O*-ethyl-*O*-[6-nitro-*m*-toly]-*N*-s-butyl phosphoramidothioate) was introduced in 1970 by Sumitomo Chemical Co. as 'Cremart' for contact pre-emergence use against annual weeds. (4) **Glyphosate** (*N*-[phosphonomethyl]glycine) a derivative of the amino acid, glycine, was introduced by Monsanto as 'Roundup' in 1971. It is used post-emergence and is rapidly absorbed by the leaves and translocated from vegetative parts to underground roots, rhizomes or stolons of perennial grass and broad-leaved weed species giving good control of both above-ground and underground organs. *Agropyron repens* is very sensitive. Glyphosate is inactivated on contact with the soil. It provides excellent weed control in pre-tillage or post-harvest treatments of annual crops or when applied as a directed spray in woody crops such as vineyards, deciduous fruit, stone fruit, rubber, coffee, citrus, tea and oil palm. It can also be used in non-agricultural areas and for bush control in forestry. (5) **Fosamine-ammonium** (ammonium ethyl carbamoylphosphonate) was introduced as 'Krenite' by Du Pont in 1974. It is a contact herbicide effective against many woody plants. Its discovery introduced a new concept for the control of susceptible brush. Application prior to leaf senescence causes little if any effect on the foliage. Normal seasonal leaf drop occurs during the autumn but bud and shoot development the following spring is either severely limited or is prevented entirely. Alder, hornbeam, hawthorn, beech, privet and many others are susceptible. The herbicide is not normally systemic affecting only those parts which it covers and can therefore be used as a chemical trimmer in that sprayed shoots are controlled without affecting the rest of the tree. It does appear however, to be translocated in some herbaceous plants such as bracken (*Pteridium aquilinum*) and *Convolvulus* sp. (6) In 1978 Murphy Chemical Co. marketed **aminophon** (*O,O*-dibutyl-1-butylamino-cyclohexyl-phosphonate) as 'Trakephon', a contact-acting herbicide, for weed control in cucumbers where it affects a wide range of dicotyledonous weeds, and as a desiccant and defoliant for use on potatoes, onions and hops. (7) **Isophos** (*O*-2,4-dichlorophenyl-*N*-isopropyl chloromethyl phosphonamidothionate) was developed at the Sheremetskaya Base of the All Union Institute for Plant Protection Chemicals. The best method of application is surface spraying 1 day before sowing. It may be used for rice, cotton, beans, sunflower and radish which are tolerant, and it will control *Chenopodium album*, *Amaranthus retroflexus* and *Agrostis stolonifera.*

Organophosphorus compounds – structural formulae

$C_6H_5-SO_2NHCH_2CH_2S-P(=S)[OCH(CH_3)_2]_2$ Bensulide

$(C_3H_7O)_2P(=S)-S-CH_2CO-N$ (2-methylpiperidine, H_3C) Piperophos

$2\text{-}NO_2\text{-}5\text{-}H_3C\text{-}C_6H_3-O-P(=S)(OC_2H_5)-NHCH(CH_3)-C_2H_5$ Butamifos

$(HO)_2-P(=O)-CH_2NHCH_2COOH$ Glyphosate

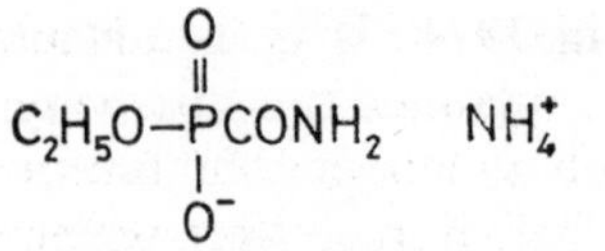

$$C_2H_5O-\overset{O}{\overset{\|}{P}}(O^-)CONH_2 \quad NH_4^+$$

Fosamine-ammonium

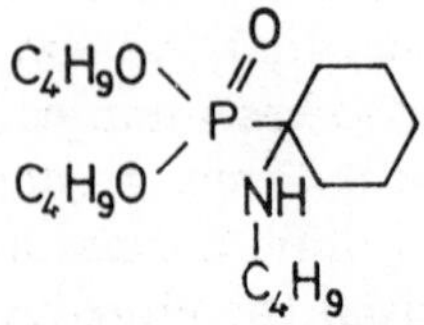

$$(C_4H_9O)_2P(=O)-C_6H_{10}-NH-C_4H_9$$

Aminophon

$$Cl-C_6H_3(Cl)-O-P(=S)(NHCH(CH_3)_2)-CH_2Cl$$

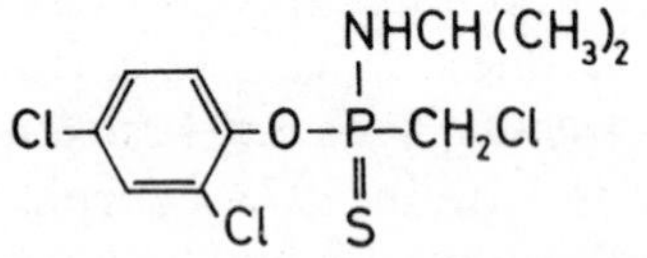

Isophos

UNCLASSIFIED COMPOUNDS

There are a number of herbicides which do not fit into any of the groups dealt with. They include: **allyl alcohol** introduced as a herbicide in 1950 and used for weeding forestry nursery beds. **Acrolein** which was introduced by Shell as 'Aqualin', an aquatic herbicide and algicide. **Metham-sodium** (methyldithiocarbamic acid) introduced as 'Vapam' by the Stauffer Chemical Co. in 1955 as a soil fumigant for the control of weeds, fungi and nematodes. Its activity in the soil is due to its decomposition to methyl isothiocyanate. **Quinonamid** (2,2-dichloro-*N*-[3-chloronaphthoquinon-2-yl]acetamide) introduced by Hoechst A.G. as 'Alginex' for the control of algae and mosses. **Dow 221** (α-2,2,2-trichloroethylstyrene) introduced by the Dow Chemical Co. in 1970. It is formulated as M3429 granules ('Tavron G') containing 2,4-D. It may be applied 3–8 days after transplanting rice and the paddy should be flooded with 2–4 inches of water at the time of application. It is absorbed primarily through the mesocotyl of young seedlings which become more resistant after the 2-leaf stage. It controls *Echinochloa crus-galli* and other mositure-loving plants. **Perfluidone** (1,1,1-trifluoro-*N*-[4-phenylsulphonyl-*o*-tolyl]-methane-sulphonamide) introduced by 3M Company in 1971 as 'Destun' for the control of some grass and dicotyledonous weeds and *Cyperus esculentus* in emerging cotton, established turf, transplanted tobacco and several other crops. **4-Methoxy-3,3′-dimethylbenzophenone** introduced in 1971 by Nippon Kayaka Co. as 'Kayametone' for use as pre-emergence for the control of many weeds in rice and vegetable crops. **Alloxydim-sodium** (methyl-3-[1-(allyloxyimino)butyl]-4-hydroxy-6,6-dimethyl-2-oxocyclohex-3-ene carboxylate) introduced in 1976 by Nippon Soda Ltd as a translocated selective herbicide and used post-emergence for the control of many annual and perennial grass weeds in sugar-beet, vegetables and other dicotyledonous crops such as clovers, sunflower, flax and hops. It has short soil persistence. **NP 55** (2-[*N*-ethoxybutyrimidoyl]-5-[2-ethylthiopropyl]-3-hydroxy-cyclohexene-1-one) introduced by Nippon Soda in 1979 as a selective grass herbicide for post-emergence application. It controls a wide spectrum of common annual and perennial grasses of field crops as well as volunteer cereals. It is tolerated by almost all broad-leaved crops.

Unclassified compounds – structural formulae

Allyl alcohol $H_2C{=}CH{-}CH_2OH$

Acrolein $H_2C{=}CH{-}CHO$

Metham-sodium $CH_3NHCSS^-Na^+$

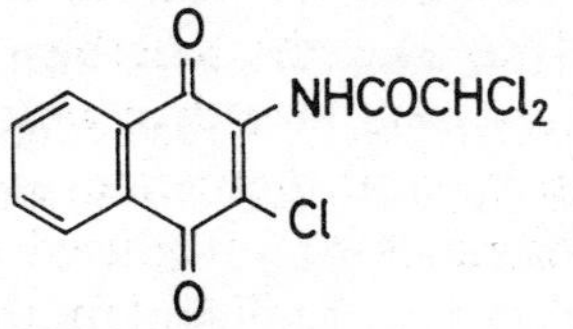

O, $NHCOCHCl_2$, Cl, O

Quinonamid

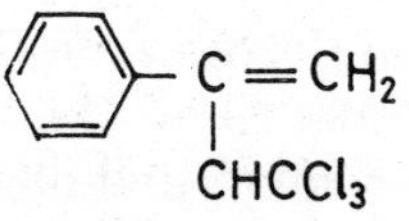

$C{=}CH_2$, $CHCCl_3$

Dow 221

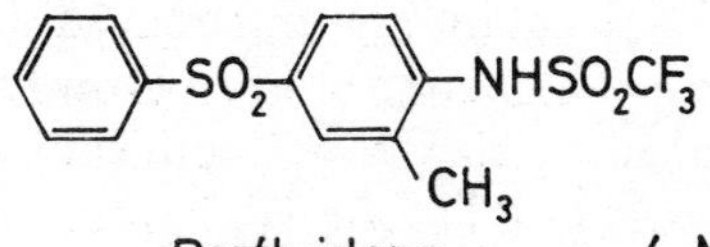

SO_2, $NHSO_2CF_3$, CH_3

Perfluidone

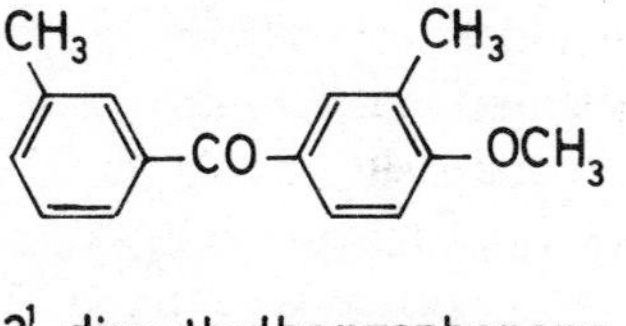

CH_3, CH_3, CO, OCH_3

4-Methoxy-3,3'-dimethylbenzophenone

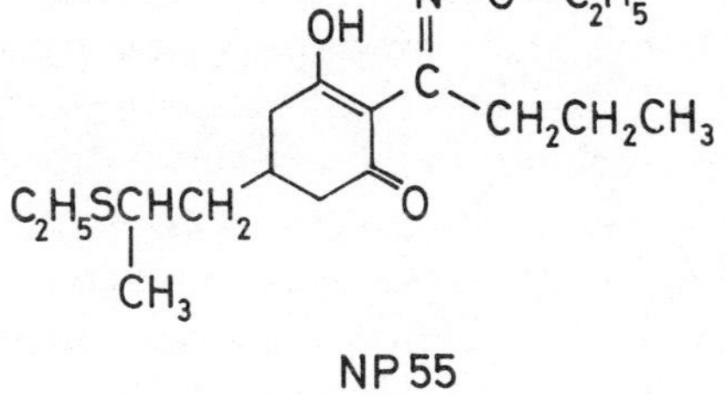

OH, $N{-}O{-}C_2H_5$, C, $CH_2CH_2CH_3$, $C_2H_5SCHCH_2$, CH_3, O

NP 55

$OCH_2CH{=}CH_2$, N, Na^+O^-, C, $CH_2CH_2CH_3$, H_3C, H_3C, O, C, O, OCH_3

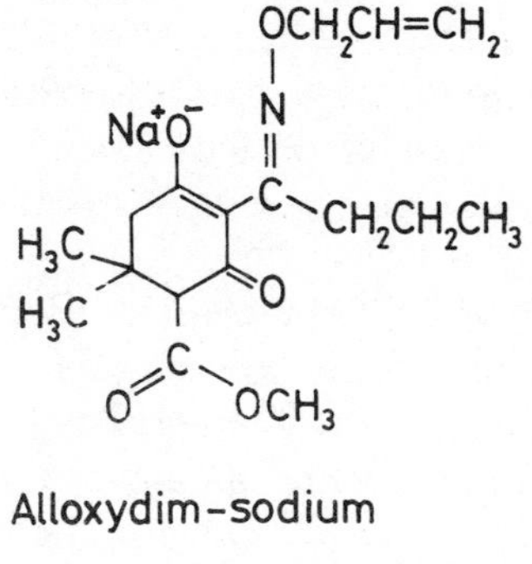

Alloxydim-sodium

HERBICIDE PROTECTANTS AND ANTIDOTES

A new development in weed herbicide usage is the use of protectants and antidotes in order to protect the crop plant from possible damage by a herbicide (Fryer, 1977). This means that it may be possible to use certain herbicides on crops that would normally be affected by the herbicide. Among the protectants that are in use are **R 27 788** (*N,N*-diallyl-2,2-dichloroacetamide) introduced by the Stauffer Chemical Co. in 1972 and marketed in mixtures with EPTC and with butylate for soil application; and **1,8-naphthalic anhydride (NA)** which is sold as a seed dressing by Gulf Oil Corp. to allow EPTC to be applied safely in crops. NA also provides some protection against the effects of several other herbicides including alachlor, ethofumesate, and epronaz. Both antidotes have resulted in a dramatic reduction in damage to maize from EPTC. Fryer also notes that NA has been used as a seed dressing on rice protecting this crop against molinate and alachlor and making possible the selective control of the important weed, red rice (*Oryza rufipogon*) in rice.

Another 'herbicide safener' **cyometrinil (CGA 43 089)** (α-[cyanomethoximino]benzacetonitrile) has been introduced by Ciba Geigy which will also allow the safe use of metolachlor in grain sorghum (*Sorghum bicolor*) and other crop species.

Herbicide protectants and antidotes – structural formulae

$Cl_2CHCON(CH_2CH{=}CH_2)_2$

R 27788

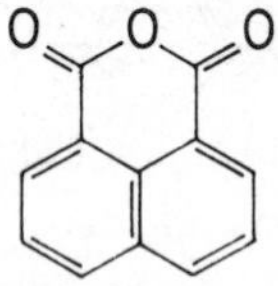

NA

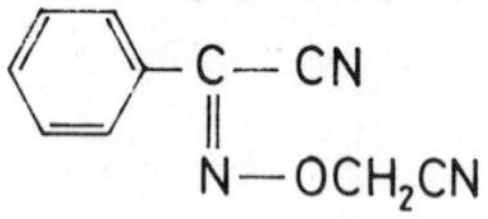

$C_6H_5{-}C({-}CN){=}N{-}OCH_2CN$

CGA 43089

(Cyometrinil)

References

ASHTON, F.M. and CRAFTS, A. S. (1973). *Mode of Action of Herbicides.* John Wiley & Sons.

ASLANDER, A. (1926). *J. Amer. Soc. Agron.*, **18**, 1101.

BLACKMAN, G. E. (1945). *Nature (London)*, **155**, 497.

BRIAN, R.C. (1976). In *Herbicides* (Ed. L. J. Audus), Vol 1, 2nd edn., Academic Press.

BUCHA, H. C. and TODD, C. W. (1951). *Science N.Y.*, **114**, 493.

BUCHA, H. C. *et al.* (1962). *Science N. Y.*, **137**, 537.

CALDERBANK, A. (1968). In *Advances in Pest Control Research* (Ed. R. L. Metcalf) Vol. 8, John Wiley & Sons, p. 127.

CALUWE, P. DE (1900). *Vereen Oudleerl. Rijks-Landbouwschool*, **12**, 103.

CASTELFRANCO, P., FOY, C. L. and DEUTSCH, D. B. (1961). *Weeds*, **9**, 580.

COTTRELL, H. J. and HEYWOOD, B. J. (1965). *Nature (London)*, **207**, 655.

CRAFTS, A. S. (1945). *Science N.Y.*, **101**, 417.

DISKUS, A. *et al.* (1976). *Proc. 1976 Br. Crop Protect. Conf. Weeds*, **2**, 717.

EVANS, S. A. *et al.* (1977). In *Weed Control Handbook* (Eds. J. D. Fryer and R. J. Makepeace), Vol. 1, 6th edn., Blackwell, p. 150.

FAWCETT, C. H., OSBORNE, D. J., WAIN, R. L. and WALKER, R. D. (1953). *Ann. Appl. Biol.*, **40**, 232.

FLETCHER, W. W. (1974). *The Pest War*, Blackwell.

FRYER, J. D. (1977). In *Ecological Effects of Pesticides* (Eds. F. H. Perring and K. Mellanby), Academic Press, p. 27.

GREEN, M. B., HARTLEY, G. S. and WEST, T. F. (1979). *Chemicals for Crop Protection and Pest Control*, Pergamon Press.

GUTHRIE, F. B. and HELMS, H. R. (1904). *Ag. Gaz. N.S.W.*, **14**, 114 and **15**, 29.

GYSIN, H. and KNÜSLI, E. (1957). *Proc. 4th Int. Cong. Crop Protect. Hamburg.*

GYSIN, H. and KNÜSLI, E. (1958). *Proc. 4th Brit. Weed Control Conf.*, 225.

GYSIN, H. and KNÜSLI, E. (1960). *Adv. Pest Control Res.*, **3**, 289.

HAMNER, C. L. and TUKEY, H. B. (1944). *Science N.Y.*, **100**, 154.

HOFFMAN, A.W. (1885). *Ber. dt. Chem. Ges.*, **18**, 2755.

HOFFMAN, O. L. and SMITH, A.E. (1949). *Science N.Y.*, **109**, 588.

MARTH, P. C. and MITCHELL, J. W. (1944). *Bot. Gaz.*, **106**, 224–32.

NUTMAN, P. S., THORNTON, H. G. and QUASTEL, J. H. (1945). *Nature (London)*, **155**, 497.

PEARLMAN, W. M. and BANKS, C. K. (1948). *J. Amer. Chem. Soc.*, **70**, 3726.

POIGNANT, P. and CRISINEL, P. (1967). *J. Etud. 4th Herbic. Conf. Columna,* 196.

POKORNY, R. (1941). *J. Amer. Chem. Soc.*, **63**, 1768.

SCHWARTZBECK, R. A. (1976). *Proc. 1976 Brit. Crop Protect. Conf. Weeds.*

SILK, J. A., DEAN, M. L., RICHARDSON, G. and TAYLOR, W. A. (1977). In *Weed Control Handbook* (Eds. J. D. Fryer and R. J. Makepeace), Vol. 1, 6th edn., Blackwell, pp. 85–154.

SLADE, R. E., TEMPLEMAN, W. G. and SEXTON, W. A. (1945). *Nature (London)*, **155**, 497.

SYNERHOLM, M. E. and ZIMMERMAN, P. W. (1945). *Contrib. Boyce Thompson Inst.*, **14**, 39.

TEMPLEMAN, W. G. and SEXTON, W. A. (1945). *Nature (London)*, **156**, 630.

TEMPLEMAN, W. G. and SEXTON, W. A. (1946). *Proc. Roy. Soc. London*, **133B**, 480.

VAN DEN BRINK, B. J. *et al.* (1969). *Proc. E.W.R. Symp. New Herbicides*, 35.

WAIN, R. L. (1955a). *Ann. Appl. Biol.*, **42**, 151.

WAIN, R. L. (1955b). *J. Agric. Fd. Chem.*, **3**, 128.

WAIN, R. L. and WIGHTMAN, F. (1955). *Proc. Roy. Soc. London*, **142B**, 525.

WOODS, C. D. and BARTLETT, J. M. (1909). *Bull. Me. Agric. Exp. Stn.*, No. 167.

WORTHING, C. R. (Ed.) (1979). *The Pesticide Manual*, 6th edn., British Crop Protection Council.

ZIMMERMAN, P. W. and HITCHCOCK, A. E. (1942). *Contrib. Boyce Thompson Inst.*, **12**, 321.

ZIMMERMAN, P. W. and HITCHCOCK, A. E. (1944). *Proc. Amer. Soc. Hortic. Sci.*, **45**, 353.

CHAPTER TWO

Plant Growth Regulators

Auxins - natural and synthetic

Although there had long been speculation about the presence of substances within plants which correlated their growth and Sachs in the latter half of the 19th century put forward the idea of 'organ-forming substances' it was the publication by Darwin of his book *The Power of Movement in Plants* in 1880 that led eventually to the isolation of a plant hormone. Darwin reported on his studies of the responses of plants to light and gravity. He observed that if the extreme tip of the coleoptile of a seedling of *Phalaris canariensis* were removed or covered with tinfoil no response took place to light from the lower regions of the coleoptile where bending normally took place. He postulated that the stimulus (light in this instance) was received at the tip and that some 'influence' was transmitted to the reacting region lower down. Similar types of experiments indicated that the response to gravity by shoots and roots might also be governed by the transmission of similar 'influences' from shoot and root tips. Some 30 years after publication of this book, Boysen Jensen (1911) showed that this 'influence' was chemical in nature when he demonstrated that it would pass through a gelatine barrier. This view was substantiated by Paál (1919) but it was not until some 7 years later using *Avena* coleoptiles that this chemical messenger was isolated when Went (1926) placed coleoptile tips upon agar blocks and allowed the chemical messenger to diffuse into the agar. He demonstrated that these blocks could act as 'physiological tips' when they were placed on decapitated coleoptiles. He undertook quantitative measurements of the chemical and showed that the curvature of the coleoptile by cell extension was directly proportional to the concentration of the hormone in the agar and that it could act at very low concentrations. It was indeed a plant hormone (phytohormone) which has been defined (Pincus and Thimann, 1948) as 'an organic substance produced naturally in higher plants, controlling growth or other physiological functions at a site remote from its place of production and active in minute amounts'. Those phytohormones which are involved in the extension growth of plant cells are given the name 'auxins' and Audus (1959) considers that the latter term may be used to include both 'natural auxins' for those produced by plants themselves and 'synthetic auxins' (of which as we shall see there are many) which have the same action as natural auxins but which do not occur naturally in plants but are synthesised in the laboratory. It is with this latter group that we shall largely be concerned. First, however, a word about the naturally occurring auxin. Because it occurs in such minute amounts in plants early attempts to determine its chemical nature failed until Kögl *et al.* (1934) isolated it from human urine and showed that it was indolyl-3-acetic acid (**IAA**).

Important though biological assays are it was desirable that IAA should be identified in plant tissues. This is now possible through the development and application of sensitive physicochemical techniques such as ultraviolet and infrared spectroscopy and more recently by mass spectrometry (Wareing and Philips, 1978).

These studies have shown that although IAA is the principle auxin nevertheless there are other chemicals such as indole-3-acetonitrile (IAN), 4-chloroindole-3-acetic acid and phenylacetic acid occurring in certain plants which have auxin activity.

It is now known that IAA is involved in a great many growth processes in plants. These include internode elongation, root growth, leaf growth, initiation of vascular tissues, cambial activity, fruit set, fruit growth and apical dominance (Wareing and Philips, 1978).

It is not surprising that synthetic auxins have found a wide application in horticulture, particularly for the stimulation of rooting, fruit set and fruit thinning. Among them are:

IBA (4-[indol-3-yl]butyric acid) first prepared by Jackson and Manske (1930) and recognised as a stimulant to root formation by Zimmerman and Wilcoxon (1935). It was therefore one of the first chemicals to be recognised as a growth regulator and it is still widely used for the rooting of cuttings in combination with 1-naphthylacetic acid and naphthylacetamide. The combination is particularly useful for the rooting of some of the conifers which are virtually impossible to root naturally.

NAA (α-naphthylacetic acid) introduced by Amchem Products Inc., I.C.I. and by May & Baker Ltd. As mentioned above it is used as a rooting agent. It is also used as a thinning agent (applied after petal fall) in apples and pears to prevent pre-harvest fruit drops in the same crop and as the ethyl ester to control regrowth of tree sprouts following pruning (Amchem Technical Data Sheet). The M & B product may be used on cotton to induce flowering and to increase yields by preventing the shedding of buds, flowers and bolls; to increase yield and quality of mango; to induce uniform flowering in pineapple; to inhibit pre-harvest fruit drop in citrus and grapes, and prevent flower shedding in chillies (*Capsicum annuum*) and as a fruit drop inhibitor in apples and pears; to prevent premature shedding of grain and rice; to increase pod weight and number in ground nuts; and to prevent flower shedding in pulse crops and soybeans (Technical Data Sheet).

NOA (2-naphthyloxyacetic acid) introduced in 1946 by Synchemicals Ltd as a fruit setting spray. It may be used on tomatoes, holly, grapes and strawberries. It has been reported that Cox's orange pippin apple trees sprayed at petal fall with a mixture of 2-naphthyloxyacetic acid, *N,N*-diphenylurea and GA_3 yielded up to 38% more fruit (Goldwin, 1978).

NAD (1-naphthylacetamide) introduced by Amchem Products Inc. as a thinning agent for apples and pears. It will also prevent pre-harvest drop. As noted above it may also be used as a rooting agent. The same company have also introduced a mixture of 2-(*m*-chlorophenoxy)-propionamide and 2-(*m*-chlorophenoxy)propionic acid for experimental use as a pineapple growth regulator to reduce crown growth and to increase the size of the fruit.

Phenoxyalkanoic acids are best known and most widely used for their selective weedkilling properties but sublethal doses of 4-CPA, 2,4-D, 2,4,5-T and fenoprop are used as growth regulators to promote fruit set and for thinning. 2,4-D has also been shown to increase the dry matter and yield of potatoes, peas, beans, corn and sugar-beet.

Auxins – structural formulae

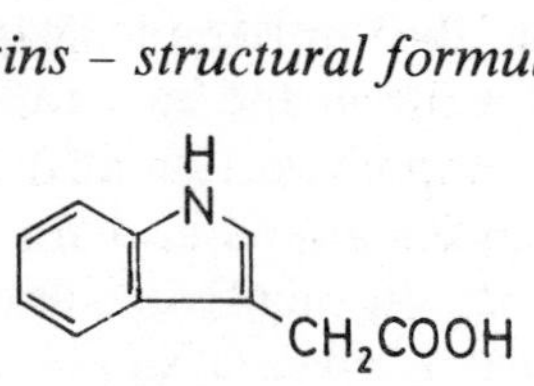

H N CH_2COOH
IAA

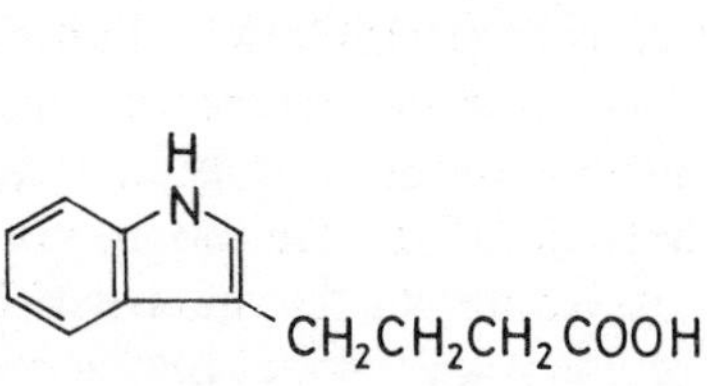

H N $CH_2CH_2CH_2COOH$
IBA

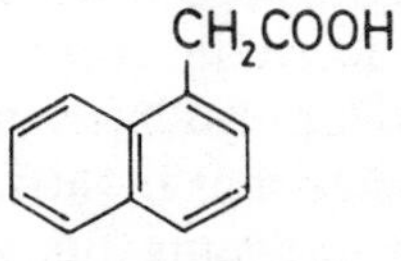

CH_2COOH
NAA

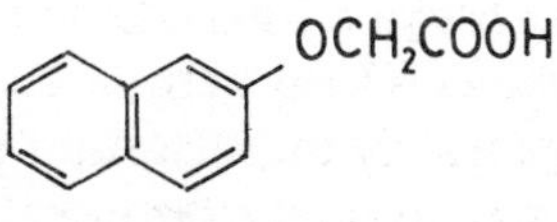

OCH_2COOH
NOA

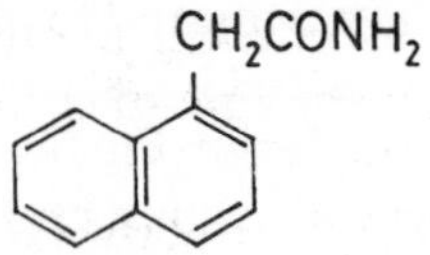

CH_2CONH_2

NAD

Gibberelins

There are more than fifty naturally occurring gibberellins. Commercial formulations are used for breaking dormancy, flower initiation, promoting vegetative and fruit growth and for the induction of parthenocarpy.

In the 1920s a Japanese physiologist, Kurosawa (1926) working in Formosa investigated the stem elongation of rice seedlings infected with the fungus *Fusarium heterosporum* (now known as *Gibberella fujikuroi*). This condition known as 'bakanae' or 'foolish seedling' disease of rice had first been noted in an agricultural book by Konishi in 1898. Kurosawa found that the fungus produced a growth-promoting substance when grown in culture and shortly afterwards the active substance was isolated and named gibberellin. This work remained unknown in the western world until Brian *et al.* (1955) obtained a *Fusarium* culture and isolated **gibberellic acid** (GA) from it. Cross *et al.* (1961) worked out its structure. They showed that GA and other gibberellins which had been isolated possessed a common feature, namely the gibbane ring skeleton.

Paleg (1965) has defined gibberellins as 'compounds having an ent-gibberellane skeleton and biological activity in stimulating cell division or cell elongation, or both, or such other biological activity as may be specifically associated with this type of naturally occurring substance'. They must have the ability to reverse stem dwarfism and be capable of inducing α-amylase activity in barley endosperm. More than fifty GAs have been isolated from the tissues of various plants and Hedden *et al.* (1978) have published their structural formulae. Rather than give these gibberellins trivial names they have been designated as GA_1, GA_2, etc. Gibberellic acid is GA_3. No plant has been found which contains all of the gibberellins but most plants – angiosperms, gymnosperms, ferns, algae, fungi and bacteria – possess several of them. Tests with a great variety of plants by workers in various parts of the world have shown that they are a very active group of growth regulators, in particular promoting the rapid expansion of plant cells, inducing flowering under non-inductive conditions, stimulating seed germination, breaking dormancy of overwintering plants and inducing fruit set which often results in the production of seedless fruit (Thomas *et al.*, 1972).

Because of the complex structure of the GAs it has not been possible to synthesise substances of comparable activity but several gibberellins can be produced commercially by large-scale fermentation and they are available for horticultural and agricultural use.

Luckwill (1976) describes how the receptacle of apple and pear flowers (which are more resistant to frost damage than the styles and ovules) can, following frost damage, be stimulated to develop parthenocarpically by the application of GA sprays. Such treatment has proved particularly valuable in saving the pear crop after spring frost damage (Luckwill, 1960; Turner, 1973). A mixture of GA with the auxin 2,4,5-T applied before and after flowering has improved the cropping of sour cherries (Wierszyllowski *et al.*, 1963). A triple hormone spray of GA, an auxin and a cytokinin can stimulate parthenocarpic fruiting in some apple cultivars (e.g. Cox's orange pippin) and increase fruit set in sweet cherries and plums (Kotob and Schwabe, 1971). Luckwill cautions that sprays with GA can depress flower initiation though this can be minimised by careful attention to concentration and time of application. Kapoor and Turner (1976) report that a combined application of 2,4-D and gibberellic acid is much more effective than 2,4-D or 2,4,5-T alone in preventing fruit drop in citrus fruits and that the GA in the mixture retards the breakdown of chlorophyll on the fruit so that the skin affords better protection against disorders and the ripe fruit can be stored on the tree and harvest delayed for several weeks. This extension of life and storage life of citrus fruits is a very important property of the GA/2,4-D mixture leading to increased profits on the sales of limes, lemons, grapefruit and oranges (Kapoor and Turner, 1976). Delay in the ripening of bananas in storage can also be effected by mixtures of GA and benomyl.

Sprays of GA increase fruit set and yield of clementines several-fold. Provided the correct concentration is used there need not be a reduction in fruit size.

GA is also very effective in increasing the yield of sugar from sugar-cane. Tanimoto and Nickell (1968) have shown that by using gibberellin sprays it has been possible to get a gain of about 5 tons of cane per acre (14.8 t/ha) under Hawaiian plantation conditions with a gain in sugar of from 0.2 to 0.5 tons per acre (0.59–1.48 t/ha).

Gibberellins are also widely used in vegetable cultivation. Thomas (1976) states that their use is recommended to induce early flowering in artichoke, increase stalk length and yield in celery, induce maleness in gynoecious lines in cucumber, give uniform bolting and thus increase seed production in lettuce, break dormancy and thus give uniform crop emergence in seed potato, reduce cold requirement and increase yield in forcing rhubarb and increase stem length during winter in watercress.

Gibberellins have also great potential use in forestry (Pharis and Ross, 1976). These authors state that with the world's forest lands rapidly declining in area there is a great need for trees that are genetically superior in growth and quality. These have to be developed from seed by the geneticist and since economically important species of conifers take 10–20 years to flower and may need another 20 years to produce significant quantities of seed, the process is a very slow one. Now, however, it has been possible to bring members of the *Cupressaceae* (cedars, cypresses, junipers) and the *Taxodiaceae* (redwoods, bald cypress) to early flowering using GAs (Kato *et al.*, 1958) and the process has been extended to economically important members of the *Pinaceae* (pine, Douglas fir, fir, spruce and hemlock). Using a mixture of GA_4 and GA_7 by injection into the stem many of these species have been induced to flower in 4–6 years (Pharis *et al.*, 1974).

Gibberellins – structural formulae

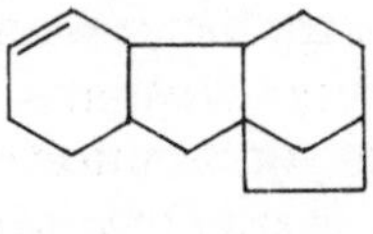

Gibbane skeleton

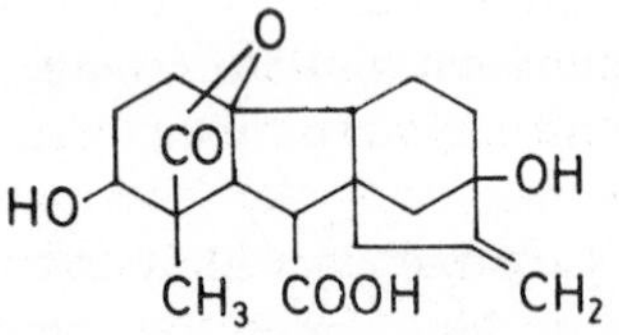

GA_3
(Gibberellic acid)

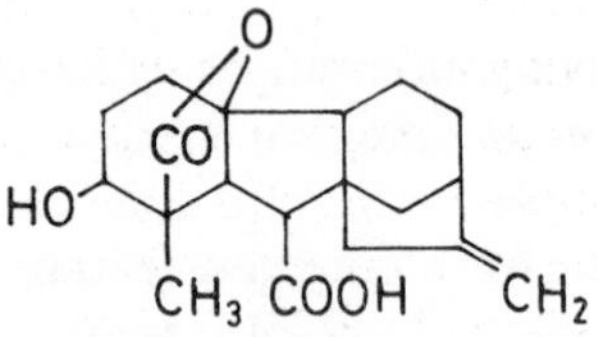

GA_7

Cytokinins

Cytokinins may be natural (zeatin) or synthetic (e.g. kinetin). Zeatin affects cell division and leaf senescence and synthetic cytokinins are used to promote lateral bud development and inhibit senescence.

The discovery of cytokinins came about as a result of investigations into the growth of plant tissue cultures. Skoog and his co-workers (Jablonski and Skoog, 1954; Miller *et al.*, 1955) found that coconut milk and malt extract, yeast extract and autoclaved DNA were all active in inducing good tissue culture growth from tobacco stems and in 1954 Miller isolated a crystalline substance from autoclaved herring sperm DNA that was able to induce cell division in tobacco cultures, at concentrations as low as one part per billion when auxin was also present in the medium. The substance was named **kinetin** andwas later found to be 6-furfurylaminopurine (6-furfuryladenine) (Letham, 1963; Letham *et al.*, 1964; Shaw and Wilson, 1964).

Later other adenine derivatives were found which had similar biological activity and they were referred to collectively as kinins, a name which was later changed to *cytokinins* because animal physiologists had a prior claim to the term kinin referring to a class of polypeptides which had quite different properties. Kinetin has never been found in plants but it is clear that some cytokinins are widely synthesised in plants. They are found particularly in young and actively dividing tissues such as embryos, seedlings, and apical meristems. Whether or not these are the sites of synthesis is still open to question since it is possible that they are synthesised elsewhere and translocated to these sites. Cytokinin activity has been detected in tomato juice, extracts of fruits and apples (*Pyrus malus*), fruits of quince (*Cydonia oblonga*), pear (*Pyrus communis*), plum (*Prunus cerasifera*), peach (*Prunus persica*) and tomato (*Lycopersicum esculentum*), cambial tissues of a number of stems, extracts of immature fruits of *Zea mays*, *Aesculus noerlitzensis*, *Juglans* sp. and *Musa* sp. (Fox, 1969). Miller (1962) found that the cytokinin in maize was adenine-substituted in the 6-amino group with an unsaturated side chain having at least one hydroxyl component. Later Letham (1963), Letham *et al.* (1964) and Shaw and Wilson (1964) showed that it was an analogue of kinetin and Letham (*loc. cit.*) who obtained it in crystalline form named it **zeatin**. It is likely that it is widely distributed in the plant kingdom (Fox, 1969). Other naturally occurring cytokinins such as N^6-methylaminopurine, N^6-dimethylaminopurine, and N^6-Δ^2-isopentenylamino)purine have also been identified (Fox, 1969).

A large number of compounds based on kinetin have been synthesised and tested for cytokinin activity. It was found that the side chain of kinetin could be replaced by a variety of substituents which yielded compounds showing this activity and some of them were more powerful than kinetin itself. Alterations in the adenine moiety however either reduced activity or rendered the compounds inactive (Strong, 1958).

A number of non-purine compounds, e.g. *N,N'*-diphenylurea, and benzimidazole have been shown to have cytokinin activity. Fox (1969) considers it to be at least a possibility that these and others may be precursors of kinetin-like compounds and suggests that 'the structure needed for cell division stimulating activity is that of adenine bearing on the amino group in the six position, a side chain of a non-polar character' or put in another way: 'Cytokinins are substances composed of one hydrophilic group of high specificity (adenine) and one lipophilic group without specificity'. Although cytokinins stimulate cell division this is not their only feature. They have also been shown to affect cell enlargement, morphogenesis, dormancy of seeds, apical dominance and senescence. These topics have been reviewed by Fox (1969). Miller and Skoog (1953) using tobacco pith cultures showed that by adjusting the proportions of IAA and kinetin it was possible to produce undifferentiated callus, or buds or roots, e.g. IAA (2 mg/l) and kinetin (0.02 mg/l) in the medium produced a mass of undifferentiated cells from the pith. Increasing the ratio of kinetin – either by lowering the auxin concentration or raising the kinetin level in the medium – resulted in the formation of buds that eventually produced complete tobacco plants. When the kinetin to auxin level was lowered the pith grew roots only. Fox notes that there are many other examples of bud formation under the influence of cytokinins in the literature but cautions that the interrelationships between IAA and cytokinins is not necessarily a simple one and that in some cases bud formation is dependent upon prior root formation.

Promotion of seed germination is another feature of cytokinins. They can substitute for the red light requirement in lettuce seed (*Lactuca sativa*) (Miller, 1956). Germination response has also been shown in tobacco and white clover and in combination with red light kinetin can overcome the effect of certain naturally occurring inhibitors of seed germination (Khan and Tolbert, 1965).

Cytokinins can also counteract apical dominance (Sachs and Thimann, 1964) though the mechanism is not clear. A most interesting feature is that they can retard senescence and it has been shown that they can postpone the disappearance of chlorophyll and the degradation of proteins in ageing leaves (Richmond and Lang, 1957). It has been shown also that they can act as mobilising agents directing the movement of a number of substances to the treated area (Möthes and Engelbrecht, 1959), but Fox (1969) observes that much more than a simple mobilising of foodstuffs is involved in that Osborne (1962) has shown that kinetin treatment augmented the ratios of RNA or protein to DNA and suggests that the action of kinetin might be the maintenance of the protein-synthesising machinery.

Cytokinins – structural formulae

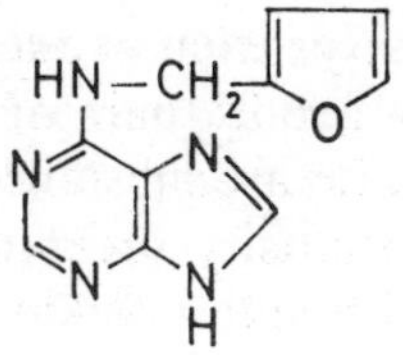

Kinetin

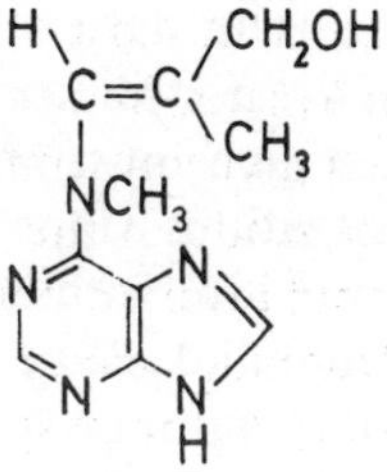

Zeatin

Ethylene

Ethylene gas is produced naturally in plant tissues often in response to environmental stress or wounding and also during ripening and abscission processes. Synthetic ethylene generators (mainly ethephon) are used to induce fruit abscission, promote flowering and fruit ripening, break dormancy, increase production of female flowers in cucumbers and melons, thin grapes and stimulate latex flow in rubber trees.

Ethylene is a rather unusual plant hormone in that it is gaseous and is a very simple organic molecule. It is however a very active one and at very low concentrations can have profound effects.

In 1924 Denny showed that 'stove gas' induced the yellowing of lemons and further that this effect could be duplicated if ethylene replaced stove gas. Gane (1935) finding that a gas from ripe apples could ripen green ones showed that the gas was ethylene and it was subsequently shown that ethylene was involved in the ripening of many fruits. Because of the small amounts involved however some workers believed that it was produced as a result of, and was not the cause of, ripening but it has now been well established that ethylene is produced by fruit and initiates the ripening process (Mapson, 1970). There are still a few fruits however such as peaches and apricots which are not affected by ethylene. The gas is remarkably active at very low concentrations and in many fruits the intercellular threshold concentration required to initiate ripening lies between 0.04 and 0.5 p.p.m. (Burg and Burg, 1965). In 1938 Kidd and West seeking to extend the storage life of apples showed that storing them in an atmosphere of increased carbon dioxide and decreased oxygen reduced the ripening process. Later research has confirmed this view and with most fruits a level of 2–5% oxygen would appear to be the best for controlling the synthesis of ethylene at temperatures between 1 and 25°C. When this is combined with increased concentrations of CO_2 maximum retardation of ripening can be achieved (Mapson, 1970) although the optimum combination of temperature, carbon dioxide and oxygen must be established for each variety. In a long series of investigations Mapson and his collaborators (Mapson, 1970) have established that within plant tissues ethylene can be produced from methionine and also from linolenic acid. Their studies on the mechanism of ethylene synthesis underline the absolute need for oxygen.

As well as being involved in the ripening of fruit ethylene has a number of other effects on plant life. At 1 p.p.m. it prevents the opening of flower buds of carnation; it can cause epinasty in leaves (the upper side of the petioles growing more rapidly and the leaves bending downwards as a result). It also interferes with the normal geotropic responses of seedlings and it has been suggested that it does so by interfering with the polar transport of auxin. In 1935 Zimmerman and Wilcoxon showed that auxin applications stimulate ethylene production from plants and it has subsequently been shown by Abeles (1966) and by Burg and Burg (1966) that several of the effects on plants that are shown when auxins are applied, such as stimulation of root initials, inhibition of root elongation and stimulation of leaf abscission, can be duplicated by exposure of the plants to ethylene. It has been suggested that many of the formative effects of auxin are really due to ethylene production resulting from auxin stimulus. It can thus be seen that ethylene has a wide range of effects from strongly stimulatory to powerfully inhibitory and according to Bidwell (1979) a real problem in the study of ethylene is the difficulty of separating its effects from those of auxins.

The introduction of **ethephon** (2-chloroethylphosphonic acid) as 'Ethrel' by Amchem Products Inc. in 1967, which when applied liberates ethylene into plant cells, has led to a wide range of commercial applications. Under slightly alkaline conditions the ethephon decomposes to produce mainly ethylene, hydrochloric acid, phosphoric acid and other unidentified metabolic by-products (Audley *et al.*, 1976).

A very important application of ethylene is in its use to increase the duration of latex flow in rubber trees. The subject has been reviewed by Dickenson (1976). Substituted phenoxyacetic acids such as 2,4-D and 2,4,5-T have been used commonly since the 1950s and it was the recognition that they acted by stimulating higher levels of ethylene in the plant tissue that led to the introduction of ethylene-generating compounds, in particular ethephon and adsorbent powders impregnated with ethylene known as 'Ethad' (Dickenson, 1973; Dickenson *et al.*, 1976) which have been shown to be highly effective as latex yield stimulants.

Ethephon is also a powerful abscission-promoting agent which is used for thinning heavy cropping cultivars of apples and peaches (Luckwill, 1976). Some apple varieties may be harvested early to benefit from early market prices by treating them about 10 days before picking with a mixture of ethephon and 2,4,5-T (Luckwill and Child, 1973). The ethylene from the ethephon accelerates ripening so that the apples improve in flavour and appearance while the 2,4,5-T counteracts the abscission effects of the ethephon. Ethephon can be used as a sugar-cane ripener to increase the sucrose content of the cane at harvest (Nickell, 1976). It may also be used to accelerate the ripening of tomatoes (Thomas, 1976) and peppers (Knavel and Kemp, 1973). It will produce a more compact soybean plant, less subject to lodging (Blomquist and Schrader, 1973). It has been noted too that ethephon is able to modify the sex expression of a number of plants including curcubits, treated plants producing a preponderance of early female flowers leading to increased, early yields (Tompkins and Smay, 1971). Because of its abscission properties ethephon is used commercially as a harvesting aid on a wide variety of fruits and nuts (Thomas, 1976). Its ability to inhibit stem elongation has meant that it can be used as a soil drench in the culture of forced narcissi producing shorter sturdier plants (Briggs, 1975). Formulated with hydriodic acid it is used as a defoliant for rose bushes prior to lifting and applied to the base it stimulates the development of basal shoots.

In addition to ethephon a number of other ethylene-generating chemicals have been introduced in recent years. They include:

Etacelasil (2-chloroethyl tris[2-methoxyethoxy]silane) introduced by Ciba Geigy (Ciba Geigy Technical Data Sheet, 1974) as an abscission-promoting chemical for use as a harvest aid in olives. It loosens the fruits by releasing ethylene.

CGA-15281 (2-chloroethyl methyl bis[benzyloxy]silane) introduced by Ciba Geigy has been used for thinning of nectarines, apples and peaches.

Glyoxime (ethane diol dioxime) was introduced by Ciba Geigy as 'Pikoff', a citrus abscission agent for pre-harvest use on oranges.

From the U.S.S.R. Rakitin and Rakitin (1977) have proposed new ethylene-producing defoliants. These are organomercury compounds containing a 2-hydroxyethyl group. They have the general formula

RCH_2CH_2HgX, where *R*=—OH or CH_3COO— and X=—OH,—ClO_4 or —CH_3COO group. They include 2-hydroxyethylmercuric hydroxide; 2-hydroxyethylmercuric perchlorate; and 2-acetoxyethylmercuric acetate. They exert their effect by releasing ethylene as they are decomposed in the plant.

Ethylene-releasing compounds – structural formulae

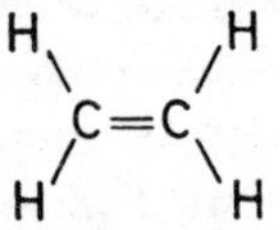

$H_2C{=}CH_2$

Ethylene

$ClCH_2CH_2P(=O)(OH)_2$

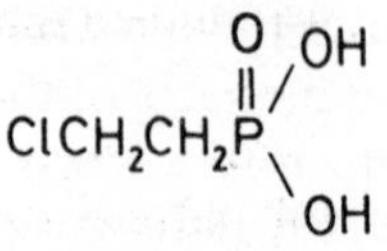

Ethephon

$(CH_3OCH_2CH_2O)_3SiCH_2CH_2Cl$

Etacelasil

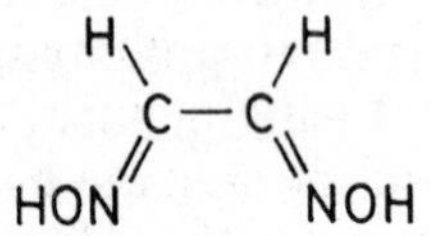

$HON{=}CH{-}CH{=}NOH$

Glyoxime

$(C_6H_5{-}CH_2{-}O)_2Si(CH_3){-}CH_2{-}CH_2{-}Cl$

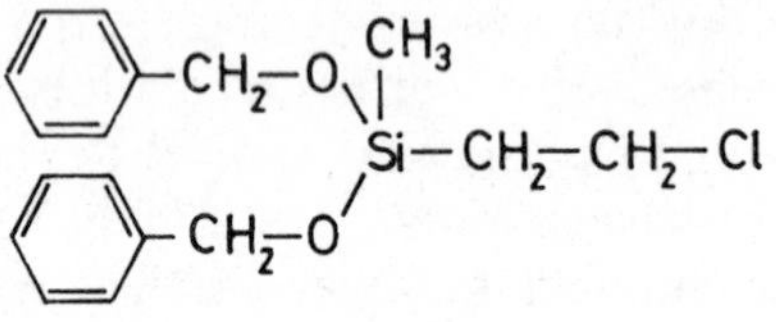

CGA 15281

Natural growth inhibitors

There are many natural inhibitors found in plants of which abscisic acid and the closely related xanthoxin are the best known: they have not found commercial application but are used as research tools.

Abscisic acid (ABA) (penta-2,4-dienoic acid-5-[1-hydroxy-2,6,6-trimethyl-4-oxocyclohex-2-en-1-yl]-3-methyl) in spite of its name, seems to be more involved in bud dormancy than in leaf shedding. The hypothesis that inhibitors are involved in bud dormancy was first put forward by Hemberg (1949). Work by Wareing (1954) using *Acer pseudoplatanus* and *Betula pubescens* led him to the conclusion that the bud inhibitor is transmitted from the leaves (maintained under short days). Subsequently Wareing and his group (Robinson *et al.*, 1963; Robinson and Wareing, 1964) isolated this inhibitor from *Acer* leaves and they named it 'dormin'. Over the same period Addicott and his group in the United States (Okhuma *et al.*, 1963, 1965) were studying a substance present in the young cotton fruits which accelerates their abscission. They called this substance 'abscisin II' and they identified it as a sesquiterpenoid. Further investigations by Cornforth *et al.* (1965a, 1965b) showed that dormin and abscisin II were identical and it was agreed that this substance should be called abscisic acid (ABA).

ABA has a wide range of physiological effects when applied externally. It will induce abscission of petiole bases in explants of cotton seedlings and induce rapid senescence of leaf discs of various species (though less so in intact plants). It acts as a growth inhibitor and antagonises the effect of auxins, gibberellins and cytokinins in certain tests. It is believed to play an important part in the dormancy of buds and certain types of seeds. Applied to leaves it causes rapid closure of the stomata and may have an important role as a potential antitranspirant though it is degraded so rapidly in plant tissues that its effects are short lived. It has been shown to induce flower production in species that otherwise require short days and inhibition of flowering in others that require long days.

Natural growth inhibitor – structural formula

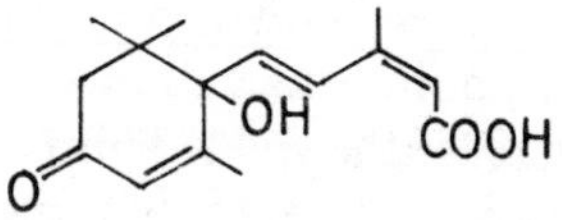

OH COOH O

Abscisic acid (ABA)

Synthetic growth inhibitors and retardants

The search for new chemicals which inhibit or retard growth is a very active field of research. Some are thought to act by inhibiting gibberellin synthesis, others by inhibiting terminal or lateral bud development affected by auxin distribution. Some are used to reduce the height of cereals and thus prevent lodging, some to retard the growth of grasses and shrubs, some to interfere with apical dominance and thus change the shape of a plant, some as defoliating agents and some to prevent sucker (axillary bud) development. In addition however to their morphological effects these inhibitors and retardants may also affect flowering, fruit development, sucrose accumulation and other physiological activities.

Maleic hydrazide (MH) is the common name for 6-hydroxy-3-(2*H*)-pyridazinone. Its plant growth activity was first described by Schoene and Hoffman (1949) and it was introduced in 1948 as 'MH-30' by the U.S. Rubber Co. Within treated plants it inhibits cell division. It is used for suppressing sprouting in stored onions, to retard the growth of hedges and grass verges and for prevention of sucker development in tomato and tobacco.

Propham and **chloropropham** (see chapter 1) are also cell division inhibitors and are used as sprout inhibitors in ware potatoes stored in buildings.

***S,S,S*-Tributyl phosphorotrithioate** was introduced by the Chemagro Corpn. in 1956 as a defoliant for use on cotton.

***S,S,S*-Tributyl phosphorotrithioite** was introduced by the Mobil Chemical Co. in 1957 as a cotton defoliant.

Chlormequat is the common name for the (2-chloroethyl)trimethylammonium ion; it is also known as CCC from the trivial name chlorocholine chloride. Its effects on plant growth were first described by Tolbert (1960a, 1960b) and it was introduced as 'Cycocel' by the American Cyanamid Co. in 1959. It is a retardant of plant growth and is used to prevent lodging of wheat and oats by shortening and thickening the lower internodes of the stem. It may also stimulate root development and increase the number of fertile tillers.

It is also used to control the growth of pot plants and flower crops such as chrysanthemums, poinsettias, lilies and azaleas. Used on apple and pear trees it may increase the yield (Thomas *et al.*, 1977). It may be used to induce early flowering in pears but may cause chlorosis of the foliage in the year of application (Luckwill, 1976). The action of growth retardants in promoting flower initiation is thought to depend on their ability to block the synthesis of endogenous gibberellins which are known to be inhibiting to flowering.

Daminozide is the common name for *N,N*-dimethylaminosuccinamic acid. Its effects on plant growth were first described by Riddell *et al.* (1962) and it was introduced as 'Alar' and 'B-Nine' by Uniroyal Inc. in the same year. It is used as a plant retardant for the vegetative growth of fruit trees and the modification of stem length and shape of ornamental plants. In Europe the use of daminozide has been restricted to apples and pears where earlier cropping, improved fruit colour, increased firmness and a reduction in pre-harvest drop can be achieved, but plums, cherries, vines and peaches are also responsive to this treatment and in the U.S.A. it is used to advance colour development and improve fruit uniformity in cherries. It enhances the fruit set of seedless grapes. It is used on bedding plants and shows promise for the promotion of rooting in cuttings of a number of ornamental species (Thomas *et al.*, 1977). On apples it reduces internode length giving a more compact type of tree and encouraging the early formation of fruit buds and its effect can be enhanced if it is combined with ethephon (Luckwill *et al.*, 1975).

Glyphosine (*N,N*-bis[phosphonomethyl]glycine), with or without surfactants (depending on the variety of cane), has been shown to give substantial (10–15%) increases in yield in cane sugar from treated plants (Nickell, 1974). It was introduced as 'Polaris' by Monsanto Co. in 1969.

Ancymidol (α-cycloprop-4-methoxy-α-[pyrimidin-5-yl]benzyl alcohol) was introduced by Elanco Products (Eli Lilly and Co.) in 1971, as 'A-rest' and 'Reduymol', as a growth regulator which causes stunting of a range of greenhouse ornamental plants, such as chrysanthemums, poinsettia, dahlia and many others, by reducing the internodal elongation. It has also been recommended for the control of stem elongation of ash and maple trees (Frank *et al.*, 1978). A related compound **EL 500** (α-[1-methylethyl]-α-[4-(trifluoromethoxy)phenyl]-5-pyrimidinemethanol) has also been produced by the same company for the regulation of turf growth.

Chlorphonium chloride (tributyl [2,4-dichlorobenzyl]phosphonium chloride) was introduced by the Mobil Chemical Co. as 'Phosphon'. It is a height retardant used as a compost additive on pot chrysanthemum where it causes a shortening of the internodes to form a more compact plant. It may also be used on rhododendrons, and on roses it has been reported to cause more laterals and increase the number of flowers, though flower size was reduced (Nanjan and Muthswamy, 1975).

Bocion *et al.* (1975) described some of the features of a new plant growth regulator, **dikegulac sodium** (2,3:4,6-di-*O*-isopropylidene-α-L-xylo-2-hexulofurnandonate) produced as 'Atrinal' by R. Maag Ltd, Switzerland as a pinching agent for ornamentals and as a growth retardant and branching agent for hedge plants. It is systemic, reduces apical dominance, encourages the formation of side branches, and increases flower bud formation on ornamentals. It has been shown to decrease the height of *Poa pratensis*, *Lolium perenne*, *Digitaria sanguinalis* and barley. Application has also increased the abscission of the petioles of *Phaseolus vulgaris*. It will also reduce the height and increase the number of shoots of fuchsia, cyclamen and begonia, increase the branching of rhododendrons and act as a pinching agent on azaleas. It will stimulate flower development in a number of pot plants. Using a number of bioassay systems Bocion and Silva (1977) have shown that dikegulac retarded the growth of wheat, pea and wild oats (*Avena fatua*) and that this effect could be overcome by GA_3. They have also found dikegulac to stimulate parthenocarpic fruit set of pears.

Morpholinium chloride (*N*,*N*-dimethylmorpholine chloride) was introduced by B.A.S.F. in 1974 as a stalk shortener for use on cereals, cotton, tomato, vines and various ornamentals (B.A.S.F. Technical Data Sheet, 1975).

Fosamine (ammonium ethyl carbamoyl phosphonate) was introduced by Du Pont for the control of woody weeds. It controls new growth without completely destroying the plant making it suitable for use in residential areas because environmental damage is minimal. It acts only on contact area and has therefore possibilities for chemical pruning (Dodel, 1975).

Mefluidide (5-[trifluoromethanesulphonamido]acet-2,4-xylidide) was introduced by 3M Co. for retarding turf and brush growth. Growth and seed head production of grasses may be suppressed for 3–8 weeks. It retards the growth of many ornamental species and controls the formation of suckers in tobacco (Fridinger *et al.*, 1975). It has given excellent post-emergence control of *Sorghum halepense* and *Sesbania exaltata* in soybeans. It has been shown to be a sugar-cane ripener, increasing sucrose levels by 20% and it reduced red winter wheat height with a consequent increase in yield by up to 54% (Miles *et al.*, 1977).

4-Methoxybenzophenones introduced by the Nippon Kayaku Co. Ltd show diverse activities against rice, barnyard grass (*Echinochloa crus-galli*), crabgrass (*Digitaria* spp.), lucerne, tomato and turnip. The 2-carboxy sodium salt derivative inhibits the roots but not the shoots of *D. adscendens*. Compounds substituted with a halogen, an amino group or a cyano group eliminate the geotropic response of tomato and other test plants and the 2-carboxy derivative disturbs the phototropic response in rice.

PP 528 (ethyl-5-[4-chlorophenyl]-2*H*-tetrazol-2-yl acetate) was introduced by I.C.I. Ltd as a chemical pinching agent. When sprayed on foliage it is translocated to the growing points of shoots which are then arrested or interrupted. It is suggested that this will have great practical use in feathering and chemical pruning of fruit trees, particularly apple, plum and cherry. Some varieties have shown increase in flowering and fruit yield. Significant increases in yield of soybeans have been reported and the use of the free acid form indicates the possibility of increases in yield of groundnuts (I.C.I. Technical Data Sheet, 1976). Cristoferi *et al.* (1978) analysing shoot apices of apple trees, 5 days after treatment, have noted a reduction in levels of auxin and gibberellin-like substances.

Piproctanyl bromide (bromide salt of 1-allyl-1-[3,7-dimethyloctyl]-piperidinium) was introduced by R. Maag Ltd in 1976. It reduces internodal elongation and is used on a variety of potted plants. It has been reported to give excellent height control of chrysanthemums when applied either as a foliar spray or as a compost drench, though much higher concentrations were required when using the latter method (Menhenett, 1977).

2-(3-Aryl-5-pyrazoyl)benzoic acids were introduced by Du Pont as a new class of auxin transport inhibitors. They reduce basipetal auxin transport in bean plants (Beyer *et al.*, 1976).

BTS 34723 (1-[*N*-2-phenoxyethyl)-*N*-propylcarbamoyl]-1*N*-imidazole) and **BTS 34442** (1-[*N*-2,4-dichlorobenzyl]-*N*-isopropylcarbamoyl-1*N*-imidazole) were introduced by Boots Co. Ltd. They are foliar-acting growth retardants which increase the sucrose concentration and juice purity of sugar-cane (Boots Co. Ltd Technical Data Sheet, 1976). They also reduce the height of wheat, barley, sorghum and rice.

UB1 P293 (2,3-dihydro-5,6-diphenyl-1,4-oxathiin) was introduced by Uniroyal Inc. as a localised growth inhibitor. It blocks cell division and expansion, although well-formed tissues are not affected and develop normally. When applied to chrysanthemum plants initiating flowers it causes chemical disbudding by stopping development of laterals but the terminal bud expands into a normal inflorescence. When applied to vegetative plants it stops development of the top 8–10 nodes but lateral shoots develop at the same rate, number and weight as those on manually pruned plants.

M & B 25.105 (propyl 3-t-butyl phenoxyacetate) was introduced by May & Baker Ltd as a plant growth regulator for temperate fruit in 1979. In apple, pear and cherry tree trials throughout Europe it was shown to suspend the activity of apical buds thus promoting the formation of lateral shoots. The subsequent growth of the leader and the lateral shoots encourages wider angled branching, leading to improved blossom counts in the second and third years, bringing trees to commercial cropping earlier. It also acts as a fruitlet thinner. It has systemic activity (Hibbit and Hardisty, 1979). It may also be used for blossom/fruitlet thinning, defoliation and cluster thinning of grapes. Trial results indicate that it may increase the yield of leguminous, sunflower and cotton crops.

Thidiazuron (*N*-phenyl-*N'*-[1,2,3-thiadiazol-5-yl]urea) was introduced by Schering A.G. in 1976 as a defoliant of cotton (Schering Information, 1977). The leaves drop while still green. It appears to be absorbed by the leaves and accelerates the formation of the natural abscission layer between plant stem and petioles causing defoliation in 7–21 days.

Mepiquat (1,1-dimethylpiperidinium chloride) was introduced by B.A.S.F. for use on cotton. It shortens internodes giving smaller cotton plants with intensification of chlorophyll synthesis. More cotton bolls may reach maturity achieving harvestability in time for the first picking. Bolls are heavier, giving increased yields and disease may be reduced (B.A.S.F. A.G., 1978). In field trials throughout the U.S.A. it was shown that when mepiquat was applied at the early flowering stage of cotton, plant height was reduced by 20–40%, the length of lateral branches was reduced by 40%, yields of seed cotton were increased by up to 18% and boll retention was increased by 17% (Willard *et al.*, 1977).

BAS 09800W ('Terpal') (mepiquat chloride + ethephon) was introduced by B.A.S.F. It is a growth retardant for winter and spring barley, rye and oats. It shortens the stem, increases its diameter and strengthens the stem wall, thus improving resistance to lodging and making it difficult for the fungus causing eye-spot and stem break (*Cercosporella herpotrichoides*) to enter the stem (B.A.S.F. Technical Data Sheet, 1977).

IZAA (5-chloroindazole-8-acetic ethyl ester) has been studied in Japan since 1971 as a thinning agent for satsuma mandarin oranges. It gave satisfactory thinning even 50 days after anthesis – compared with 30 days with NAA (Hirose *et al.*, 1977).

MON 8000 (contains glyphosate (*N*-[phosphonomethyl]glycine) sodium 60% a.i.) was introduced by Monsanto as a sugar-cane ripener. According to Brown (1978) results were equal to or better than those obtained with 'Polaris' (glyphosine-*N,N*,bis-[phosphonomethyl]glycine) producing maximum response 4–6 weeks after application while 'Polaris' needs 6–10 weeks. It is able to induce enhancement of sucrose under conditions (e.g. irrigated cane when drying off) which render 'Polaris' ineffective.

DOWCO 242 (tetraisopentylammonium bromide) was introduced by the Dow Chemical Co. Wang and Huang (1978) investigating its effect on soybean plants have shown that root growth unlike stem growth is not affected by the chemical and that the older the plant the less effect there is on stem growth. Leaf number and leaf area are not affected. They state that the chemical acts by reducing the GA_3 and IAA content in the plants.

Twelve new **quaternary ammonium iodides** have been synthesised and their growth-retarding activities examined by Cho *et al.* (1978) at the Institute Phys. Chem. Res. in Japan. *N,N,N*-Trimethyl-1-methyl-3-(3,3,5-trimethylcyclohexyl)- and *N,N,N*-trimethyl-1-methyl-3-

(3,3,5,5-tetramethylcyclohexyl)-2-propenylammonium iodides were most effective in suppressing growth of rice and cucumber and according to the authors far stronger than those of any growth retardants hitherto known.

Synthetic growth inhibitors and retardants – structural formulae

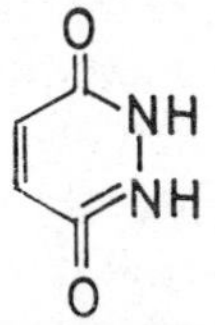

Maleic hydrazide (MH)

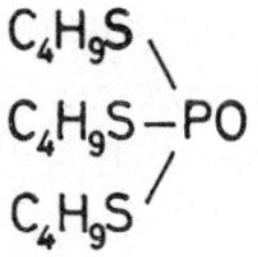

$(C_4H_9S)_3PO$

S,S,S-Tributyl phosphorotrithioate

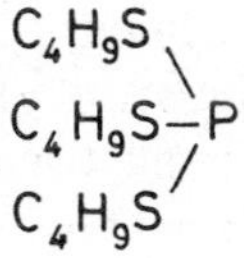

$(C_4H_9S)_3P$

S,S,S -Tributyl phosphorotrithioite

$ClCH_2CH_2N^+(CH_3)_3 Cl^-$

Chlormequat chloride

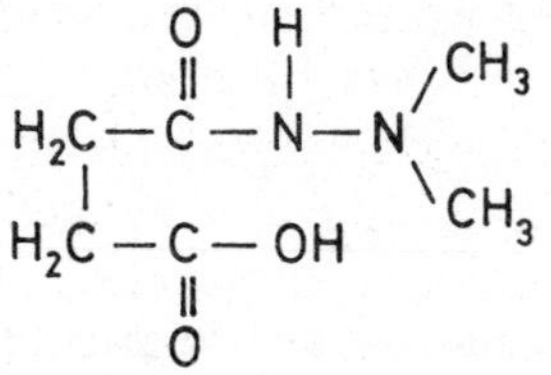

$H_2C(CONHN(CH_3)_2)-H_2C-COOH$

Daminozide

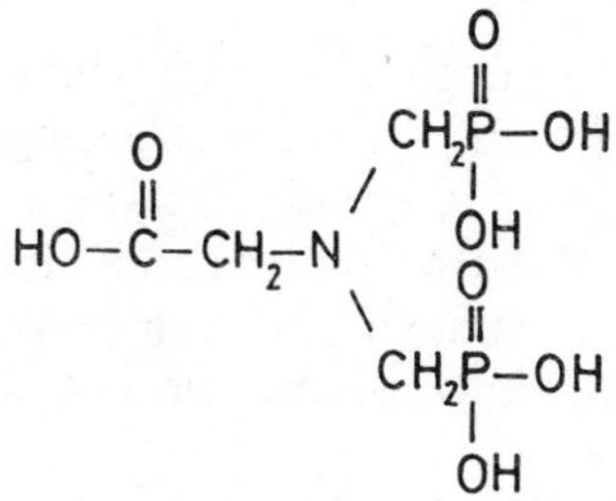

$HO-CO-CH_2-N(CH_2PO(OH)_2)_2$

Glyphosine

$N-NCH_2COOC_2H_5$, $N{=}N$, Cl

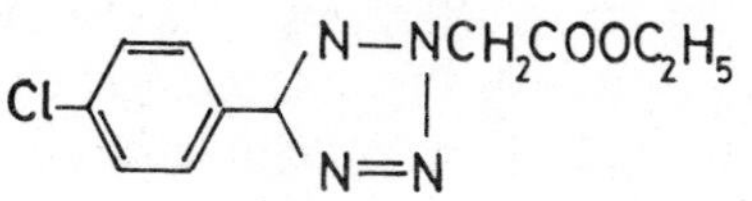

PP 528

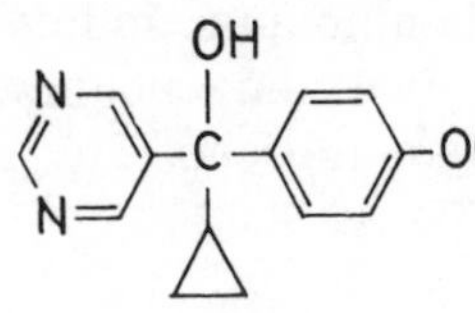

OH
N N C OCH_3
Ancymidol

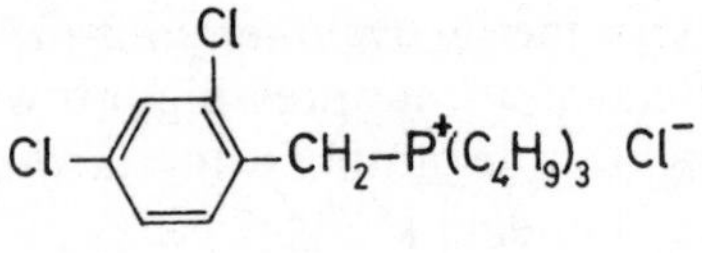

Cl
Cl $CH_2{-}P^{+}(C_4H_9)_3$ Cl^-
Chlorphonium chloride

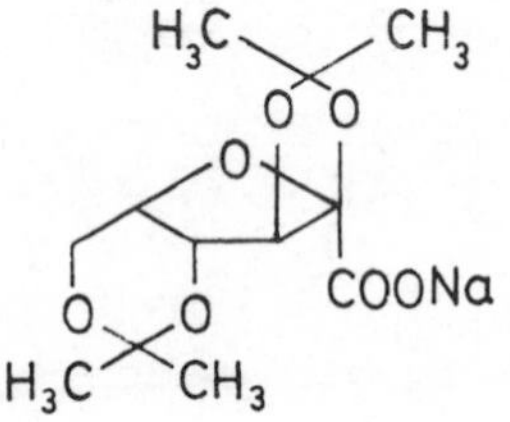

H_3C CH_3
O O
O
COONa
O O
H_3C CH_3
Dikegulac sodium

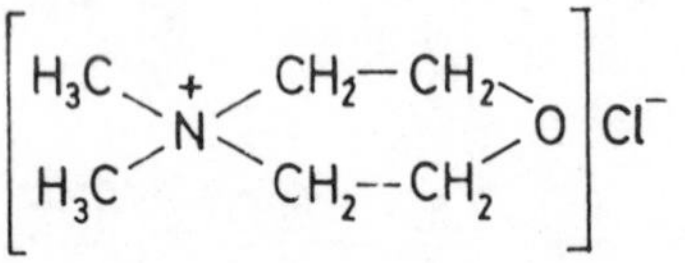

H_3C + $CH_2{-}CH_2$
N O Cl^-
H_3C $CH_2{-}CH_2$
Morpholinium chloride

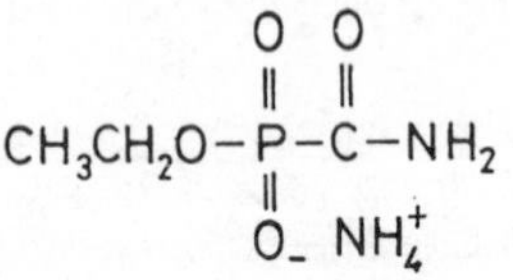

O O
$CH_3CH_2O{-}P{-}C{-}NH_2$
O_- NH_4^+
Fosamine

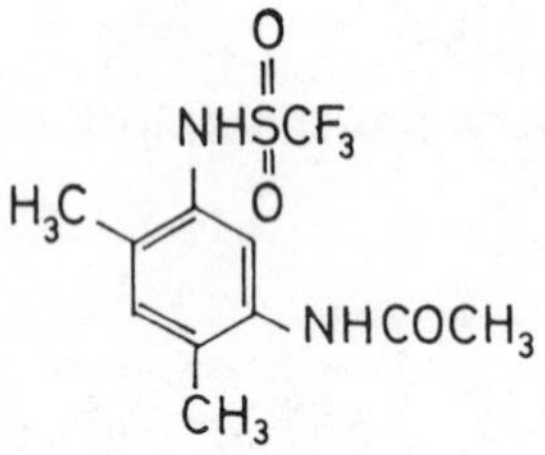

O
$NHSCF_3$
H_3C O
$NHCOCH_3$
CH_3
Mefluidide

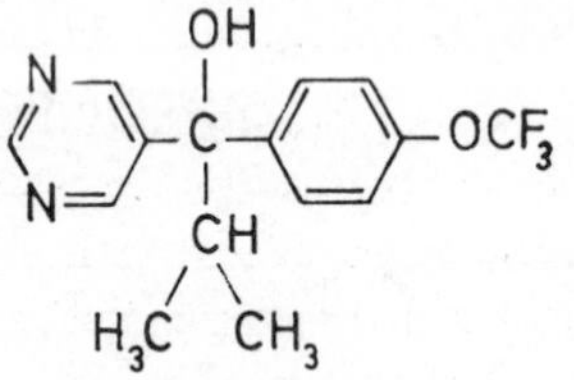

OH
N C OCF_3
N CH
H_3C CH_3
EL 500

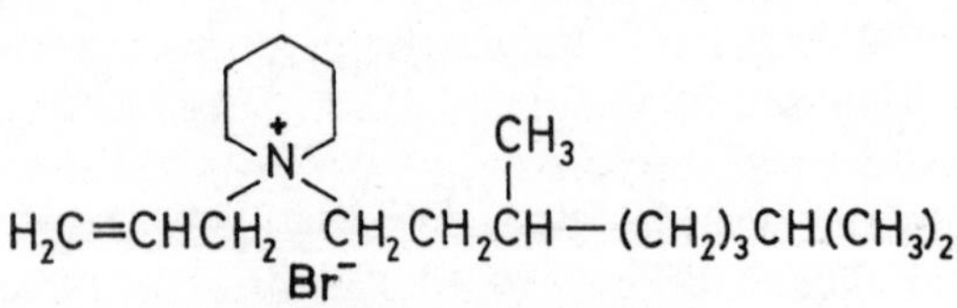

+
N CH_3
$H_2C{=}CHCH_2$ $CH_2CH_2CH-(CH_2)_3CH(CH_3)_2$
Br^-
Piproctanyl bromide

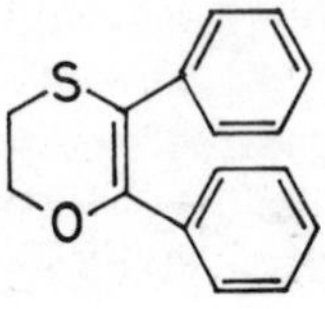

S
O
UB1 P 293

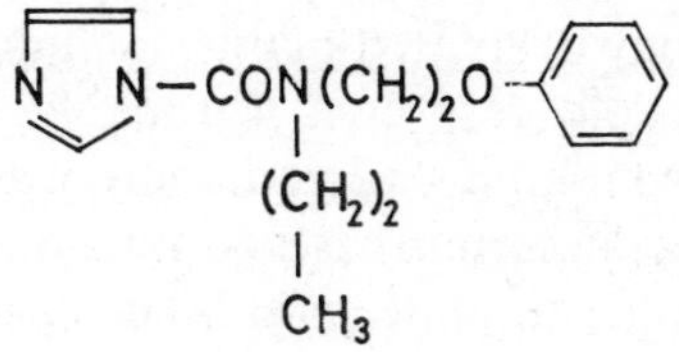

BTS 34 723

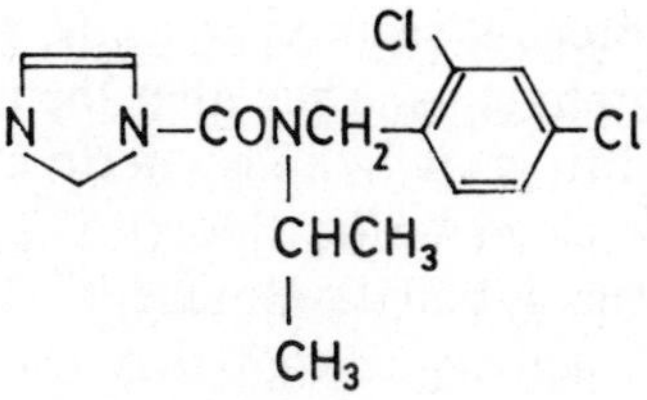

BTS 34 442

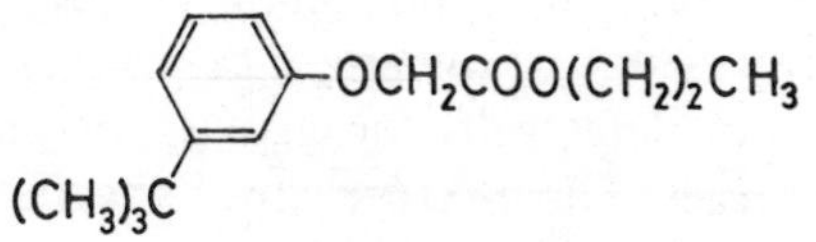

M & B 25.105

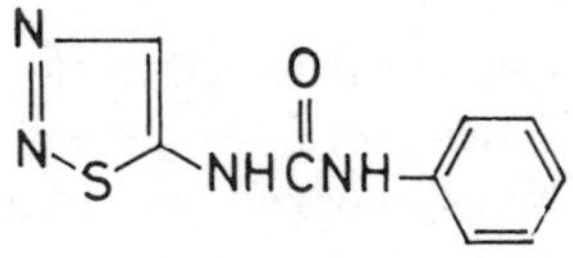

Thidiazuron

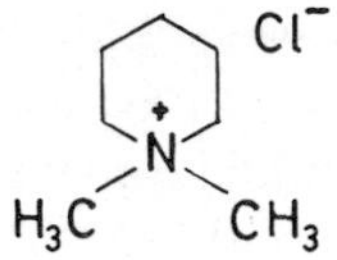

Mepiquat

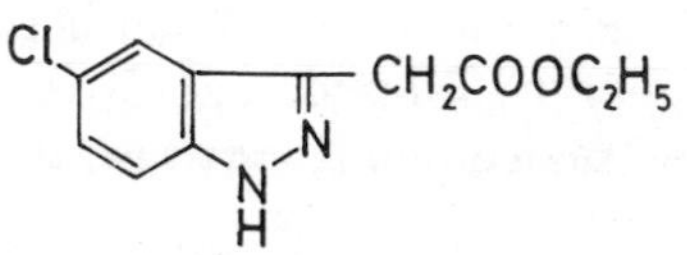

IZAA

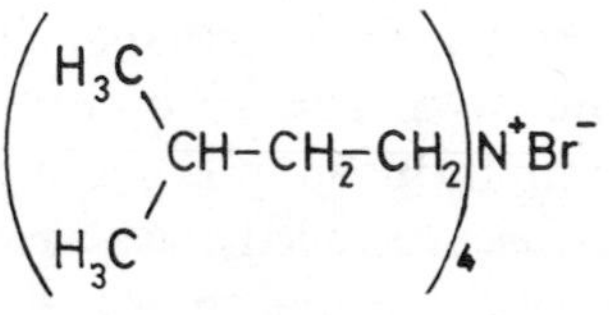

DOWCO 242

Morphactins

These are synthetic morphogenetically acting substances which are potent inhibitors of auxin transport, thus affecting tropic responses, reducing apical dominance and promoting lateral growth. Chlorflurecol is the best known but there are others. They are widely used to suppress growth in ornamentals.

This is an interesting group of synthetic growth regulators developed first in Germany by Schneider (1964, 1965, 1970) as a result of his investigation into the growth-regulatory activities of derivatives of fluorene-9-carboxylic acids. It was he who first proposed the name morphactins. They alter the morphology of treated plants, including apical growth and interfere with tropic responses (probably by affecting IAA transport). They may also inhibit seed germination and dormant bud development. They appear in certain cases to antagonise GA activity in that they prevent the 'bolting' of vernalised rosette plants. They have low toxicity, show a high degree of selectivity, are not too persistent and are active over a wide range of concentrations.

Chlorflurecol-methyl (chlorflurenol-methyl) (methyl 2-chloro-9-hydroxyfluorene-9-carboxylate) was introduced by E. Merck in 1965. It inhibits apical growth, thus stimulating lateral branching. It may be used with maleic hydrazide to restrain the growth of grasses and dicotyledons on roadside verges. The mixture should be applied when the grasses are growing actively but before the flower heads appear. Most grasses and other weeds are suppressed for 10–14 weeks but cow parsley (*Anthriscus sylvestris*) and hogweed (*Heracleum sphondylium*) are controlled.

Flurecol-butyl (butyl 9-hydroxyfluorene-9-carboxylate) introduced by E. Merck in 1964 is an interesting morphactin in that although it does cause inhibition of plant growth (being absorbed by the leaves) it is mainly used in conjunction with the phenoxyalkanoic acid herbicides, whose action it potentiates, for the control of weeds in cereals.

TIBA (2,3,5-tri-iodobenzoic acid) is used in the United States on soybeans. Treated plants have a compact form and can be planted in close rows. Due to improved plant shape light is distributed better within the crop, pod set increased and maturity accelerated. Striking increases in yield have been obtained and treated plants are resistant to lodging (Thomas *et al.*, 1977).

Morphactins – structural formulae

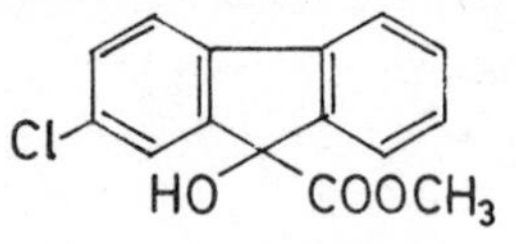

Cl
HO $COOCH_3$

Chlorflurecol-methyl

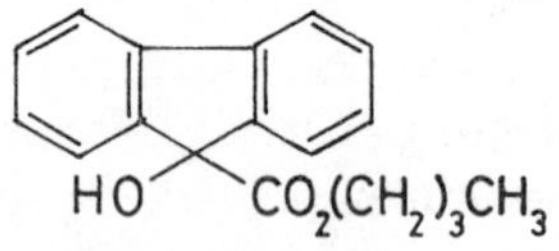

HO $CO_2(CH_2)_3CH_3$

Flurecol-butyl

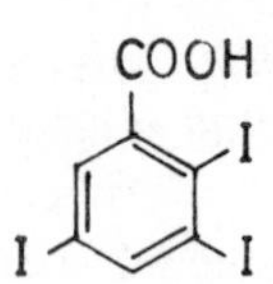

COOH
I
I I

TIBA

Gameticides

The inhibition of pollen production or the inducing of pollen sterility has an important part to play in cereal breeding programmes. Research in the United States and elsewhere has demonstrated that greater yields, and other advantages, may be got from hybrid wheats compared with pure line varieties. Seed production for hybrid wheats depends of course on male sterility in the seed-producing plant. It is not easy to control self-pollination and the use of chemical gametocides would have obvious advantages.

According to Batch (1978) two compounds with known gametocide activity are **RH 531** (sodium 1-[4-chlorophenyl]-1,2-dihydro-4,6-dimethyl-2-oxonicotinate) and **DPX 3778** (3-[4-chlorophenyl]-6-methoxy-1,3,5-triazine-2,4-dione triethanolamine).

Johnson and Brown (1976) found that foliar sprays of DPX 3778 reduced or prevented anther dehiscence in several wheat and oat varieties. The effect was more pronounced in wheat and they consider that the chemical has potential use in wheat breeding programmes. The Canadian Agricultural Research Institute Report (1975–79) records that the use of the gametocide DPX 3778 on soft white winter wheat in the field gave some 50–100% male sterility.

Hanna (1977) found that the same chemicals prevented anther dehiscence in pearl millet. Wang and Lund (1975) reported that under greenhouse conditions the application of RH 531 applied several days before meiosis in the pollen mother cells induced complete male sterility in three cultivars of barley. The pollen sterility was apparently related to an absence or decrease of carbohydrates. RH 531 inhibited anther growth leading to a degeneration of the pollen grains or a reduction of the carbohydrates in the pollen grains.

Miller (1976) found that RH 531 induced sufficient male sterility for the production of hybrid grain. But it also reduced female fertility and induced morphological changes in the plant.

Rajendra *et al.* (1978) applied RH 532 to the leaves of *Triticum aestivum* and found that many pollen abnormalities developed particularly at the meiotic stages resulting in multisporate pollen and abnormal exine differentiation. Anthers exhibited a persistent tapetum and failed to dehisce.

It is not only the chemicals mentioned above that induce sterility in cereals. Reich and Martin (1976) reported that ethephon sprays applied to two varieties each of winter barley and winter wheat induced male sterility in three of the varieties though not in the fourth – Barsoy.

It is not, however, only in cereals that gametocides have been used and it is not only the chemicals that have been mentioned above that affect sexuality in plants.

Orth and Leuchs (1961) found that 2,3-dichlorobutyric acid inhibited mitosis in the anther filaments of *Tussilago farfara* thus preventing seed set and this chemical has been successfully used to prevent the spread of the weed.

Rudich and Halevy (1974) found that GA_4+GA_7 increased male flower formation in a gynoecious line of cucumber but that this could be offset by application of abscisic acid (ABA). They suggest that ABA antagonises the activity of GA in promotion of maleness thus enhancing female sex expression. The Annual Report of the Institute for Horticulture and Plant Breeding (1979) states that GA_4+GA_7 gave complete male sterility in onions but it reduced seed setting. It also induced male sterility in Brassicas.

Chailakhyan and Khryapnin (1977) found that when they put the roots of seedlings of hemp (*Cannabis sativa*) in 25 mg/1 GA for 24 h and then allowed the plants to grow normally, then the percentage of male flowers increased from 28% to 84% while the percentage of female and hermaphrodite flowers fell.

Varna (1974) reported that **TH 6241** (5-methyl-7-chloro-4-ethoxy-carbonylmethoxy-2-[1,3-benzothiazole]) has similar effects to that of GA on plants in that it induced staminate flowers in a strongly gynoecious line. The effect persisted into the second generation suggesting a mutagenic effect. Hirose and Fujime (1973) found that sodium salts of dalapon suppressed anther dehiscence in pepper. Ghosh and Sen (1975) whilst showing that foliar applications of GA and MH enhanced maleness in papaya (*Carica papaya*) and, interestingly enough, that the animal sex hormone, testosterone propionate increased the proportion of male plants, found that NAA and chlormequat applied by the same route led to a higher female:male ratio.

Chailakhyan and Khryapnin (1977) showed that treatment of the leaves of *Cannabis sativa* with IAA or 6-benzylaminopurine gave plants with no male flowers, slightly more female flowers and appreciably more hermaphrodite flowers. Treatment with ABA also increased the percentage of hermaphrodite flowers and reduced the percentage of male flowers without affecting the percentage of female flowers.

It has been reported that caprylic acid, pelargonic acid, capric acid, Δ^2-decanoic acid and β-hydroxydecanoic acid (myrmicacin) inhibit pollen germination and pollen tube elongation in *Camellia sinensis* and mitotic division of the generative nucleus in *Ornithogalum virens* (Iwanami and Iwadare, 1979). The authors propose the name 'myrmic acids' (MYA) for these new inhibitors.

Many pollens are capable of inducing allergies in humans and the use of gametocides may have a role in preventing pollen production particularly in cases where it is not desirable to destroy the plant. Grigsby (1945) found that spraying ragweed (*Ambrosia* spp.) with 2,4-D prevented the release of pollen but did not harm the plant. Ambrosia is useful in certain situations for the prevention of soil erosion.

Gametocides – structural formulae

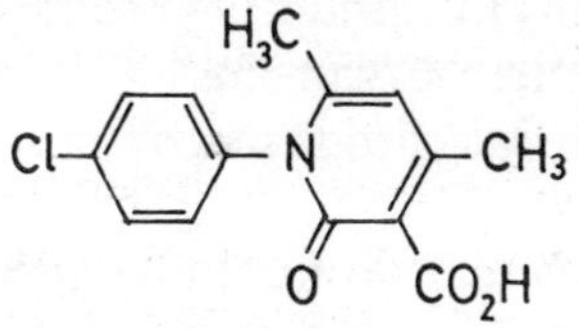

RH 531

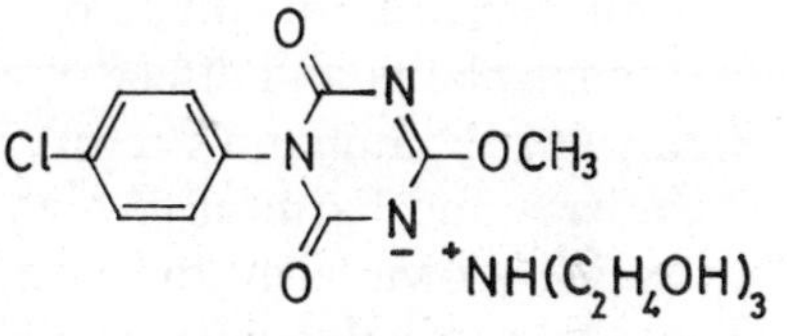

DPX 3778

Chemical pinchers

'Pruning' mixtures, the active ingredients of which are methyl esters of the 9- and 10-carbon chain fatty acids, are used to kill shoot tips, thus inhibiting apical or lateral shoot tip development. They are not translocated. They have been used on chrysanthemums and to control primary and secondary sucker development in tobacco. Mixtures of the methyl esters of C_6, C_8, C_{10} and C_{12} fatty acids applied at concentrations of 2–3% in mid-summer are generally effective on apple. pear and plum trees but not on sweet cherry (Quinlan and Preston, 1973). The chain length of such chemical pruning agents is critical since they act on the basis of differential penetration through young and old cuticles (Luckwill, 1976).

Allelopathins

De Candolle (1832) was apparently one of the first scientists to suggest that some plants may excrete something from their roots which is injurious to other plants observing that thistles (*Cirsium*) in fields injure oats; *Euphorbia* and *Scabiosa* injure flax; and rye plants (*Lolium*) injure wheat. The term allelopathy was coined by Molisch

(1937) to include both detrimental and beneficial biochemical interactions between plants at all levels of complexity, including micro-organisms, although it is derived from two Greek words meaning mutual harm. It is concerned with the *addition* of a chemical compound to the environment. Grümmer (1955) suggested that the following special terms be adopted for the chemical inhibitors involved in allelopathy, based upon the type of plant producing the inhibitor and the type of plant affected:

1. *antibiotics* for chemical inhibitors produced by micro-organisms and effective against micro-organisms
2. *phytoncides* (a term coined by Waksman) for inhibitors produced by higher plants and effective against micro-organisms
3. *marasmins* (Gaumann's term) for chemicals produced by micro-organisms and harmful to higher plants
4. *kolines* (his own term) for chemical inhibitors produced by higher plants and effective against higher plants.

It is recognised that there may be some overlaps between the groups, e.g. many kolines may inhibit the growth of micro-organisms and many phytoncides may inhibit the growth of higher plants.

The subject of allelopathy as it relates to phytoncides and kolines has been reviewed by Rice (1974) whose book contains more than 600 references. Determining the chemical nature of allelopathins is a difficult problem as they appear to be produced in small amounts, but Rice has listed them as falling into the following groups:

1. Simple water-soluble organic acids, straight-chain alcohols, aliphatic aldehydes and ketones
2. Simple unsaturated lactones
3. Long-chain fatty acids
4. Naphthoquinones, anthroquinones and complex quinones
5. Terpenoids and steroids
6. Simple phenols, benzoic acid and derivatives
7. Cinnamic acid and derivatives
8. Coumarins
9. Flavonoids
10. Tannins
11. Amino acids and polypeptides
12. Alkaloids and cyanohydrins
13. Sulphides and mustard oil glycosides

14. Purines and nucleosides
15. Miscellaneous, e.g. ethylene, agropyrene and some unidentified inhibitors.

Rice (*loc.cit.*) lists the parts of plants that are known to contain inhibitors. They are: *stems* (in some instances the chief source of toxicity); *leaves* (the most consistent source of inhibitors); *roots* (fewer and less potent or at least smaller amounts than leaves); *flowers and inflorescences* (only a few have been investigated but it is now known that many flowers contain high concentrations); *fruits* (generally neglected but toxins are present in many of them); *seeds* (widely assayed for presence of inhibitors of seed germination which many contain).

The following account relates to some of the work that has been done since 1974. It is not a comprehensive coverage but it does deal with some of the more interesting findings. Work prior to 1974 has of course been comprehensively covered by Rice (loc.cit.). Many allelopathins affect the germination and early growth of other species. Thus Zaki and Tewfik (1974) found that root exudates from flax stimulated the germination of seeds of two *Orobanche* spp. and Krishnamurthy and Chandwani (1975) reported that the germination of seeds of *Orobanche cernua* (which is a parasite of tobacco in India) was increased from less than 1% (control) to over 50% in the presence of chilli (*Capsicum*), sorghum and cowpea (*Vigna unguiculata*) plants. On the other hand, Delui (1974) found that when seedlings of wheat were grown in the presence of seeds of any of a number of root hemiparasites such as *Odonites rubra*, *Pedicularis palustris*, *Rhinanthus romelicus*, *Melampyrum arvense* and *Euphrasia stricta* then the elongation of roots and coeoptiles was reduced by 21–68%. He concluded that inhibitors can diffuse from seeds and affect hosts before forming appresoria. Fabian *et al.* (1974) found that the presence of the seeds of red fescue (*Festuca rubra*) inhibited the germination and seedling growth of red clover (*Trifolium pratense*) and conversely the presence of the legume inhibited the germination and growth of red fescue.

It is not, however, only germination that is affected. There are many examples on record of a direct effect on other plants. The following are but a few of the recent reports.

Hozumi *et al.* (1974) found that sowing mixtures of rice and barley together reduced the growth of the rice plants and Sagar and Ferdinandez (1976) demonstrated that the rhizomes and roots of *Agropyron repens* had a severe repressive effect on the root system of wheat. Friedman *et al.* (1977) reported that *Artemisia herba* inhibits the growth of annual plants in the Negev and Caussanel *et al.* (1976)

showed that maize growth is inhibited in the presence of *Chenopodium album*.

Newman and Rovira (1975) grew eight common grassland species individually in sand culture in buckets. The leachate was allowed to pass through holes in the bottom of the bucket, collected and applied to pots in which the eight species were growing separately. They were allowed to grow for eight months and then harvested. All leachates were inhibitory effecting a 7–15% reduction in dry matter. Some, e.g. *Lolium perenne*, *Plantago lanceolata*, and *Trifolium repens*, grew more slowly when supplied with their own leachate. Others, e.g. *Holcus lanatus, Anthoxanthrum odoratum* and *Cynosurus cristatus*, grew faster. Only *Rumex acetosa* was unaffected. The authors suggest that these results could help to explain why some species are strongly dominant in grassland whereas others are interspersed with other species.

Numata (1977) believes that all weeds in arable lands probably show allelopathy and that they excrete substances which maintain their temporary predominance.

There can be considerable variation in the amount of allelopathic substances released by individual plants within the same species. Thus Putnam and Duke (1974) sowed seeds from over 500 lines of cucumber from forty-one countries, in pots together with either *Brassica hirta* (=*Sinapis alba*) or *Panicum miliaceum* as indicator species. One cucumber seedling inhibited *P. miliaceum* growth by 87% and twenty-five seedlings inhibited growth of one or two other indicator species by 50% or more. Two of these cucumber accessions inhibited the germination of red root pigweed (*Amaranthus retroflexus*), barnyard grass (*Echinochloa crus-galli*) and pros millet (*Panicum miliaceum*). Fay and Duke (1977) assessing the allelopathic potential in *Avena* germ plasm screened 3000 accessions of the U.S.D.A. world collection of *Avena* sp. for their ability to exude scopoletin (an allelopathin exuded by many plants). They found that twenty-five accessions exuded more blue-fluorescing materials than the standard oat cv. Garry. Four accessions exuded up to three times as much scopoletin as Garry. Massantini *et al.* (1977) grew soybean lines from several countries in hydroponic culture to examine their effect on *Helminthia echoides* and *Alopecurus myosuroides*. Out of the 141 lines tested two severely inhibited the growth of *H. echoides* while one promoted the growth of both weeds.

There are many reports of extracts of plants and plant litter having toxic effects. The following are among some of the most recent.

Arines *et al.* (1974) found that acid extracts of *Erica vagans* flowers inhibited the germination of *Trifolium pratense* and *Phleum pratense* seed and reduced hypocotyl and radicle growth of *T. pratense* and coleoptile and radicle growth of *P. pratense* by 90–100%. Acid extracts of leaves, stems and roots had no effect on the germination of both species and little effect on the hypocotyl and radicle growth of *T. pratense*. On the other hand the coleoptiles and radicles of *P. pratense* were totally inhibited. Bendall (1975) reported that water and ethanol extracts of the roots and foliage of *Cirsium arvense* inhibited the germination of its own seed and of the seed of subterranean clover (*Trifolium subterraneum*). They also inhibited the growth of its own seedlings and those of three annual thistles, perennial ryegrass (*Lolium perenne*), subterranean clover and *Hordeum distichon*. Hoffman and Hazlett (1977) found that water-soluble and aromatic compounds extracted from *Artemisia* spp. had an adverse effect on the germination of a number of grass seeds and Friedman *et al.* (1977) noting that *Artemisia herba* inhibits the growth of annual plants in the Negev found that both water-soluble and volatile substances from this plant inhibited the germination of a number of species, including *Helianthemum ledifolium* and *Stipa capanensis*, while a number of others were unaffected.

It would seem that plant extracts can affect not only the growth but also the chemical composition of affected species. Gopalakrishnan and Siracar (1973) treated seedlings of rice with root extracts from water hyacinth (*Eichhornia crassipes*) and found that after 10 days asparagine had disappeared from the shoot and there were only traces in the roots. Plant litter has been shown to have toxic effects on other plants. Rietveld (1975) found that phytotoxic grass residues of *Festuca arizonica* and *Multenbergia montana* reduced the germination of the seeds of *Pinus ponderosa*. Eussen and Soerjani (1976) found that the growth of cucumber plants was inhibited in soil mixed with fresh or dried leaves of alang alang (*Imperata cylindrica*) and Lodhi (1978) states that the amount of toxin released by the dominant species in a lowland forest corresponded with the amount of litter and the rate of decay. He has isolated fourteen phytotoxins from leaf litter and soil.

The nature of many of the allelopathic substances released by plants remains obscure. Reynolds (1975) reviewing aggressive chemicals in plants reminds us that protein-destroying enzymes can be got from paw paw (papain) and from pineapple (brolemin), insecticides from *Derris eliptica* (rotenone) and *Chrysanthemum cinerariafolium* (pyrethrum) and cytotoxic agents from several plants including *Sorbus aucuparia*, *Anemone pulsatilla* and *Catharanthus* spp.

Oats and other plants exude scopoletin (β-hydroxy-7-methoxy-coumarin), a toxic phenolic coumarin derivative, which is a powerful root inhibitor. Fay and Duke (1974) exposed seedlings of weeds, lucerne and oats, growing in nutrient solution, to scopoletin at various concentrations and found that root growth was inhibited in all of them.

Strigol is another growth regulator which is found in the root exudate of certain plants. It has been identified and synthesised (Eplee, 1975; Heather *et al.*, 1976). Agrostemmin, an extrametabolite of *Agrostemma githago*, was applied to a pasture in Yugoslavia (Gajic, 1974). Some species, e.g. *Potentilla zlatiborensis* and *Ranunculus acer*, were considerably reduced; other species, e.g. *Alectorolophus major*, *Plantago lanceolata* and *Trifolium pratense*, were stimulated and agrostemmin favoured especially the spread of *Cynosurus cristatus*. Agrostemmin doubled the tryptophan content of hay and considerably increased the total nitrogen content. It is of interest to note that Vrbaski *et al.* (1977) found extracts from seeds of *Agrostemma gigatho* stimulated or inhibited the growth of 4-day-old wheat plants depending on the solvent used. Substance 'A' which was separated from the other components of the ethyl alcohol extract had the greatest stimulatory powers. Analytical studies suggested that it was a gibberellin. Lucena (1974) found that radicle growth of a number of species, including sorghum, oats and soybean, when exposed to extracts from the underground parts of *Cyperus rotundus* were either stimulated (low concentrations) or inhibited (high concentrations). Paper chromatographs of the extracts indicated that the active substances are phenols.

Vancura and Stotzky (1976) analysed the gaseous and volatile exudates from the germinating seeds and seedlings of *Phaseolus vulgaris*, sweet corn, cotton and peas. They found that all liberated ethanol and most methanol, formaldehyde, acetaldehyde, formic acid, ethylene and propylene. Propionaldehyde and acetone was also evolved by cotton and peas. Any or all of these could obviously play an important part in allelopathy. Popovici (1974) notes that seeds and seedlings of maize exposed to volatile substances from the root of horseradish had their actively dividing cells blocked at metaphase.

Exudates from the leaves of *Robinia pseudoacacia* and sumac (*Rhus* spp.) which had fallen in autumn had a strong adverse effect on the germination and seedling growth of a number of herbaceous plants. These substances were identified as miasmins and saprolins (Matveev *et al.*, 1975). Lodhi (1975) isolated toxins from the soil beneath trees of *Celtis laevigata* and identified them as ferulic acid, caffeic acid and *p*-coumaric acid. These chemicals adversely affected the germination of brome grass (*Bromus* sp.). It is of interest to note that Naqvi (1976) also isolated caffeic and ferulic acids, as well as syringic and vanillic acids, from leaf extracts of Italian ryegrass (*Lolium italicum*). These adversely affected the germination of lettuce seed.

References

ABELES, F. B. (1966). *Pl. Physiol., Lancaster*, **41**, 585.

Annual Rept. Inst. for Hortic. Plant Breeding (1979). Wageningen, Neth. 1979.

ARINES, J., VIEITEZ, E. and MANTILLA, J. L. F. (1974). *Anales de Edafologia y Agrobiologia*, **33**(7/8), 689.

AUDLEY, B. G, ARCHER, B. L. and CARRUTHERS, I.B. (1976). *Arch. Environ. Contam. Toxicol.*, **4**, 183.

AUDUS, L. J. (1959). *Plant Growth Substances,* Leonard Hill.

B.A.S.F. A.G. (1975). *Technical Data Sheet: Morpholinium Chloride.*

B.A.S.F. A. G. (1977). *Technical Data Sheet: BAS 09800 W.*

B.A.S.F. A.G. (1978). *Agrochemicals of Our Time*, 8 pp.

BATCH, J. J. (1978). *Brit. Crop. Prot. Monograph*, No. 21, p. 33.

BENDALL, G. M. (1975). *Weed Res.*, **15**, 77.

BEYER, E. M., JOHNSON, A. L. and SWEETSER, P. B. (1976). *Pl. Physiol., Lancaster*, **57**(6), 839.

BIDWELL,R. G. S. (1979). *Plant Physiology*, 2nd edn., MacMillan Publ. Co.

BLOMQUIST, R. V. and SCHRADER, I. E. (1973). *Crop Sci.*, **13**, 23.

BOCION, P. T. and SILVA, W. H. (1977). *Proc. 9th Int. Conf. Plant Growth Substances*, 1976.

BOCION, P. T. *et al.* (1975). *Nature (London)*, **258**(5513), 142.

Boots Co. Ltd (1976). *Technical Data Sheet: BTS 34.442.*

BOYSEN JENSEN, P. (1911). *K. Danske Vidensk Selsk*, **3**, 1.

BRIAN, P. W., ELSON, G. W., HEMMING, H. G. and RADLEY, M. (1955). *J. Sci. Fd. Agric.*, **5**, 602.

BRIGGS, J. B. (1975). *Acta Hort.*, **47**, 287.

BROWN, D. A. (1978). *Proc. Sugar Cane Ripener Seminar.*

BURG, S. P. and BURG, E. A. (1965). *Science*, **148**, 1190.

BURG, S. P. and BURG, E. A. (1966). *Proc. Nat. Acad. Sci.*, **55**, 262.

Canad. Agricultural Res. Inst. Report (1975–1979).

CAUSSANEL, J. P. *et al.* (1976). *4e Coll. Int. Ecol. et Biol. des Mauvaise Herbes*, 240.

CHAILAKHYAN, M. K. and KHRYAPNIN, V. N. (1977). *Dokl. Akad. Nauk SSSR*, **236**(1), 268.

CHO, K. Y. *et al.* (1978). *Agric. and Biol. Chem.*, **42**(7), 1389.

Ciba Geigy (1974). *Technical Data Sheet: Etacelasil.*

CORNFORTH, J. W., MILLBORROW, B. V. and RYBACK, G. (1965a). *Nature (London)*, **206**, 715.

CORNFORTH, J. W., MILLBORROW, B. V., RYBACK, G. and WAREING, P. F. (1965b). *Nature (London)*, **205**, 1269.

CRISTOFERI, G., FILITI, N. and SANSAVINI, S. (1978). *Acta Hort.*, No. 80, 199.

CROSS, B. E. *et al.* (1961). *C. Adv. Chem. Ser.*, **28**, 3.

DARWIN, C. (1880). *The Power of Movement in Plants*, London.

DE CANDOLLE, M. (1832). *Physiologie Vegetale III*, Becket Jeune Lib. Fac. Med., Paris.

DELUI, C. (1974). *Cont. Bot. Univ. 'Babis-Bolyai' din Cluj-Napoca*, 197.

DENNY, F. E. (1924). *J. Agric. Res.*, **27**, 757.

DICKENSON, P. B. (1973). *Brit. Patent*, No. 1315131.

DICKENSON, P. B. (1976). *Outlook on Agric.*, **9**(2), 88.

DICKENSON, P. B., SIVUKUMARAN, S. and ABRAHAM, P. D. (1976). *Proc. Int. Rubber Conf. Kuala Lumpur*, 1975.

DODEL, J. B. (1975). *Compt. Rend. 8e Conf. du Columa*, 123.

EPLEE, R. E. (1975). *Pl. Growth Reg. Bull.*, **3**(2), 24.

EUSSEN, J. H. H. and SOERJANI, M. (1976). *Compt. Rend. Ve Coll. Inter. sur L'Ecologie et Biol. des Mauvaises Herbes*, 451.

FABIAN, A., CHIRCA, E. and CARAULEAN, I. (1974). *Cont. Bot. Grad Botanica Univ. 'Babis-Bolyai' din Cluj. Romania*, 201.

FAY, P. K. and DUKE, W. B. (1974). *Abstr. Weed Sci. Amer.*, 66.

FAY, P. K. and DUKE, W. B. (1977). *Weed Sci.*, **25**(3), 224.

FOX, J. E. (1969). 'The cytokinins' in *Physiology of Plant Growth and Development* (Ed. M. B. Wilkins), McGraw Hill.

FRANK, J. R. *et al.* (1978). *Hort. Sci*, **13**(4), 434.

FRIDINGER, T. L. *et al.* (1975). *Abstr. 170th Nat. Mtg. Amer. Chem. Soc.*

FRIEDMAN, J., ORSHAM, G. and ZIGER-CFIR, Y. (1977). *J. Ecol.*, **65**(2), 413.

GAJIC, D. (1974). *Fragments Herbologica Jugoslavica*, **46**, 12 pp.

GANE, R. (1935). *Rept. Fd. Invest. Board*, 1934, 142.

GHOSH, S. P. and SEN, S. P. (1975). *J. Hort. Sci.*, **50**(2), 9.

GOLDWIN, G. K. (1978). *Acta Hort.* **80**, 115.

GOPALAKRISHNAN, S. and SIRACAR, S. M. (1973). *Ind. J. Agric. Sci.*, **43**(11), 1012.

GRIGSBY, B. H. (1945). *Science. N.Y.*, **52**, 99.

GRÜMMER, G. (1955). *Die gegenseitige Beeninflussung hoherer Pflanzen-Allelopathie*, Fischer, Jena.

HANNA, W. W. (1977). *Crop. Sci.*, **17**(6), 965.

HEATHER, J. B., MITTAL, R. S. D. and SIH, C. J. (1976) *J. Amer. Chem. Soc.*,**98**(12), 3661.

HEDDEN, P., MACMILLAN, J. and PHINNEY, B. O. (1978). *Ann. Rev. Pl. Physiol.*, **29**, 149.

HEMBERG, T. (1949). *Physiol. Plant*, **2**, 24 and 37.

HIBBIT, C. J. and HARDISTY, J. A. (1979). *Med. Fac. Landboww. Rijk.*, **44**(2), 835.

HIROSE, K. *et al.* (1977). *Abstr. 1333 Okitsu Branch, Fruit Tree Res. Sta., Japan.*

HIROSE, T. and FUJIME, Y. (1973). *J. Japan Soc. Hort. Sci.*, **42**(3), 235.

HOFFMAN, G. R. and HAZLETT, D. L. (1977). *J. Range Management*, **30**(2), 134.

HOZUMI, Y., NAKAYAMA, K. and YOSHIDA, K. (1974). *J. Cent. Agric. Exp. Sta.*, **20**, 87.

I.C.I. (1976). *Technical Data Sheet: PP 528.*

IWANAMI, Y. and IWADARE, T. (1979). *Bot. Gaz.*, **140**(1), 1.

JABLONSKI, J. R. and SKOOG, F. (1954). *Physiol. Plant*, 7, 16.

JACKSON, R. W. and MANSKE, R. F. (1930). *J. Amer. Chem. Soc.*, **52**, 5029.

JOHNSON, R. R. and BROWN, C. M. (1976). *Crop Sci.*, **16**(4), 584.

KAPOOR, J. K. and TURNER, J. N. (1976). *Outlook on Agric.*, **9**(2), 52.

KATO, Y., MIYAKE, T. and ISHIKAWA, H. (1958). *J. Japan For. Soc.*, **40**, 35.

KHAN, A. A. and TOLBERT, N. E. (1965). *Physiol. Plant*, **18**, 41.

KIDD, F. and WEST. C. (1938). *Gt. Brit. Dept. Sci. Ind. Res. Food Invest. Bd. Rept.*, 1937.

KNAVEL, D. E. and KEMP, T. R. (1973). *Hort. Sci.*, 403.

KÖGL, F., HAAGEN SMIT, A. J. and ERXLEBEN, H. (1934). *Hoppe-Seylz*, **228**, 90.

KOTOB, M. A. and SCHWABE, W. W. (1971). *J. Hort. Sci.*, **46**, 89.

KRISHNAMURTHY, G. V. G. and CHANDWANI, G. H. (1975). *PANS*, **21**(1), 64.

KUROSAWA, E. (1926). *J. Nat. Hist. Soc. Formosa*, **16**, 213.

LETHAM, D. S. (1963). *Life Sci.*, **8**, 569.

LETHAM, D. S., SHANNON, J. S. and MCDONALD, I. R. (1964). *Proc. Chem. Soc. (London)*, 230.

LODHI, M. A. K. (1975). *Amer. J. Bot.*, **62**, 618.

LODHI, M. A. K. (1978). *Amer. J. Bot.*, **65**(3), 340.

LUCENA, J. M. (1974). *Revista Comalfi*, **1**(2), 40.

LUCKWILL, L. C. (1960). *Rept. Long Ashton Res. Sta. for 1959*, 59.

LUCKWILL, L. C. (1976). *Outlook on Agriculture*, **9**(2), 46.

LUCKWILL, L. C. and CHILD, R. D. (1973). *Exp. Hortic.*, **25**, 1.

LUCKWILL, L. C., CHILD, R. D., ATKINS, H. and MAGGS, J. (1975). *Rept. Long Ashton Res. Sta. for 1974*, 36.

MAPSON, L. W. (1970). *Endeavour*, **29**, 29.

MASSANTINI, F., CAPORALI, F. and ZELLINI, G. (1977). *Proc. E.W.R.S. Symp. on Diff. Methods of Weed Control and Their Integration*, **1**, 23.

MATVEEV, N. M., KRISANOV, G. N. and LYZHENKO, I. I. (1975). *Nauchyne Dokl. Vysshei Sbikoly Biol. Nauki*, **10**, 80.

MENHENETT, R. (1977). *Ann. Appl. Biol.*, **87**(3), 451.

MILES, H. E. *et al.* (1977). *Proc. 4th Ann. Mtg. Plant Growth Reg. Working Group.*

MILLER, C. O. (1956). *Pl. Physiol., Lancaster*, **31**, 318.

MILLER, C. O. (1962). *Pl. Physiol., Lancaster*, **37**, (suppl. XXXV).

MILLER, C. O. and SKOOG, F. (1953). *Amer. J. Bot.*, **40**, 768.

MILLER, C. O. *et al.* (1955). *J. Amer. Chem. Soc.*, **77**, 1239.

MILLER, J. F. (1976). *Diss. Abstr. Int. B.*, **36**(12), 5900 B.

MOLISCH, H. (1937). *Der Einfluss einer Pflanze auf die ander-Allelopathie*, Fischer, Jena.

MÖTHES, K. and ENGELBRECHT, L. (1959). *Monatsber. Deutsch. Akad., Wiss. Berlin*, **1**, 367.

NANJAN, K. and MUTHSWAMY, S. (1975). *South India Hort.*, **23**(3/4), 99.

NAQVI, H. H. (1976). *Pakistan J. Bot.*, **8**(1), 63.

NEWMAN, E. I. and ROVIRA, A. D. (1975). *J. Ecol.*, **63**(3), 727.

NICKELL, L. G. (1974). *Bull. Pl. Growth Regulators*, **2**, 51.

NICKELL, L. G. (1976). *Outlook on Agriculture*, **9**(2), 57.

NUMATA, M. (1977). *Proc. 6th Asian Pacific Weed Sci. Soc. Conf.*, **1**, 80.

OKHUMA, K., LYON, J. L., ADDICOTT, F. T. and SMITH, O. E. (1963). *Science (N. Y.)*, **142**, 1592.

OKHUMA, K. *et al.* (1965). *Tetrahedron Letters*, **29**, 2529.

ORTH, H. and LEUCHS, F. (1961). *Compt. Rend. Ie Conf. du Columa, Paris*, 314.

OSBORNE, DAPHNE, J. (1962). *Pl. Physiol., Lancaster*, **37**, 595.

PAÁL, A. (1919). *Jb. Wiss. Bot.*, **58**, 406.

PALEG, L. G. (1965). *Ann. Rev. Pl. Physiol.*, **16**, 291.

PHARIS, R. P. and ROSS, S. D. (1976). *Outlook on Agric.*, **9**(2), 82.

PHARIS, R. P., ROSS, S. D. and WAMPLE, R. L. (1974). *Proc. 3rd NA for Biol. Workshop Colorado State Univ.*

PINCUS, G. and THIMANN, K. V. (1948). *The Hormones*, Academic Press, New York.

POPOVICI, N. (1974). *An. Stiin. ale Univ. 'Al I Cuza' Romania*, **20**(1), 35.

PUTNAM, A. R. and DUKE, W. B. (1974), *Science N. Y.*, **185**(4148), 370.

QUINLAN, J. D. and PRESTON, A. P. (1973). *Acta Hort.*, **34**, 123.

RAJENDRA, B. R., MUJEEB, K. A. and BATES, L. S. (1978). *Agron. Abstr.*

RAKITIN, YU. V. and RAKITIN, V. YU. (1977). *Fiz. Rast.*, **24**(5), 1004.

REICH, V. H. and MARTIN, J. T. (1976). *Agron Abstr.*

REYNOLDS, T. (1975). *Chem. and Ind.*, **14**, 603.

RICE, E. L. (1974). *Allelopathy*, Univ. of Oklahoma Academic Press.

RICHMOND, A. and LANG, A. (1957). *Science N. Y.*, **125**, 650.

RIDDELL, J. A. *et al.* (1962). *Science N. Y.*, **136**, 391.

RIETVELD, W. J. (1975). *USDA Forest Service Res. Paper*, No. RM-153, 15 pp.

ROBINSON, P. M. and WAREING, P. F. (1964). *Physiol. Plant*, **17**, 314.

ROBINSON, P. M., WAREING, P. F. and THOMAS, T. H. (1963). *Nature (London)*, **199**, 874.

RUDICH, J. and HALEVY, A. H. (1974). *Pl. Cell Physiol.*, **15**(4), 635.

SACHS, T. and THIMANN, K. V. (1964). *Nature (London)*, **201**, 939.

SAGAR, G. R. and FERDINANDEZ, D. E. F. (1976). *Ann. Appl. Biol.*, **83**(2), 341.

Schering Information (1977).

SCHNEIDER, G. (1964). *Naturwissenschaften*, **51**, 416.

SCHNEIDER, G. (1965). *Nature (London)*, **208**, 1013.

SCHNEIDER, G. (1970). *Ann. Rev. Pl. Physiol.*, **21**, 499.

SCHOENE, D. L. and HOFFMAN, O. L. (1949). *Science N. Y.*, **109**, 588.

SHAW, G. and WILSON, D. V. (1964). *Proc. Chem. Soc. (London)*, 231.

STRONG, F. M. (1958). *Topics in Microbial Chemistry*, John Wiley & Sons.

TANIMOTO, T. T. and NICKELL, L. G. (1968). *Rept. Hawaii Sug. Technol. 1967*, 137.

THOMAS, T. H. (1976). *Outlook on Agric.*, **9**(2), 62.

THOMAS, T. H. *et al.* (1972). In *Weed Control Handbook* (Eds. J. D. Fryer and R. Makepeace) Vol. 2, 7th edn., Blackwell.

THOMAS, T. H. *et al.* (1977). In *Weed Control Handbook* (Eds. J. D. Fryer and R. Makepeace) Vol. 1, 6th edn., Blackwell.

TOLBERT, N. E. (1960a). *Pl. Physiol., Lancaster*, **35**, 380.

TOLBERT, N. E. (1960b). *J. Biol. Chem.*, **235**, 475.

TOMPKINS, D. R. and SMAY, S. E. (1971). *Arkansas Farm Res.*, 20.

TURNER, J. N. (1973). *Acta Hort.*, **34**, 287.

VANCURA, V. and STOTZKY, G. (1976). *Can. J. Bot.*, **54**(5/6), 518.

VARNA, S. P. (1974). *Diss. Abstr. Int. B*, **34**(8), 358 B.

VRBASKI, M. M., GRUJIC-INJAC, B. and GAJIC, D. (1977). *Biochem. Physiol. Pflanzen*, **171**(1), 69.

WANG, R. C. and LUND, S. (1975). *Crop Sci.*, **15**(4), 550.

WANG, S. M. and HUANG, C. Y. (1978). *Taiwania*, **23**, 128.

WAREING, P. F. (1954). *Physiol. Plant*, **7**, 261.

WAREING, P. F. (1969). *Physiology of Plant Growth and Development* (Ed. M. Wilkins), McGraw Hill.

WAREING, P. F. and PHILIPS, I. D. (1978). *The Control of Growth and Differentiation in Plants*, 2nd edn., Pergamon Press.

WENT, F. W. (1926). *Proc. Kon. Nederl. Akad. Wetensck, Amsterdam*, **30**, 10.

WIERSZYLLOWSKI, J., REBANDEL, Z. and BABILAS, W. (1963). *Bull. Acad. Polonaise des Sciences*, Cl.V., **11**(4), 191.

WILLARD, J. *et al.* (1977). *Proc. 4th Ann. Mtg. Plant Growth Regulator Group.*

ZAKI, M. A. and TEWFIK, M. S. (1974). *Egypt. J. Bot.*, **17**(2–3), 179.

ZIMMERMAN, P. W. and WILCOXON, F. (1935). *Contrib. Boyce Thompson Inst. Pl. Res.*, 7, 209.

CHAPTER THREE

The Absorption and Translocation of Foliage-applied Herbicides

The cuticle barrier

The activity of a foliage-applied translocated herbicide depends largely on factors which govern the amount of active ingredient reaching the sites of action. Thus the efficiency of cuticle retention and penetration, absorption from the symplast, short- and long-distance transport, metabolism and the degree of immobilisation at metabolically non-active sites may determine activity and selectivity (fig. 3.1). The activity of contact herbicides is largely influenced by the first two or three of these factors.

The leaves and stems of plants are covered by a lipoidal, non-cellular, non-living membrane called the cuticle which minimises water loss from the plant and at the same time acts as a barrier to penetration of exogenous materials, particularly those of a polar nature. The physicochemical properties of the cuticle in relation to herbicide absorption have been reviewed by several workers (Hull, 1970; Martin and Juniper, 1970; Kirkwood, 1972; Kirkwood, 1978).

The gross morphology of the plant plays an important role in retention and penetration of herbicides (Blackman *et al.*, 1958; Boize *et al.*, 1976). The stage of development, the shape and lamellar area of the leaf, the angle and orientation of the leaves, the presence of veins, cell size and arrangement and the presence of hairs or trichomes influence retention. There is evidence to show that water retention is greater on leaves having an 'open' as opposed to 'closed' trichome pattern (Challen, 1962). The open pattern enhances wetting due to capillary action, while the closed pattern depresses it by the trapping of air beneath the spray droplets.

Droplet size is of considerable importance in determining the effectiveness of herbicide application (e.g. McKinlay *et al.*, 1972).

The cuticle membrane is composed of two integrated regions, the epicuticular wax and the cutin, the latter being continuous with the intercellular middle lamella and the cellulose layer which is actually the outer wall of the epidermal cell (fig. 3.2). The plasma membrane lies to the inside of this wall and its penetration presumably constitutes

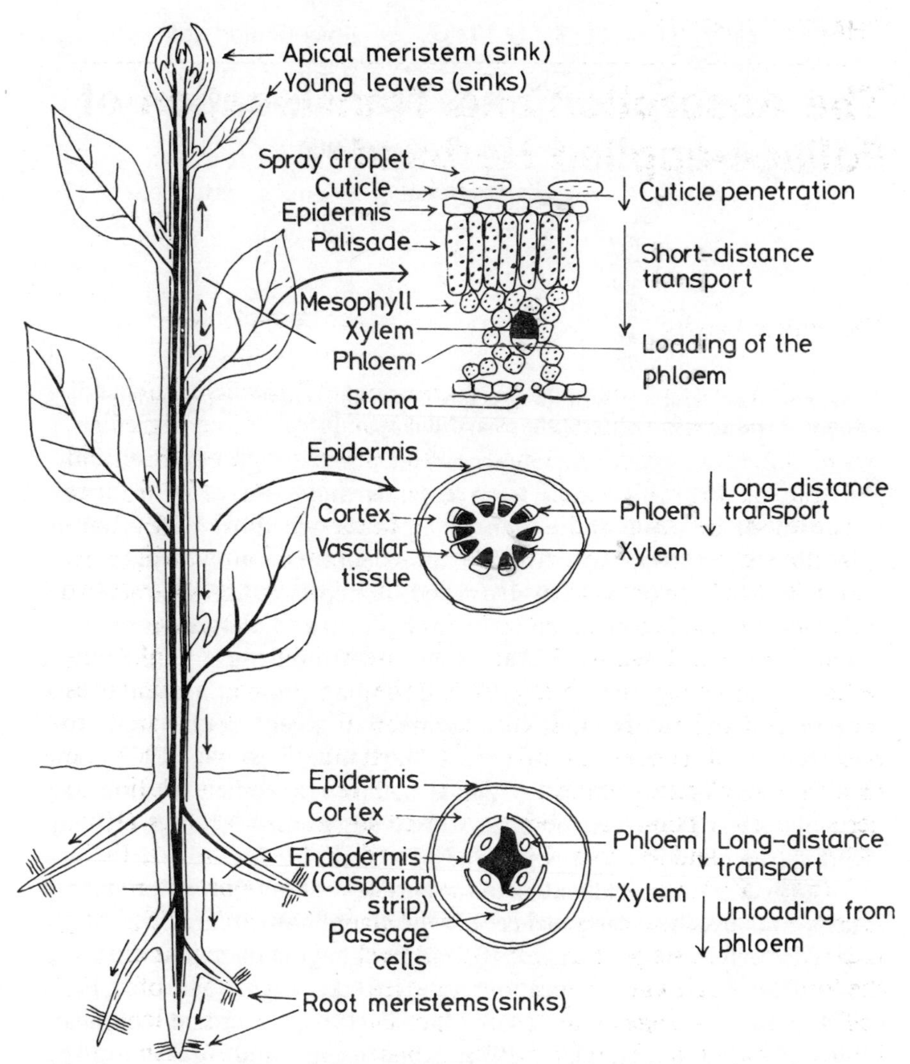

Fig. 3.1 A schematic view of the probable stages and route of uptake and translocation of a foliage-applied systemic herbicide.

absorption into the symplast or cytoplasmic continuum which includes the phloem system.

The epicuticular waxes of most plants contain a wide diversity of complex constituents though certain classes of compound are common to the leaf waxes of many species (e.g. Mazliak, 1968; Kolattukudy, 1970; Martin and Juniper, 1970; Kolattukudy and Walton, 1973). Generally they are composed of complex mixtures of long-chain alkanes, primary and secondary alcohols, ketones, aldehydes, fatty and hydroxy fatty acids and esters. Those with a high

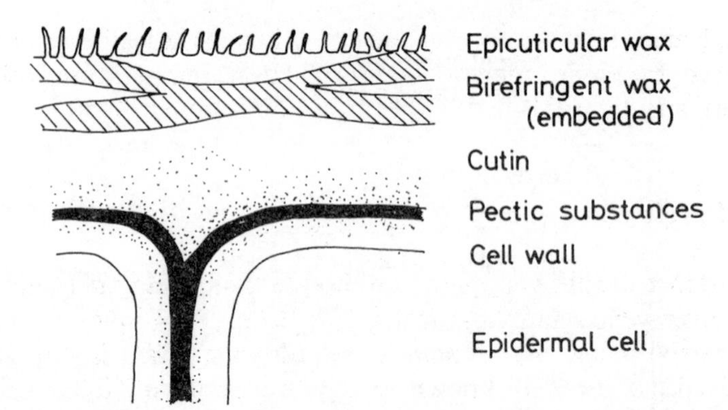

Fig. 3.2 A schematic view of the structure of the cuticle of the upper epidermis of a leaf of *Pyrus communis* (after Norris and Bukovac, 1968).

alkane or ketone content form crystals and tend to be relatively hydrophobic, while those rich in alcohols are most easily wetted (Holloway, 1969a, b, 1970). The orientation and structure of epicuticular waxes is often deposited in a specific and characteristic fashion which is determined by their chemical and physical properties and not by a mechanism of wax extrusion through micropores (Jeffree *et al.*, 1975; Chambers *et al.*, 1976). The quantity of epicuticular wax present varies widely and in a number of weed species examined ranged from <10 to 20 $\mu g\ cm^{-2}$ (Baker and Bukovac, 1971). The surface wax features have been classified into (a) rodlet, (b) granular, (c) platelet, (d) layered or crust, (e) aggregate or (f) liquid or semi-liquid (Amelunxen, Morgenroth and Picksak, 1967).

Cutin, the structural matrix of the cuticle is a polyester of long-chain fatty and hydroxy fatty acids. The principal constituents are 10,16-dihydroxyhexadecanoic acid, 10,18-dihydroxyoctadecanoic acid and 9,10,18-trihydroxyhexadecanoic acid (Baker and Martin, 1963; Eglington and Hunneman, 1968; Mazliak, 1968; Baker and Holloway, 1970; Holloway and Dees, 1971; Kolattukudy and Walton, 1973). While the exact structure of the cutin polymer is unknown, there is evidence that ester linkages between the hydroxy and carboxyl groups dominate and some peroxide and ether linkages may be present (Heinen and Brand, 1963; Crisp, 1965; Kolattukudy and Walton, 1973). Cutin has hydrophilic characteristics due to —OH and —COOH groups and lipophilic properties due to $—CH_2$ and $—CH_3$ groups (van Overbeek, 1956). It behaves like a highly cross-linked, high capacity ion exchange resin of a weak organic-acid type (Schönherr and Bukovac, 1973). In water, cutin is believed to swell separating the wax platelets and thus increasing permeability.

The physicochemical properties of the cuticle vary according to species, leaf maturity and surface and position on the leaf; they are also influenced by environmental factors such as temperature, relative humidity and light (Hull, 1970; Martin and Juniper, 1970).

Sites of preferential entry

The existence of sites of preferential entry (e.g. stomata, trichomes, the cuticle over veins, anticlinal walls, leaf bases, etc.) involves certain controversial issues. For example, the role of stomata is still largely unresolved. It is well known that absorption of foliar-applied chemicals is generally greater through the abaxial (lower) leaf surface of dicotyledons (e.g. Currier and Dybing, 1959; Sargent and Blackman, 1962; Hull, 1970) and this has been partly attributed to the greater number of stomata on that surface.

Stomata may act as specialised portals of entry. For example, the spray solution, under certain conditions, may move in mass through the stomatal pore (Dybing and Currier, 1961; Schönherr and Bukovac, 1972; Greene and Bukovac, 1974), or preferential penetration may occur through the cuticle over the guard or accessory cells which may be more permeable than over the epidermal cells (e.g. Franke, 1964a; Neumann and Jacob, 1968). The question of mass movement of spray into the leaves through the stomatal pores is controversial and has yet to be conclusively proved. The mechanism of pore penetration is complex and a number of factors must be taken into account including the surface tension and contact angle of the droplet, the morphology and chemistry of the pore wall. Penetration of the pore can be obtained if the surface tension of the spray droplets is equal to or less than the critical surface tension of the plant surface (Fox and Zisman, 1950), i.e. if the droplets form zero contact angles on the surface. Otherwise, pore penetration can only be expected if the contact angle formed by the spray droplet is lower than the minimum wall angle of the stomatal pore (Schönherr and Bukovac, 1972). A critical surface tension of many cuticular surfaces is approximately 30 dyn cm^{-1}, but few commonly used surfactants will reduce the surface tension below this figure. Thus it is uncertain whether penetration of the stomatal pore occurs in view of the difficulty in achieving the very low surface tension which would be necessary. Even if this does occur it would still be necessary for the herbicide to penetrate the thin cuticle which surrounds the substomatal cavity (Martin and Juniper, 1970). Other workers believe that preferential penetration of the guard cells is more probable than entry via the pore (e.g. Sargent and Blackman, 1962).

It has been suggested that preferential penetration of the cuticle over the guard cells, basal cells of trichomes, anticlinal walls of epidermal cells, etc., may be due to the relatively high numbers of ectodesmata (or 'ectocythodes') to be found at these sites (Franke, 1961, 1964a, b, 1967, 1970, 1971).

Ectodesmata are not protoplasmic strands *per se* but rather well-defined cell wall structures rich in reducing substances (Franke, 1964b). They have been demonstrated by mercury precipitates formed during the fixation process (Schnepf, 1959). However, doubts have been cast on their existence as definable cell wall structures since the characteristic distribution pattern and number can be altered by brushing the leaf surface or removing the epicuticular wax with chloroform (Schönherr and Bukovac, 1970). These authors produced evidence indicating that distribution is determined by the cuticle and not by structures in the cell wall. They conclude however, that there are areas in the cuticle which are preferentially permeable to polar compounds, possibly serving as 'polar bridges' across the cuticular membrane.

Cuticle penetration

The importance of epicuticular and cuticular wax as a barrier to penetration has been demonstrated by the enhanced uptake of herbicides following the addition of kaolin (Eveling and Elisa, 1976), pretreatment of plants with sodium trichloroacetate (TCA) (Siebert and Köcher, 1972a, b), di-allate (Still *et al.*, 1970; Davis and Dusbabek, 1973; Wilkinson, 1973, 1974; Wilkinson and Smith, 1973), dalapon, carbamate and other herbicides (Siebert and Köcher, 1972a, b; Flore and Bukovac, 1974, 1975). Cuticle thickness is reduced following treatment with those compounds, reflecting their interference with lipid biosynthesis (Still *et al.*, 1970; Davis and Dusbabek, 1973; Wilkinson, 1973; Wilkinson and Smith, 1973; St. John and Hilton, 1973).

The exact pathways involved in cuticle penetration are still subject to conjecture but it is generally believed that non-polar compounds follow a lipoidal route and polar materials an aqueous route (Crafts, 1956a, b). It seems likely that penetration would involve four major steps: (a) sorption into the cuticle, (b) movement across the cuticular membrane, (c) desorption into the apoplast and (d) uptake by the underlying cells. It is believed that transmission across the cuticular membrane is a physical process which may be directly affected by such factors as pH, particle size, cuticle thickness and indirectly by the rate of uptake, translocation and metabolism within the underlying tissues;

these latter events would influence the concentration gradient across the cuticular membrane.

Penetration of the cuticle is believed to be a physical process which is directly influenced by a number of factors including surface tension of the carrier, polarity and chemical structure of the active chemical. Polarity of the applied herbicide appears to be a critical factor, non-polar compounds being more readily sorbed than polar compounds. For example, 2,4-D is more readily sorbed as the undissociated molecule than as the anion and progressive chlorination of phenoxy-acetic acid (POA) results in a corresponding increase in lipid solubility and sorption (Bukovac *et al.*, 1971). Again, there is evidence that ionic compounds become associated with the cuticular surface to a lesser degree than non-dissociated molecules (Yamada *et al.*, 1965, 1966).

FACTORS INFLUENCING CUTICLE PENETRATION

The rate of cuticle penetration tends to remain linear, provided that large changes in concentration are avoided (Norris and Bukovac, 1968; Bukovac *et al.*, 1971). Absorption, however, often shows an initial rapid phase followed by a sustained slower phase (Barrentine and Warren, 1970; Singh *et al.*, 1972; Sands and Bachelard, 1973a). This progressive reduction in penetration with time may be associated with the rate at which the droplet evaporates (Luckwill and Lloyd-Jones, 1962). Penetration may be prolonged if the chemical is hygroscopic or if a humectant is added to the spray solution (Pallas, 1960), or if the residue is rewetted (Bukovac, 1965).

In addition, the rate of penetration may be influenced by a number of factors including the following.

Leaf age

Sargent and Blackman (1972) examined the comparative patterns of penetration of 2,4-D into the leaves of *Phaseolus vulgaris*, *Pisum sativum*, *Beta vulgaris*, *Helianthus annuas*, *Zea mays* and *Gossypium hirsutum*. With the exception of the last two species, the rates of penetration into both surfaces declined as the leaf matured.

Surface of application

While both surfaces of the leaf are generally readily penetrated by herbicides, in dicotyledons the lower surface (abaxial) is often more permeable, possibly due to increased frequency of stomata and trichomes (Hull, 1970) or less highly oriented cuticular wax (Norris and Bukovac, 1968). In monocotyledons, penetration is generally greater at the basal rather than at the distal end of the leaf, probably reflecting differences in the structure and amount of wax between these regions.

Herbicide concentration

Generally, penetration is linearly related to the external concentration (e.g. Sargent and Blackman, 1972), but at high concentrations physiological changes may be effected altering the rate of subsequent penetration (Hull, 1957; Hai and Hung, 1975).

pH

The pH of the herbicide solution often plays an important role in the penetration of weak organic acid type herbicides (Sargent, 1965), increased pH resulting in reduced uptake (e.g. Baur, Bovey and Riley, 1974; Zetterberg *et al.*, 1977). Generally the undissociated molecule is more lipid soluble and penetrates more readily than the anion (Sargent and Blackman, 1962). In the case of esters or amides the hydrogen ion concentration has little effect on penetration (e.g. Robertson *et al.*, 1971; Greene and Bukovac, 1971). The effect of pH appears to be primarily on the penetrating herbicide, though dissociable groups at the cuticle surface or within the cuticle may be influenced (Norris and Bukovac, 1971).

Molecular structure

Modifications of molecular structure which result in increased lipid solubility generally enhance foliar penetration. For example, progressive chlorination increased penetration of POA, but not benzoic acids, into bean leaves (Sargent *et al.*, 1969). Similar results were obtained with isolated tomato fruit cuticles (Bukovac *et al.*, 1971). This difference in the activity of the two chemicals is explained by the difference in effect on pH of chlorination of these compounds. Chlorination of POA has little effect on its p*K*, while chlorination of benzoic acid depresses the p*K*. Thus at a given pH value, the chlorinated POA is more lipid soluble and the chlorinated benzoic acid less lipid soluble than the corresponding parent acid. The importance of lipophilic surface activities has been examined by Kennedy (1974) with reference to the adsorption of benzoic acid and some of its chlorine-substituted derivatives at an alkane/H_2O interface. He found that the relative surface affinities depended upon the number and position of the substituted chlorine atoms, substitution of the chlorine in the 2-position conferring greatest affinity for a surface lipophilic site. Penetration of benzoic acid type herbicides into leaf discs of *Phaseolus vulgaris* (in the dark) was related to the mole fraction of the dissociated acids at the lipophilic leaf surface/water interface.

Additives

The addition of surfactant to polar compounds may have a marked effect on herbicide activity, though the role of the surfactant is complex and little understood. Perhaps the most important effect is on surface tension. In general as the concentration of surfactant is increased so the surface tension is reduced to a point (critical micelle concentration – CMC) at which further addition of surfactant has no effect. For most efficient surfactants the CMC lies between 0.01 and 0.5% (Osipow, 1964), but penetration is often maximal at concentrations much higher than those needed for wetting action (Foy and Smith, 1965) suggesting that effects additional to surface tension, such as solubilisation of the waxes (Furmidge, 1959; Parr and Norman, 1965), may be operating. Three classes of surfactants are available, anionic, cationic and non-ionic, the latter being the most commonly used. Jansen (1973) examined the enhancement of herbicides by silicone-glycol surfactant and found that the threshold for enhancement in susceptible and resistant species was 0.01% and 0.1% respectively. In *Cyperus esculentus*, a silicone-glycol surfactant at 0.5% was most successful.

Formulation with certain oil carriers has been reported to enhance the absorption of certain herbicides (e.g. 2,4-D and linuron in *Cyperus rotundus*, Burr and Warren, 1972; Scifres, Baur and Bovey, 1973) probably as a result of increased wetting of the leaf and the increased lipophilicity of the herbicide–oil complex (Wrigley, 1973). Other additives such as ammonium nitrate (NH_4NO_3), ammonium sulphate [$(NH_4)_2SO_4$]; ammonium citrate and ammonium chloride (NH_4Cl) (Turner and Loader, 1972) and ascorbic acid (Baur, Bowman and Bovey, 1973) are also capable of increasing foliage absorption of herbicides. Pretreatment with certain herbicides or growth regulators may also be effective, possibly reflecting direct action on membrane permeability or indirect effects due to sink stimulation (Kirkwood, 1978).

Environmental factors

The effects of environmental factors on foliage absorption are very subtle acting either directly or indirectly due to effects on the plant, especially cuticle development. Penetration is generally enhanced by light (e.g. Sargent and Blackman, 1965; Greene and Bukovac, 1971), though the reverse was reported for diquat and paraquat (Brian, 1967). Inhibitors of the Hill reaction negated light-enhanced uptake of 2,4-D (Sargent and Blackman, 1969) and NAA (Greene and Bukovac, 1972) suggesting that uptake may be limited by ATP-dependent processes.

Penetration is also temperature dependent (Sargent, 1965; Hull, 1970; Wilson and Norris, 1973; Richardson, 1976), Q10 values of 1.5–3.0 being obtained, further suggesting that the uptake process is at least partly metabolically governed. Sharp increases in the Q10 values, however, could also indicate changes in the physical nature of the fatty substances in the plasma membrane. Norris and Bukovac (1968) studied the effect of temperature on the penetration of NAA through isolated cuticular membranes and suggested that marked change in cuticular permeability between 15 and 25° C may be associated with a phase change in the cuticle.

High humidity generally favours foliar absorption of herbicides possibly reflecting the increased drying time of spray droplets (Prasad *et al.*, 1967), enhanced stomatal opening (Prasad *et al.*, 1967) and phloem transport (Pallas, 1960). On the other hand, drought conditions may cause reduced absorption (and translocation) of herbicide (Brady, 1974). In addition to these direct effects, the environmental conditions under which plants develop may affect cuticle development and indirectly affect subsequent foliar penetration (e.g. Templeton and Hurtt, 1972; Hull, Morton and Wharrie, 1975). Leaves developing under full sunlight, for example, form a thicker cuticle than those expanding in the shade (Skoss, 1955). Predisposing plants to relatively high temperature and humidity resulted in greater herbicide penetration than under low temperature and humidity conditions (Westwood and Batjer, 1958; Edgerton and Haesler, 1959; Donoho *et al.*, 1961). Conversely, moisture stress reduces foliar absorption (Skoss, 1955). These environmental factors markedly affect the quantity, fine structure and chemistry of the epicuticular wax (Hull, 1958; Juniper, 1960; Baker, 1972; Whitecross and Armstrong, 1972). Contrary to these findings however, Darwent and Behrens (1972) found that pretreatment conditions of light, temperature and humidity had a relatively minor influence in determining the response of peas and *Abutilon theophrasti* to subsequent treatment with 2,4-D.

Short- and long-distance transport

Subsequent to penetrating the cuticle, lipophilic herbicides must partition into the apoplast (or cell wall continuum including the xylem). Ester formulations which are strongly lipophilic are apparently hydrolysed and thereafter partition into the aqueous phase. Some compounds, such as the substituted ureas and triazines when foliage applied are apparently unable to penetrate the symplast and move only in the apoplast within and not out of the treated leaf; such compounds are normally root absorbed and xylem transported (see chapter 4). Compounds which do penetrate the symplast and are phloem tranported appear to be capable of cellular accumulation and yet are not irreversibly bound. This incubation of leaf sections of tobacco and maize with certain ^{14}C-labelled herbicides revealed that whereas 2,4-D, aminotriazole and picloram were accumulated but not irreversibly bound, accumulated fractions of atrazine and diuron were readily desorbed by water (Lumb, 1972).

Short-distance transport of sugars and apparently some foliage-applied herbicides, appears to involve cell to cell movement by way of cytoplasmic connections (plasmodesmata) (Robards, 1975), with active accumulation occurring in the companion cells associated with the sieve tubes. Studies with ATP and dinitrophenol (DNP) indicate the involvement of an energy-requiring process for both phloem loading and subsequent translocation of sugars in *Beta vulgaris* (Sovonick *et al.*, 1974). It appears that in this species, sugars can be actively accumulated in the phloem directly from the apoplast, the process involving membrane thiol groups (Giaquinta, 1976). Phloem-translocated herbicides appear to follow the same route as assimilates and possibly involve the same mechanisms. The route of long-distance transport is the sieve tube system and there is evidence to suggest that movement occurs from 'source' (the region of sugar synthesis) to 'sink' (the region of sugar demand). The mechanism involved is still a matter of controversy and its solution appears to hinge on the nature of the sieve pores and whether they are sufficiently open (Fisher, 1975) to allow a pressure flow of solution (Fellows and Geiger, 1974), or filled with strands (Evert *et al.*, 1969) or tubules (Thaine *et al.*, 1975; De Maria *et al.*, 1975) implying a more complex mechanism such as intercellular streaming (Thaine, 1962; Canny, 1962).

Certain structural criteria for phloem transport of insecticides have been proposed by Crisp (1972) and these appear to have relevance to the transport of herbicides. These criteria include the presence of a free carboxylic group or functional groups which form conjugates with sugars or amino acids or alternatively the presence of azino structures in the heterocyclic ring which can resonate. The remaining portion of the molecule attached to the acid must have a hydrophilic/lipophilic balance (HLB) which does not particularly favour water or lipid. There are many examples in the literature of compounds which appear to bear out Crisp's proposals. For example, most of the phenoxyalkanoic compounds which possess a carboxylic group are readily phloem translocated, the exception being more lipophilic compounds such as MCPB (Kirkwood *et al.*, 1972) or 4-(2,4-DB) (Linscott, 1964) which are relatively lipophilic and may not partition readily from the cuticle wax.

Factors influencing the efficiency of herbicide translocation in the phloem have been reviewed by Kirkwood (1978). Translocation is apparently dependent on a number of factors including the efficiency of absorption, certain 'plant' characteristics including the presence of C_4 as opposed to C_3 characteristics, leaf age and the relative position of 'source' to 'sink' and environmental factors including light, temperature and humidity conditions and time of day. Sink stimulation follows treatment with certain growth regulators such as BA, ethephon, or 2,4-D (Kirkwood, 1979), or nitrogen supply or capitation (Hunter and McIntyre, 1974).

There is evidence to suggest that transmission of certain herbicides may be inhibited due to their effect on energy-requiring steps associated with loading or long-distance transport. Some confirmation is found in the results of studies on the effect of ATP, sucrose or metabolic inhibitors on herbicide translocation by excised plant tissues (Sharma and Vanden Born, 1973). A further factor which may restrict phloem transport is the possibility of some form of fixation at metabolically non-active sites due to adsorption or binding to protein, lipid or polysaccharide compounds of leaf tissues. Indeed, there is evidence that selectivity may be partly due to differential binding (Ezerzha *et al.*, 1971; Matlib *et al.*, 1971; Camper and Moreland, 1971; Dexter *et al.*, 1971; Hallmén and Eliasson, 1972; Hilton and Christiansen, 1972; Harvey and Muzik, 1973; King, 1973).

Again, non-translocation may result from disruption of cell and organelle membranes by absorbed herbicides (Anderson and Thomson, 1973), with a consequent disruption of the normal photosynthetic and translocation processes.

It must be emphasised that while our knowledge of the pathways and mechanisms involved in absorption and translocation has increased in the last two decades, much still remains to be resolved. It is clear, however, that complex interrelationships exist between the various factors involved in absorption and translocation of foliage-applied herbicides. Understanding of the nature of the factors involved can only lead to an improvement in the efficiency of these compounds, reducing the size of the applied dose, thus reducing the costs of material and possible environmental impact of treatment.

Uptake and translocation characteristics of foliage-applied herbicides

HALOALKANOIC ACIDS

While **TCA** is generally soil applied and root absorbed, **dalapon** is predominantly foliage applied. Studies with *Lemna minor* have indicated that penetration of dalapon into the fronds (and roots) is initially rapid and the subsequent decline in the rate of absorption is believed to reflect a metabolically governed process involving thiol groups (Prasad and Blackman, 1965). Foliage absorption of dalapon has also been reported in a number of other species including cotton (Foy, 1961), sorghum (Foy, 1961), maize (Blanchard *et al.*, 1960; Foy, 1962), sugar-beet (Anderson *et al.*, 1962), yellow foxtail (*Setaria glauca*) (Anderson *et al.*, 1962) and dallis grass (*Paspalum distichum*) (Smith and Davies, 1965).

The rate of absorption of dalapon is influenced by a number of factors including plant age, surfactant, temperature, light intensity and relative humidity. For example, Prasad and Blackman (1965) found that uptake of dalapon by *Lemna minor* and *Salvinia natans* increased with a rise in temperature from 20 to 30° C and by an increase in humidity from 28 or 60% to 88%; pretreatment at high humidity also increased absorption. They concluded that the degree of cuticle hydration and rate of drying of the herbicide were involved.

Dalapon appears to be capable of both symplastic and apoplastic transport (Blanchard *et al.*, 1960; Foy, 1961; Smith and Dyer, 1961; Anderson *et al.*, 1962) and may move from one system to another (Ashton and Crafts, 1973). Its application to the leaves of maize resulted in diffusion into the minor veins and larger vascular bundles (Foy, 1962) and its presence in the xylem and phloem has been detected by histoautoradiography (Pickering, 1965). The efficiency of translocation of dalapon is influenced by a number of factors including plant age, light intensity, temperature and relative humidity. For example, Wilkinson (1956) found that it was exported from leaves of barley up to 2 weeks old, but not 3-week-old plants. Again, translocation of dalapon from unshaded leaves of couch grass (*Agropyron repens*) was greater than from shaded leaves. This was attributed to reduced water flow in the xylem resulting in diminished removal of the compound from phloem to xylem (Sagar, 1960). McIntyre (1962), however, found no evidence to support this view and suggested that a natural metabolite may be formed in light which complexes with dalapon inhibiting its translocation from the treated leaf.

Translocation of dalapon is greater at high (43° C) than low (25° C) temperature and at high (88%) as opposed to low humidity (60 or 28%) (Prasad *et al.*, 1967). Generally similar results were obtained by McWhorter and Jordan (1976) who studied the effects of adjuvants and environment on the toxicity of dalapon to Johnson grass (*Sorghum halepense*). Without surfactant, dalapon (1.2 kg/ha) was more active at lower temperature and high humidity (r.h. 100%) as opposed to low humidity (r.h. 35%), but, incorporation of surfactant (1% nonoxynl 9.5) and/or 0.1 M KH_2PO_4 altered the response to temperature and humidity. Studies using ^{14}C-dalapon confirmed that translocation of dalapon was greater at high than at low humidity (100% r.h. as opposed to 35%), while the addition of 0.5% surfactant increased translocation under all the test environmental conditions. Incorporation of 0.1 M KH_2PO_4 to the dalapon/surfactant solutions increased translocation at high temperature (38° C), especially at low humidity (35%) (McWhorter and Jordan, 1976).

PHENOXYALKANOIC ACIDS

The mechanism and factors influencing the absorption and translocation of foliage-applied phenoxy acid herbicides have been reviewed by Robertson and Kirkwood (1969, 1970). While there is little doubt that the cuticle acts as a barrier to movement of herbicide molecules, there is some controversy as to its importance in herbicide selectivity. It has been suggested that differences in the rate of cuticle penetration of **MCPA, 2,4-D** and **2,4,5-T** play no more than a minor role in the selectivity of these compounds (Holly, 1956). On the other hand, differences in the susceptibility of certain weed species to 2,4-D have been attributed to differences in the rate of penetration of cuticles of differing structure and thickness (Repp, 1958; Hamilton, 1977).

Cuticle penetration of **POA** depends on a number of factors which include molecular structure, formulation and environment. Molecular structure is very important and a suitable balance of the polar–apolar groups of the molecule seems to be essential for efficient penetration. Crafts (1956a) showed that 2,4,5-T was absorbed by *Phaseolus* leaves more slowly than 2,4-D and suggested that the more lipophilic 2,4,5-T was less likely to partition from the lipid phase of the leaf surface into the aqueous medium of the cell. Increased chlorination of the POA molecule favoured penetration into the lipoid phases of leaf discs of *Phaseolus vulgaris* (Sargent, 1964; Sargent and Blackman, 1969) and isolated tomato cuticle (Sargent, 1966). However, high lipid solubility may result in cuticle wax binding, immobilisation and inactivation of the chemical. The resistance of broad bean (*Vicia faba*) to **MCPB**, for example, has been attributed, in part, to the relatively high levels of binding in the cuticle wax; in contrast MCPA partitions relatively easily out of the wax and is relatively mobile in this species (Kirkwood *et al.*, 1972). Penetration of both compounds is enhanced at low pH presumably reflecting a higher proportion of molecules in the undissociated form. A similar situation has been reported by Wathana *et al.*, (1972) who found that **2,4-DB** was immobilised in the treated leaves of soybean and cocklebur (*Xanthium pennsylvanicum*) whereas 2,4-D was transported to actively growing regions, particularly in cocklebur. Similar findings have been reported for ^{14}C-2,4-DB in *Rumex crispus* (susceptible) and *Plantago lanceolata* (resistant) (King and Bayer, 1972) and in peanut and prickly sida (*Sida spinosa*) (Corbin *et al.*, 1972). Kennedy and Harvey (1972) compared the binding of 2,4-D, **2,6-D**, 2,4,5-T and **IAA** to lecithin vesicles in solution at several pH values and found that the unionised form of these acids is bound more strongly than the ionised form. These findings tend to agree with those of Smith (1972a, b) who found that lecithin-treated parenchyma tissue (potato tuber) accumulated more 2,4-D and 2,4,5-T than did untreated tissue indicating possible binding of the herbicide to the lecithin moiety.

In commercial practice, phenoxy acid herbicides are normally formulated as esters, amines, or sodium salts. Hydrolysable esters of biologically active weak acids are used to enhance cuticle penetration, heavy esters being apparently preferable to light since the latter may kill the leaves by contact injury, destroying the mechanism of translocation. Successful absorption (and translocation) has been achieved with the isopropyl ester of 2,4-D in soybean and corn plants (Hauser, 1955) and the butoxyethanol esters of 2,4-D and 2,4,5-T (Crafts, 1956b); such heavy esters formed by alcohols have both lipid and fat solubility. Hydrolysis of the ester to the free acid appears to be necessary for the compound to be herbicidally active. Crafts (1960), for example, reported that in barley leaves the acid moiety of the isopropyl ester of 2,4-D was translocated leaving the alcohol moiety in the treated leaves.

Environmental factors may influence the rate of uptake of phenoxy compounds. Increased temperature enhances cuticle penetration of 2,4-D (Silva Fernandes, 1965) and high light intensities influence the penetration of 2,4-D into *Phaseolus* leaf discs removed from the abaxial surface of young expanding leaves (Sargent and Blackman, 1965).

While penetration of the cuticular membrane is dependent on

physical factors, there is evidence that the process of absorption into the leaf may involve a metabolically governed rate limiting step, possibly concerned with the movement of molecules into or across the plasmalemma into the symplast. Absorption of ^{14}C-2,4-D and ^{14}C-2,4,5-T by mesquite (*Prosopis juliflora*) was increased threefold by ascorbic acid (10^{-2} mol dm^{-3}) and by increase in temperature in the range 20–40°C (Baur *et al.*, 1973). Similarly, the penetration of isolated cuticle and leaf discs of bean (*V. faba*) by ^{14}C-MCPA and ^{14}C-MCPB was enhanced by increased temperature (Kirkwood *et al.*, 1972).

The involvement of energy-requiring processes in the phloem translocation of phenoxy compounds probably associated with vein loading, is indicated by the normally inhibitory effect of compounds which act as uncouplers or inhibitors of oxidative phosphorylation. Pretreatment of *V. faba* with MCPA or MCPB inhibits translocation of these ^{14}C-labelled herbicides (Kirkwood *et al.*, 1968). Similarly, 2,4-D significantly reduces the amount of ^{14}C-2,4-D absorbed and translocated by purple nutsedge (*C. rotundus*) (Bhan *et al.*, 1970), picloram inhibits the translocation of 2,4-DB and its metabolites in *P. vulgaris* and normal rates of MCPA, dinoseb, ioxynil and bromoxynil reduce the translocation of ^{14}C-MCPA (Fyske, 1975). A comparative study of the movement of 2,4-D and assimilates in wheat (cv. Koga II) reveals that at the 5–6 leaf stage relatively less 2,4-D moves into the roots than would have been predicted from knowledge of assimilate movement (Olunuga, Lovell and Sagar, 1977). On the other hand, translocation of MCPA in *Alnus incana* and *Betula odorata* are unaffected by application of ioxynil and bromoxynil along with MCPA, and indeed high concentrations of the nitrile increased movement of MCPA (Lund-Høie, 1973), perhaps reflecting effects on membrane permeability. Translocation of ^{14}C-2,4-D (or ^{14}C-glyphosate) by hemp dogbane (*Apocynum cannabinum*) was unaffected by pretreatment with that herbicide (Schultz and Burnside, 1979).

Disruption of cell or organelle membranes may inactivate photosynthetic and translocation systems and such damage has been reported for 2,4-D (1.6×10^{-3} mol dm^{-3}) on primary leaves of bean (*P. vulgaris*) (Hallam, 1970), 2,4-D (and picloram) on needles of *Pinus radiata* (Bachelard and Ayling, 1971), and on leaves of tobacco (White and Hemphill, 1972). The loss of herbicide from the roots of treated plants (e.g. Sanad and Müller, 1973) may reflect membrane damage. On the other hand, treatment of hypocotyl segments with 2,4,5-T (10^{-6}–10^{-4} M) increases the influx and efflux of ^{14}C-urea while at 10^{-3} M, 2,4,5-T reduces both the influx and efflux (Feng, 1976), emphasising the importance of concentration level. Disruption of the phloem may also contribute to the tolerance of certain species. The resistance of soybean to 2,4-DB (0.56 kg/ha) has been attributed to obliteration of the conducting phloem of the metaphloem and early secondary phloem (Regehr and Pizzolato, 1979),and a similar loss in *Glycine max* treated with 2,4-DB was compensated for by accelerated differentiation of the secondary phloem (Pizzolato and Regehr, 1979).

The low efficiency or absence of translocation of certain herbicides may reflect adverse effects on membrane permeability, and/or the inhibitory effect of the herbicide on the energy-requiring processes associated with 'loading' of the minor veins, or long-distance transport of assimilates and the herbicide *per se*. Additionally it would appear that fixation in the leaf tissues may result from adsorption or binding and differentials in the degree of binding may partly explain differences in the efficiency of translocation (and activity). In studies of differential binding to BSA (bovine serum albumin), 7 moles of MCPB per mole of BSA were bound in contrast to 5 moles MCPA per mole BSA (Matlib *et al.*, 1971). Binding of 1-^{14}C-2,4-D by proteins of maize leaf increases with increasing herbicide concentration (0.01–0.5%), temperature (10–35°C) and light (Zemskaya, 1971). Phenoxy derivatives with a methyl radical located at position 4 will bind to a hydrophobic radical of protein (Zemskaya *et al.*, 1975).

Further support for the hypothesis that the formation of complexes counteracts herbicide toxicity comes from the results of Hallmén (1975). He found that when root-applied 2,4-D (or picloram) was applied to susceptible species, the herbicide was recovered in an uncomplexed form. In herbicide-tolerant species however, the absorbed 2,4-D (or picloram) was converted into water-soluble or TCA-insoluble complexes. Most of the complexes released free herbicides on hydrolysis with NaOH or HCl.

There is some evidence to suggest that the rate, degree and direction of transport of certain herbicides including the phenoxyalkanoic compounds may be influenced by 'sink' stimulation induced by certain growth regulators. Thus abscisic acid (ABA) and gibberellic acid (GA) increase translocation of 2,4,5-T in bean (*P. vulgaris*) seedlings though the effects of GA are nullified by the protein synthesis inhibitor cyclohexamide (Basler, 1973). Blanton and Basler (1976) found that ABA enhances and inhibits basipetal translocation of 2,4,5-T in persimmon, the effect being both time and concentration dependent; stimulatory effects were optimal at 5 μg ABA per plant. Ethephon increases foliar absorption of ^{14}C-labelled **dicamba** and ^{14}C-labelled 2,4-D in *Cirsium arvense* (Carson and Bandeen, 1975) and penetration and translocation of 2,4-D in field bindweed (*Convolvulus arvensis*) (Goss, 1978). These growth regulators have been found to stimulate both absorption and translocation of ^{14}C-2,4-D in leaf explants of *Vicia faba*, the effect being attributed to sink stimulation in the growth regulator treated leaflet (Kirkwood, 1979).

The transport characteristics of a herbicide may be altered by changes in growth pattern, for example, by being induced by shoot decapitation. McIntyre, Fleming and Hunter (1978) observed a reduction in the basipetal transport of ^{14}C-2,4-D in *Cirsium arvense* resulting from apical decapitation. This response was attributed to the inhibiting influence of subapical lateral shoots on basal branching and root bud growth. When apical decapitation was combined with removal of the axillary buds in the subapical region, basal branching and root bud growth were promoted and ^{14}C translocation into the roots significantly increased.

The efficiency of translocation of a herbicide may also be influenced by combination with another toxic compound, the effects often being concentration and species dependent. For example, application of picloram with 2,4,5-T tended to decrease the concentration of 2,4,5-T in the phloem of honey mesquite (*Prosopis juliflora* var. *glandulosa*), whereas in most instances the concentration of picloram in the phloem was increased by 2,4,5-T (Davis *et al.*, 1972). However, the application of a combination of 2,4,5-T and picloram to young gorse plants did not improve their translocation (Field and Phung-Hong-Thai, 1979) and the lack of efficient control of gorse by 2,4,5-T is largely attributed to its inadequate uptake and lack of true systemic translocation (Phung-Hong-Thai and Field, 1979). The uptake and translocation of 2,4,5-T esters by gorse has been studied using gas liquid chromatography (g.l.c.) by Zabkiewicz and Gaskin (1979). There was generally little movement from the treated portion into the roots/lateral shoots, though the uptake/translocation trends varied according to selective uptake, ester hydrolysis and interaction with additive chemicals.

The effect of various mixtures of herbicides on growth of soybean (cv. Lee) has been examined and antagonistic inhibition or competitive inhibition reported (Barrentine and Frans, 1973). For example, high concentrations of IAA antagonised the action of 2,4-D, while low concentrations of IAA acted by competitive inhibition with 2,4-D. Non-toxic concentrations of 2,4-D frequently increased the absorption (or toxicity) of prometryne, whereas toxic concentrations of 2,4-D decreased the absorption of the latter (Diem and Davis, 1974). Further evidence of complex interactions between herbicides comes from the work of Basler (1978) who examined the effects of dichlobenil and alachlor on the translocation of 2,4,5-T in bean (*Phaseolus vulgaris*) seedlings. The dichlobenil and alachlor were applied to the nutrient solution at concentrations of 2.5 and 12.5

p.p.m. one day before injection of ^{14}C-2,4,5-T (0.5 μg, 54 mCi/mM) into the stem tissue. Dichlobenil enhanced acropetal translocation of 2,4,5-T to the young shoots (particularly in the presence of 2.5 p.p.m. GA^3), but inhibited basipetal movement. Alachlor also enhanced translocation of 2,4,5-T to the young shoots in the presence of GA_3, but had little effect on basipetal translocation.

The phenoxy aliphatic herbicides are also readily root absorbed and translocated in the xylem. Studies by Shone and Wood (1972) indicate that uptake and movement is markedly increased at low pH and high temperature conditions and reduced by treatment with sodium azide, suggesting that metabolic involvement occurs in the uptake process. In a subsequent study, barley plants were treated with ^{14}C-labelled 2,4-D, POA and other herbicides. 2,4-D had more influence on membrane permeability than POA and a high transpiration stream concentration factor resulted from the leakage of material previously accumulated in the roots (Shone and Wood, 1972).

Singh and Müller (1979a) examined the uptake, distribution and efficiency of 2,4-D and other herbicides in water hyacinth (*Eichhornia crassipes*). Absorption of ^{14}C-2,4-D from the culture solution by the roots and floats was followed by translocation into the meristematic foliar parts; 2,4-D amine (0.75–2.25 kg a.i./ha), or glyphosate (2–6 kg a.i./ha) gave complete control of this species. Foliar application of 2,4-D resulted in only source to sink movement, with translocation into the daughter plants while both xylem and phloem transport of asulam and aminotriazole occurred (Singh and Müller, 1979b). Titova (1978) found no correlation between resistance to 2,4-D by certain higher aquatic plants and 2,4-D accumulation in these plants.

NITRILES

The hydroxybenzonitrile (HBN) compounds, such as **ioxynil** and **bromoxynil** are foliar-applied contact herbicides whose selectivity apparently depends on differences in retention (Davies *et al.*, 1967) and other factors (Somerville, 1968). The basis of differential phytotoxicity was examined by Davies *et al.* (1967) using mustard, pea and barley as test species, ioxynil salt formulated with 1% Tween 20 being applied by dipping or painting. Selective phytotoxicity between barley and mustard was found to be based largely on differential retention, supported by small differences associated with plant morphology and differentials in the rate of uptake and metabolism of ioxynil. Ester formulations penetrated more readily than did salts (Somerville, 1968) and in the field were relatively unaffected by environmental conditions (Savory, 1968).

Although the HBNs are essentially foliage-applied contact herbicides, some translocation may occur. Foy (1964) demonstrated translocation using ^{14}C-ring-labelled ioxynil, autoradiographs revealing acropetal and basipetal migration of ^{14}C from the treated area; movement was extremely slow in both cases and was attributed to diffusion, followed by leakage from successively injured cells. Davies *et al.*, (1968) using autoradiography, found that 5 h after application of ^{14}C-ioxynil to the leaves of mustard, cotton, dwarf bean, barley and pea, some movement of ^{14}C occurred to the base of the petiole and acropetally to the leaf tip.

Schafer and Chilcote (1970) investigated the translocation of ^{14}C-bromoxynil octanoate in winter wheat (*Triticum aestivum* var. Nugaines) (resistant) and coast fiddleneck (*Amsinckia intermedia*) (susceptible). Limited movement of the label into the leaf sheaths and other tillers occurred in wheat and into adjoining leaves and root systems of coast fiddleneck. After 4 days, however, appreciable amounts were found in leaves of intermediate age, suggesting that symplastic movement had occurred; the parent herbicide was more mobile in coast fiddleneck than wheat.

NITROPHENOLS

The rapid contact action of this group of compounds is influenced by many environmental factors which influence the rate of absorption such as high temperature, sunlight, high humidity, and under these conditions herbicidal action is rapid and effective. While this contact action destroyed the tissues and mechanisms involved in apoplastic or symplastic transport, young rye plants readily absorbed root-applied **DNOC** and it was subsequently transported in the xylem (Bruinsma, 1967).

CARBAMATES

Unlike propham and chlorpropham, a number of carbamate herbicides are active when foliage applied. Foy (1961) found that absorption and translocation of ^{14}C increased with time following application of ^{14}C-**barban** to the adaxial surface of the first leaf of barley, oat and wild oat; treatments applied to the leaf axil were particularly effective. Selectivity did not appear to be due to differences in absorption and translocation of these species. However, Shimabukuro, Walsh and Haerauf (1976) suggest that selectivity of foliar-applied barban in wheat and wild oat (*Avena* sp.) grown in culture solution, is dependent on differences in penetration of barban through succeeding layers of coleoptile and leaf sheaths surrounding the apical meristem of grasses.

The basis of selectivity of root-applied barban has also been studied by Ishizuka, Kobayashi and Mashimo (1975) using susceptible oat and wild oat (*Avena fatua*) and resistant wheat and barley. The markedly higher absorption of barban by the roots of the susceptible species was thought to be a major factor determining selectivity; neither differential translocation nor metabolism appeared to be important. Kobayashi and Ishizuka (1977) found little difference in the rate of translocation or metabolism of foliage-applied barban in oats (susceptible) or wheat (resistant). They attributed selective activity to the greater rate of absorption by intact shoots of oats compared to wheat.

Rapid foliage absorption of ^{14}C- or ^{3}H-labelled **phenmedipham** has been reported following leaf application to beet and certain weed species, 20–50% of the herbicide being absorbed within 4 h depending on species (Kassenbeer, 1969). Translocation occurred in the xylem within 8 h and was particularly rapid in *Sinapis arvensis*. Koch *et al.* (1969) found that foliar absorption was initially rapid, thereafter continuing slowly; the rate of uptake was enhanced by increase in light intensity (2–90 lx) and temperature (10–40°C).

The retention and penetration of phenmedipham may be influenced by certain other herbicides. For example, pretreatment with certain halogen-substituted carboxylic acid and thiocarbamate herbicides inhibited the synthesis of cuticular wax thus enhancing the foliage penetration of phenmedipham (Siebert and Köcher, 1972a). The rate of uptake of phenmedipham by *Sinapis alba* and sugar-beet was increased by TCA but not by triallate or cycloate, while addition of TFP (sodium 2,2,3,3-tetrafluoropropionate) increased both retention and penetration of phenmedipham (Siebert and Köcher, 1972b).

Asulam has proved to be a most useful foliage-applied herbicide for the control of certain perennial weed species. The retention, penetration and translocation of this compound in a number of perennial weeds has been examined by Catchpole and Hibbitt (1972). Addition of a wetter enhances retention and activity of asulam in *A. repens*. and *Sorghum halepense*, probably due to the higher quantity of herbicides reaching the leaf axils. However, addition of surfactant fails to increase retention or herbicidal activity in docks (*Rumex obtusifolius*) or bracken (*Pteridium aquilinum*), though application of ^{14}C-asulam to the leaf axils of *A. repens* or dock leaves results in high levels of translocation.

The penetration of asulam into bracken is enhanced by surfactant (1.6% w/v Tween 20) (Babiker and Duncan, 1974) and into broad bean (*Vicia faba* cv. Maris Bead) (Babiker and Duncan, 1975). In the latter, high humidity also influences asulam penetration and this is further enhanced in the presence of Tween 20, glycerol and urea. Foliage application of asulam (4.4 kg/ha) to field bracken reduced frond density by 97, 88 and 79%, 1, 2 or 3 years respectively after treatment (Veerasekaran, Kirkwood and Fletcher, 1976). Autoradiography of greenhouse plants grown from sporelings or fragments and treated with ^{14}C-ring-labelled asulam reveals that uptake is rapid and progressive with time, translocation being generally basipetal and showing a source–sink pattern. Optimum penetration and basipetal translocation occurs when asulam is applied to almost fully expanded fronds while application to immature fronds results in predominantly acropetal movement. Basipetal translocation to the rhizome system is positively correlated with total uptake, this being greater under high humidity and temperature conditions, when application is made to the abaxial (as opposed to adaxial) surface and by incorporation of surfactant (0.1% Tergitol 7) (Veerasekaran *et al.*, 1977).

Foliar penetration and translocation of ^{14}C-asulam in flax (resistant) and wild oat (*A. fatua*) (susceptible) has been studied by several workers. Sharma and Vanden Born (1978) reported that foliar penetration of ^{14}C-asulam in the presence of a surfactant is rapid in both species and continues for at least 72 h, being increased with temperature (10–30°C). Root uptake is also rapid though upward movement is slow. The basis of selectivity of asulam in these species was also examined (Sharma, Vanden Born and McBeath, 1978). Wild oats retains considerably more spray solution (commercial formulation) than does flax, though penetration into flax is more rapid than into wild oats; penetration of the latter is enhanced by surfactant (0.01–0.3%) and high relative humidity. While translocation of radioactivity was extensive throughout the wild oat plants, translocation in flax was limited, probably due to the contact injury caused by the herbicide; there were no detectable differences in herbicide metabolism. They concluded that differentials in spray retention, foliar penetration and translocation all play an important role in the selectivity of asulam in these species.

TRIAZINES

Field observations have indicated that post-emergence application of certain triazine herbicides can be very effective (e.g. Thompson and Slife, 1974). Foliage absorption of these herbicides appears to depend upon a number of factors including cuticle integrity, water solubility of the applied compound and the incorporation of surfactant. For example, little absorption of **simazine** occurred into the leaves of maize, cotton or cucumber unless the cuticle was damaged (Davis *et al.*, 1959). A study of foliar absorption and translocation of a number of triazines in several crop species indicated that cuticle penetration (and acute toxicity) was directly correlated with water solubility of the compound and was enhanced by incorporation of a surfactant (Foy, 1964). The activity of **atrazine** used as a post-emergence contact herbicide to control weeds in maize and sorghum, was also greatly enhanced by the use of additives (Ilnicki *et al.*, 1965). Similar results were obtained by Dexter *et al.* (1966) using a range of surfactants, and by Thompson and Slife (1969) with non-phytotoxic oils or surfactants. The latter authors concluded that complete control of giant foxtail (*Setaria faberii*) depends partly on foliage absorption and partly on root absorption of the herbicide washed from the foliage into the soil.

The effect of oils on the foliar uptake and activity of various triazine herbicides has also been examined (e.g. Coats, 1971; Almodovar-Veja, 1971; Coats and Foy, 1974). Coats (1971) studied the influence of phytobland oils on the activity of foliar-applied atrazine in a number of species. He found that differences in oil type and viscosity influenced uptake and translocation of atrazine by maize leaves, but had no significant effect on photosynthesis or transpiration. Naphthenic oil types were generally more effective penetrating oils than were paraffinic types; the 100 second viscosity oils of both types were, however, more effective than vegetable oils, though Almodovar-Veja (1971) found that vegetable–mineral oil formulations were generally effective adjuvants for atrazine. Using an isoparaffinic oil carrier, Burr (1971) obtained a six-fold increase in the activity of atrazine (and dinoseb) against *Ipomoea hederacea*, but no effect when applied to *Agropyron repens* or *Cyperus rotundus*; however linuron or 2,4-D formulated with undiluted oil as a carrier was effective on *Cyperus rotundus*. Oil treatment also increased the uptake of atrazine applied to leaves of *Setaria glauca* and *Chenopodium album* (Smith and Nalewaja, 1972). Promotion of foliar absorption of ^{14}C-labelled **prometryne** by soybean is also enhanced by a phytobland oil (2.5% Sun 11E) but only during the early hours of treatment; in contrast 0.5% Multi-Film X-77 surfactant promotes a continuous absorption up to 20 h after treatment (Singh *et al.*, 1972). Older soybean leaves absorb more prometryne in surfactant solutions whereas young leaves absorb more herbicide in oil solutions.

When triazine herbicides are applied to the foliage, they penetrate the cuticle, but remain within the leaf to which they are applied (Davis *et al.*, 1959; Foy, 1964; Crafts, 1964; Burt, 1972; Singh, 1972; Bergmannova and Zemänek, 1973; Hilton *et al.*, 1974), moving acropetally to the tip or margin (Crafts, 1959; Foy, 1964; Burt, 1972;

Singh, 1972; Hilton *et al.*, 1974) but not basipetally from the leaf. Foy (1964) has suggested that the lack of movement can be attributed to (1) negligible penetration of the cuticle with consequently low levels of material available for entry into the conducting tissue, (2) interference by the chemical with assimilate (and thus herbicide) translocation in the phloem, due to accumulation of sub-lethal levels in the leaf tissues.

These hypotheses have been examined by Olunuga, Lovell and Sagar (1977) using atrazine. Absorption of substantial quantities of ^{14}C-atrazine occurred suggesting that cuticle penetration was not a limiting factor. They believe that the major obstacle to basipetal movement is the inability of the compound to reach or enter the phloem system. Peterson and Edgington (1976) believe, however, that the important property which allows a pesticide to be transported in the symplast is not its ability to penetrate the plasmalemma but rather its ability to be retained by the symplast after entry.

There appears to be a limited role for foliage- as opposed to root-applied triazines because activity is relatively low. This is illustrated by the results of Dorozhkina *et al.* (1975) who found that about three times as much atrazine was required with foliar than with root application to obtain an 80% inhibition of growth. In contrast to the foliage-applied situation where 72% of the applied dose remained in the leaf after 5 days, root-applied atrazine was translocated to all parts of the plant within hours.

Other herbicides in this group include **metribuzin** and **metamitron**. Metribuzin can be applied either pre- or post-emergence and is effective for the control of several annual weeds, while the latter is efficiently foliage absorbed but not basipetally translocated from the leaves (Müller and Sanad, 1975).

PYRIDINES

Early observations on the field performance of the bipyridylium herbicides indicated that these herbicides were more effective when applied in the late afternoon or evening rather than the morning. The explanation lies in the effect of light on the distribution of the herbicide. For example, when tomato plants treated with ^{14}C-**diquat** were maintained in the dark for 24 h, movement out of the treated leaf was inhibited, but extensive movement throughout the shoot occurred when the plant was subsequently placed in light for 5 h. No movement occurred when the treated plants were placed in light immediately after treatment. Steam ringing of the treated leaf petiole did not prevent the transportation of diquat from the leaf, suggesting that diquat was transported in the xylem. Similar results were obtained with a number of other species and paraquat behaved in the same fashion (Baldwin, 1963).

Translocation of **paraquat** from the leaves of purple nutsedge (*Cyperus rotundus*) and yellow nutsedge (*Cyperus esculentus*) was similarly enhanced by a 24 h dark period prior to placing the plants in light (Wood and Gosnell, 1965). While movement within the leaf was believed to be largely apoplastic Wood and Gosnell noted that some movement occurred into the new growth of rapidly growing plants suggesting some symplastic movement. However, Slade and Bell (1966) found that paraquat was transported out of young developing as well as mature leaves and steam-ringing had no effect on movement, suggesting that xylem transport was involved. They suggested that light-induced damage was a prerequisite for significant movement into the remainder of the shoot, though this damage appeared to prevent further entry of the herbicide into the xylem. The role of light on the distribution of diquat was studied by Smith and Sagar (1966). They concluded that to allow penetration darkness was necessary only in the area of application. After entry, the desiccation which follows death enables the transfer of water containing herbicide in the xylem to the other leaves; they believe that phloem transport is not involved. Xylem transport of paraquat has also been demonstrated in dallis grass (*Paspalam distichum*) (Smith and Davies, 1965). The influence of environmental factors on the action of bipyridylium herbicides has been discussed by Brian and Headford (1968), Putnam and Ries (1968), Brian (1969) and the properties of these compounds reviewed by Akhavein and Linscott (1970).

Although paraquat and diquat are normaly foliage-applied herbicides, root uptake has been reported to occur from culture solutions or from soil if the herbicide is leached into the rooting zone. For example, uptake of paraquat and diquat following pre-emergence application to sand or loamy sand has been observed when the herbicide is leached to depths greater than 1.25 cm. Apparently binding of the compound to clay or peat does not preclude the possibility of absorption. Thus absorption of paraquat by cucumber seedlings has been reported despite tight binding to montmorillonite or kaolinite clay particles (Weber and Scott, 1966); absorption from peat by a number of species has also been found (Damonakis *et al.*, 1970). Under these circumstances the activity of paraquat is restricted to the roots presumably reflecting limited translocation to the shoots.

Leaf absorption by the bipyridylium herbicides is influenced by their cationic and polar properties and may take place very rapidly. For example, application of diquat to a beet leaf resulted in 33% of the applied dose being absorbed within 30 seconds. Washing failed to remove the diquat and the effect was attributed to an ionic attraction between cations of the herbicide and negative charges associated with the leaf surface. The influence of the polar properties of these compounds is evident in a study of diquat uptake by *Elodea canadensis* in which a non-reversible rapid initial phase was followed by a relatively slower accumulation phase (Davies and Seaman, 1968); the rapid initial phase is not reversible.

The effect of diquat and several surfactants on membrane permeability in red beetroot tissue has been examined by Sutton and Foy (1971). Combinations of 10^{-5} or 10^{-3} M diquat and 0.1% or 1.0% Tween 20 increase the efflux of betanin from root-tissue discs of red beet, there being evidence of additive effects. Sutton and Foy suggest that the effect of this non-ionic surfactant may be due to a combination of its hydrophilic properties and its specific structure which may alter its energy relationships at membrane surfaces.

The influence of surfactants on the uptake and movement of paraquat has been studied at Jealott's Hill (Imperial Chemical Industries). Using mainly cocksfoot (*Dactylis glomerata*), Brian (1972) studied the action as wetters of alcohols, nonylphenol and amine oxides condensed with 2–30 mol ethylene oxide at 0.1–0.5% (w/v). In the case of nonylphenol and alcohol ethylene oxide condensates, paraquat uptake and percentage movement were inversely related, the greatest uptake being associated with lowest movement. Bland and Brian (1972) found that where wetters partitioned largely into wax, paraquat movement was high and where immobilisation in the wax was low, movement also was low. However, paraquat movement within the plant was adversely affected by the presence of wetter within the leaf tissues, the concentration there being dependent on the degree of immobilisation within the wax. Bland and Brian (1975) concluded that surfactants are essential components of a paraquat formulations in order to wet the leaf surface and increase penetration, but when the surfactant also penetrates into the leaf, the mobility and efficiency of paraquat are reduced.

There is evidence that the relative humidity at the time of application affects cuticle penetration of the bipyridyliums. For example, Brian (1966) found that two to five times more diquat than paraquat was absorbed at high humidity the effect being attributed to hydration of the cuticle. Similar effects have been found by Brian and Ward (1967) and Thrower *et al.* (1965); the latter attributing the beneficial effect of high humidity to a slower drying rate of the spray droplets.

Using perennial ryegrass (*Lolium perenne*), Harvey, Muldoon and Harper (1978) found that there was no difference in uptake or distribution of ^{14}C-paraquat within leaf tissue in susceptible and tolerant cultivars; no metabolites of paraquat were detected in either susceptible or tolerant plants. They concluded that paraquat tolerance in perennial ryegrass is unlikely to depend upon reduced uptake enhanced metabolism or altered translocation of the herbicide. Previously Brian (1972) had found comparable levels of uptake and movement of ^{14}C-morfamquat in resistant graminaceous plants and susceptible dicotyledons. He suggested that selectivity could be due to (1) accumulation of herbicide in the epidermal cells, preventing access to the chlorophyll cells, (2) selective detoxication by grasses or selective activation by dicotyledons perhaps by conversion to paraquat, (3) relatively free movement of herbicide from the cytoplasm into the chloroplasts in the cells of dicotyledons as compared to monocotyledons.

Picloram is used both as a root- and foliage-applied herbicide. It is a growth-regulating herbicide, many of its effects being similar to those associated with POA. It is not surprising, therefore, that in their studies several workers have compared picloram and 2,4-D. The foliar absorption and translocation of ^{14}C-picloram and ^{14}C-2,4-D have been compared in the quaking aspen (*Populus tremuloides*) and 2,4-D has been found to be more efficiently absorbed than picloram (Sharma and Vanden Born, 1970). The effect of plant maturity on the uptake and translocation of these compounds has been investigated using 5- and 7-week-old seedlings and 16-week-old vegetatively propagated plants of field bindweed (*Convolvulus arvensis*). After 48 h, more picloram has been absorbed by the 16-week-old plants than by the seedlings, which is perhaps surprising since the cuticle wax would be expected to be more highly developed in the older plants; in this case less 2,4-D than picloram was absorbed (Hallmén and Eliasson, 1972). The influence of phenoxy herbicides on foliar uptake and phytotoxicity of picloram in *Phaseolus vulgaris* was examined by Hamill, Smith and Switzer (1972). The combination of picloram with 2,4-D or mecoprop resulted in synergistic effects on fresh and dry weight reduction of beans. Picloram inhibited movement of 2,4-DB, while the latter increased the distribution of picloram.

There is evidence that the climatic conditions prevailing at the time of application may influence absorption and presumably the activity of picloram. Absorption of picloram (and 2,4-D) by *Populus tremuloides* was enhanced by increased temperature and humidity (Sharma and Vanden Born, 1970), while uptake of picloram (and 2,4,5-T or triclopyr) by certain woody species was also increased under relatively high temperature and long-day conditions (Radosecvich and Bayer, 1979).

There is evidence that picloram or 2,4-D may form complexes with plant components, thus counteracting translocation and toxicity of auxin herbicides. Eliasson and Hallmén (1973) found that considerable quantities of both herbicides were retained in the treated leaves of *Populus tremula*, only slight translocation occurring into the shoot tip and roots. Leaf retention was partly due to the incorporation of picloram or 2,4-D into complexes from which the active compounds could be released. Differences in metabolism between 2,4-D and picloram could not explain differences in their toxicity. Subsequent study of the uptake, translocation and fate of ^{14}C-picloram and ^{14}C-2,4-D in seedlings of rape (*Brassica napus* cv. Nilla) and sunflower showed that complex formation was particularly marked in rape (Hallmén, 1974). Investigation of the translocation and metabolism of ^{14}C-picloram in *Prosopis ruscifolia* also showed that most of the herbicide remained in the treated leaves, though transformation into a water-soluble compound resulted in extensive translocation into untreated leaves, stem and roots 48 h after treatment (Prego, Maroder and Sonvico, 1976). In contrast, foliar application of ^{14}C-picloram (and ^{14}C-glyphosate and 2,4-D) to leafy spurge (*Euphorbia esula*) resulted in extensive translocation throughout the plant, particularly in an acropetal direction; however, the translocation of 2,4-D was more basipetal than acropetal (Bybee, Messersmith and Gigax, 1979).

Additives have been found to stimulate the absorption of picloram. For example, uptake of picloram and 2,4-D was enhanced by 'Atlox 210' (1%) (Sharma and Vanden Born, 1970). Sands and Bachelard (1973a) found that uptake by leaf discs of *Eucalyptus viminalis* and *E. polyanthemos* was markedly increased by a number of surfactants. In the latter case uptake depended largely on the degree of wetting and the stomata play a major role in the absorption of picloram in these species (Sands and Bachelard, 1973b). Formulation of a picloram–carrier complex composed of 50% dimethyl sulphoxide (DMSO), 25% glycerol, 15% isoparaffinic oil and 10% water resulted in enhanced foliar absorption and translocation in leaves and stems but not in roots (Hull, Morton and Martin, 1973). The use of ammonium sulphate [$(NH_4)_2SO_4$] (0.1–10%) also increased the magnitude and rate of absorption of ^{14}C-picloram by detached leaves of strawberry guava (*Psidium cattleianum*). Similar effects have been obtained with ammonium nitrate, ammonium chloride, ammonium dibasic phosphate (pH 7.7), ammonium monobasic phosphate (pH 4.5) and potassium monobasic phosphate (pH 4.6) (Wilson and Nishimoto, 1974). Wilson (1974) suggested that $(NH_4)_2SO_4$ has a direct physical effect on the absorption pathway of picloram through the leaves. Later work shows considerable increases in absorption of ^{14}C-picloram by leaves of strawberry guava (*Psidium cattleianum*) (fivefold), guava (*P. guajava*) (fourfold) and dwarf beans (*Phaseolus vulgaris*) (fourfold) treated with $(NH_4)_2SO_4$ (0.5 and 10%) (Wilson and Nishimoto, 1975).

Picloram has been successfully used in the control of certain difficult weeds such as bracken and gorse. Farnworth and Davies (1974) studied the absorption of ^{14}C-picloram by bracken (*Pteridium aquilinum*) grown from fragments of rhizome. The ^{14}C was readily taken up by all organs (frond laminae, rhizome apices, frond buds and roots) and freely translocated, except in the case of frond laminae from which translocation was very poor. They concluded that this may explain the poor control of field bracken sprayed with picloram in July. The absorption, translocation and metabolism of ^{14}C-picloram applied with 2,4,5-T (unlabelled) to gorse (*Ulex europaeus*) plants grown from cuttings was examined by Rolston and Robertson (1975). Picloram was readily translocated from the treated stem to the roots although it did not appear to be redistributed to other treated areas. Absorption was increased by addition of non-ionic surfactants, by lowering the pH and by removal of cuticular wax (Rolston and Robertson, 1976).

HETEROCYCLIC NITROGEN COMPOUNDS (UNCLASSIFIED)

Early studies using chlorosis as an indicator produced evidence of foliage (and generally root) absorption and translocation of **aminotriazole** in cotton (Hall *et al.*, 1954), purple nutsedge (*Cyperus rotundus*) and Johnson grass (*Sorghum halepense*) (Hauser and Thompson, 1954). The availability of ^{14}C-labelled aminotriazole made possible more detailed examination of the uptake/transport characteristics. For example, absorption and extensive translocation of ^{14}C was evident within 3 days of application of ^{14}C-aminotriazole to soybean, Canada thistle (*Cirsium arvense*) and Johnson grass (*S. halepense*), small quantities of ^{14}C being detected in the rhizome and roots of *C. arvense* to a depth of about 30 cm (Bondarenko, 1957). Confirmation of the rapid and efficient translocation of aminotriazole or its metabolites in a number of plant species is evident from reports by a number of workers (Ashton and Crafts, 1973).

The effectiveness of aminotriazole is enhanced by formulation with ammonium thiocyanate (NH_4SCN), though the underlying mechanism is somewhat controversial. Application of this additive prior to or along with aminotriazole did not influence the absorption of ^{14}C-aminotriazole by couch grass (*Agropyron repens*) but did increase the translocation of ^{14}C, though application of NH_4SCN one day after the herbicide had no effect (Donnalley and Ries, 1964). They suggest that NH_4SCN reduces the binding of aminotriazole at the treatment site. Forde (1966) reporting that NH_4SCN inhibits the translocation of aminotriazole in couch grass suggested that the immobilising effect of the additive is exerted on the treated leaf rather than on long-distance transport *per se.* Van der Zweep (1965) reported that increasing concentrations of NH_4SCN enhance shoot inhibition and increase the dispersal of ^{14}C from ^{14}C-aminotriazole within the treated leaf; similar but greater effects were noted with the alternative addition of 6-*N*-benzyladenine.

The translocation of assimilates and aminotriazole and its metabolites has been examined in *Agropyron repens* at three stages of shoot development (Fiveland, Erickson and Seely, 1972). At the 2- and 3-leaf stage and at the 3- and 4-leaf stages, ^{14}C assimilates were translocated from the treated shoot to other shoots and rhizomes but at the 5-leaf stage much less ^{14}C material was translocated into the untreated shoots. At all growth stages, more ^{14}C assimilates were translocated to the roots than to untreated shoots and the translocation patterns of ^{14}C assimilates and ^{14}C-aminotriazole and its metabolites were similar. The effect of the stage of shoot development on the translocation pattern of ^{14}C-aminotriazole in *Agrostis gigantea* has been reported by Fiveland (1974). ^{14}C-Aminotriazole was metabolised in the treated leaf and two major metabolites (A and B) were found, A being slowly metabolised into B; aminotriazole *per se* was not translocated out of the treated leaf. In *A. repens*, however, ^{14}C-aminotriazole made up at least 50% of the activity in the treated leaf after 144 h and was translocated out of the leaf. Fiveland suggests that these differences probably account for the selective toxicity of low rates of aminotriazole in the two species.

The mechanism and efficiency of penetration of aminotriazole has interested a number of workers. Lund-Høie (1972) compared the uptake (and metabolism) of aminotriazole applied to the adaxial or abaxial surfaces of *Pteridium aquilinum*, *Cirsium arvense*, *Phaseolus vulgaris* and *Avena sativa* and found that more ^{14}C-aminotriazole was taken up by the abaxial surface; *Avena sativa* was exceptional in that uptake was almost equal from both surfaces. His results suggest a possible relationship between uptake of aminotriazole and stomatal densities.

The effect of certain adjuvants on the penetration of bean leaves (*P. vulgaris* cv. Canadian Wonder) by aminotriazole has been studied by Cook, Babiker and Duncan (1977). Penetration was increased by Tween 20, dimethyl sulphoxide (DMSO) and glycerol, the ratio of herbicide to surfactant appearing to be of central importance to the process of penetration. High humidity conditions enhanced penetration, but at low humidity they found that addition of polyethelene 20 sorbitan monolaurate (Polysorbate 20) to the spray always increased penetration of aminotriazole into the plant.

The selective action of foliage-applied **difenzoquat** in *Avena fatua* (susceptible) and barley (cv. Conquest) (tolerant) could not be accounted for by differences in the efficiency of absorption, translocation or metabolism (Sharma *et al.*, 1976). High levels of absorption were noted for both species, but little basipetal translocation was evident, movement being mainly acropetal; there was no evidence of metabolism in either species.

ORGANOARSENIC COMPOUNDS

Cacodylic acid is generally regarded as a contact herbicide, but when applied to cuts around the base of trees it is translocated in the apoplast (Ashton and Crafts, 1973). While there was no evidence of translocation of DSMA (disodium methylarsonate) in Johnson grass (McWhorter, 1966), it is apparently phloem translocated in purple nutsedge (*Cyperus rotundus*), the growing tubers acting as active sinks (Holt *et al.*, 1967). The translocation of DSMA and the amine salt of methylarsonic acid (AMA) (amine methane arsonate) was further studied in purple nutsedge (Duble, Holt and McBee, 1968) and evidence of acropetal and basipetal movement found in single leaves; chromatographic tests suggested that a DSMA–plant extract conjugate may have formed. Actively growing terminal tubers in a series accumulated arsenic while intermediate or dormant tubers did not, confirming that movement in a tuber depends upon its sink activity.

AMA was found to be both xylem and phloem mobile in Johnson grass and cotton, apoplastic movement being more rapid than symplastic (Sckerl and Frans, 1969). As in the case of DSMA above, chromatography of the extracts from Johnson grass indicated possible complexing with sugars and/or organic acids; there was some evidence to suggest a complex with histidine or one of its analogues.

MSMA (monosodium methylarsonate) also appears to be phloem translocated. For example in cotton as much as 25% of the detected ^{14}C activity was found in the untreated leaves following foliage application of MSMA (2.24 kg/ha) (Wilkinson and Hardcastle, 1969). Symplastic movement of ^{14}C from leaf-applied ^{14}C-MSMA has also been reported in Johnson grass (Kempen, 1970) and bean plants (Sachs and Michael, 1971), though Kempen found MSMA to be principally xylem mobile.

Application of ^{14}C-MSMA, MAA or DSMA to cotton seedlings resulted in little translocation of ^{14}C from the cotyledons to developing leaves except when treatment was carried out at 13°C (Keeley and Thullen, 1971). These herbicides were also applied to purple and yellow nutsedge (*Cyperus* spp.), more ^{14}C being absorbed and translocated by the susceptible yellow species than by the relatively resistant purple species (Keeley and Thullen, 1971).

Foliage application of ^{14}C-cacodylic acid to seedlings of French beans and green ash (*Fraxinus pennsylvanica*) resulted in exudation of ^{14}C-cacodylic acid into the nutrient solution within 24 h of treatment. The cumulative loss for beans was 19.1% after 9 days compared with 9.6% for ash after 16 days, presumably evidence of differences in the efficiency of basipetal movement. Translocation of MSMA (10,000–40,000 p.p.m. into the rhizome and roots) in *Sorghum halepense* resulted in growth inhibition, sequential applications being more effective than a single application of the same dose (Kogan and Araya, 1974). The distribution (and metabolic fate) of ^{14}C-MSMA in wheat (*Triticum aestivum* var. Waldon) was investigated 2, 4 and 13 weeks after foliage application (Domir *et al.*, 1976). While some apoplastic movement occurred, translocation was mainly symplastic with evidence of exudation from the roots into the soil (20% of the applied dose).

ORGANOPHOSPHORUS COMPOUNDS

Glyphosate is one of the most interesting and successful herbicides to be developed in the last decade. It is effective against a large range of species (e.g. Baird, Upchurch and Selleck, 1972), its activity being affected by environmental factors such as temperature, humidity, wind, rain (Caseley, 1972) and shade (Upchurch and Baird, 1972).

The movement of glyphosate (and bentazone) in maize, soybean and several weed species was examined by Sprankle *et al.* (1973, 1975). Greater movement occurred in plants treated at the 3-leaf as opposed to 5-leaf stage, different translocation characteristics being evident in several species. For example, glyphosate moved both acropetally within the treated leaf and basipetally into the roots and developing leaves of maize and *Cyperus esculentus*. In soybean and *Amaranthus retroflexus*, glyphosate was phloem translocated to all active sinks (Sprankle *et al.*, 1973). In *Agropyron repens*, glyphosate was readily absorbed with some acropetal movement to the leaf tips; however a large proportion of the absorbed ^{14}C was translocated in the phloem to new shoots developing from the rhizomes (Sprankle *et al.*, 1975).

The growth stage at the time of application affects the activity of glyphosate against *A. repens*, treatment at the 4-leaf, being more effective than at the 2-leaf stage (Rioux, Bandeen and Anderson, 1974). Some explanation may be evident in the results of Claus and Behrens (1976) who examined the effect of rhizome length and foliar height on translocation of ^{14}C-glyphosate (0.28 kg/ha) in *A. repens*. They found that bud kill was greater when glyphosate was applied to foliage 45 cm high, treatment at 25 cm and 13 cm being less effective. The greatest accumulation in the nodes occurred near the rhizome tips and was least in the nodes adjacent to the mother shoot, no doubt reflecting the relative sink intensities. Similar findings have been obtained in studies of the absorption and translocation of ^{14}C-glyphosate in *Cyperus rotundus* (Pulver and Romero, 1976). Penetration of the cuticle took twice as long in mature compared with immature plants, but translocation of ^{14}C-glyphosate was more rapid in mature plants, being complete in 5 days compared with 15 days in immature plants. The effectiveness of glyphosate was reduced by certain formulations with ametryne, atrazine, 2,4-D amine or ester, linuron or terbacil. Zandstra and Nishimoto (1977) also found that translocation of ^{14}C-glyphosate into the tubers of *C. rotundus* increased with time and plant age, there being no evidence of metabolism of glyphosate in this species.

The basis of selectivity of glyphosate in Canada thistle (*Circium arvense*) (susceptible) and leafy spurge (*Euphorbia esula*) (resistant) was investigated by Gottrup *et al.* (1976). They found that the pattern of absorption and distribution was similar in both species, glyphosate being readily absorbed and translocated (apoplastic and symplastic) particularly under high humidity conditions or with a surfactant. However, retention was greatest in Canada thistle and this may account for the difference in sensitivity of these two species. Marquis, Comes and Yang (1979) were unable to pinpoint the basis of selectivity of ^{14}C-glyphosate in creeping red fescue (*Festuca rubra* var. rubra) (resistant) and reed canary grass (*Phalaris arundinacea*) (susceptible). The mechanism of resistance appeared not to involve differentials in retention, absorption, translocation or metabolism though they observed that the tolerance of creeping red fescue appeared to be related to its ability to regenerate shoots and roots from the crown of the plant. Lund-Høie (1976) concluded that the tolerance of Norway spruce (*Picea abies*) to glyphosate was due to limited uptake, restricted distribution and rapid detoxification; he found that uptake of ^{14}C-glyphosate was only ½ to ⅓ of that in *A. repens* (susceptible). However, stage of growth seems to be important, uptake being four to five times greater when active shoot growth is taking place as opposed to the period of no shoot growth.

Various aspects of foliar entry of glyphosate have been examined, particularly the relative efficiency of penetration of various sites. Coupland, Taylor and Caseley (1978) reported that the performance of glyphosate on *Agropyron repens* (and certain herbicides on *Avena fatua*) was increased when the herbicide was applied towards the leaf or plant base; treatment of the youngest, fully expanded leaf giving the best performance. They believe that variations in the amount of epicuticular wax between different areas on the leaf could be an important factor determining performance. King and Radosevich (1978) examined the penetration of glyphosate (10^{-5} M) (and triclopyr) into leaf discs of *Lithocarpus densiflorus*, and found that only a small proportion (4%) of the applied glyphosate was absorbed, penetration being correlated with certain leaf characteristics including cuticle thickness, stomatal density and pubescence. Uptake of ^{14}C-glyphosate by isolated leaf cells of hemp dogbane (*Apocynum cannabinum*) was only significant during the first 15 min of exposure and leaf cell absorption was thought to be a limiting factor in foliage absorption (Richard, 1978).

Richard and Slife (1979) reported that in the absence of additives penetration of detached leaves of *A. cannabinum* by ^{14}C-glyphosate was negligible after 30 min. Additives increased the amount of glyphosate absorbed initially but did not extend the period of absorption. Isolated leaf cells absorbed 0.1% of applied glyphosate in a 2 h period of incubation, and binding studies *in vitro* and *in vivo* indicated that the absorbed ^{14}C was not tightly bound but was probably free to be translocated out of the cells. They concluded that cellular absorption represents the major barrier in the foliage absorption of glyphosate in *A. cannabinum*. A further study of the uptake mechanism was carried out by Leonard and Shaner (1979) using protoplasts isolated from cultured tobacco cells. ^{14}C-Glyphosate was readily absorbed at a linear rate for several hours over a range of concentrations (10^{-6}–10^{-2} M); uptake appeared to be energy dependent and they produced evidence suggesting that the amino acid transport system may be involved.

It is well established that translocation of glyphosate occurs in the phloem, movement following a typical source to sink pattern with accumulation in the meristematic areas (e.g. Haderlie, Slife and Butler, 1978; Schultz and Burnside, 1980). Sanderberg, Meggitt and Penner (1978) found that foliar application of ^{14}C-glyphosate to *Ipomoea purpurea*, *Convolvulus arvensis*, *Calystegia sepium*, *Polygonum convolvulus* and *Cirsium arvense* resulted in phloem translocation of ^{14}C throughout the plant with accumulation in the meristematic tips of shoot and root; translocation was almost complete within 3 days. Likewise Kells and Rieck (1978) reported that application of ^{14}C-glyphosate to Johnson grass (*Sorghum halepense*) resulted in ^{14}C accumulation in the growing points and other meristematic areas. Light intensity had no effect on ^{14}C absorption but increased ^{14}C translocation, significantly greater accumulation in roots and rhizomes occurring in plants exposed to full light (Kells and Rieck, 1979). Translocation of ^{14}C-glyphosate in cotton (*Gossypium hirsutum*) was also greatly influenced by relative humidity; an increase in r.h. from 40% to 100% at a constant temperature resulted in a three- to sixfold increase in translocation of ^{14}C (Wills, 1978). Relative humidity and temperature also exercised a striking influence on the quantities of ^{14}C-glyphosate absorbed and translocated in *Betula verrucosa* and *Fraxinus excelsior* (Lund-Høie, 1979). Higher temperatures (30 or 25° C) also resulted in higher translocation of ^{14}C-glyphosate in hemp dogbane (*A. cannabinum*) and higher light intensity increased movement into untreated regions of the leaf (Schultz and Burnside, 1980); these workers warned against the use of lanolin rings in foliage-application studies due to high levels of adsorption of ^{14}C-glyphosate (and ^{14}C-2,4-D) in lanolin (Schultz and Burnside, 1980).

The literature associated with the brush control agent **fosamine** (ammonium ethyl carbamoylphosphonate) is relatively sparse. Fosamine provides excellent growth suppression of brushwood when applied in the spring; late summer or early autumn applications have little immediate effect but they prevent growth of most perennial broad-leaved weeds in the following year.

Autoradiography of ^{14}C-fosamine-treated plants shows that the compound may have difficulty in penetrating leaves, but does enter young stems; leaf penetration is significantly increased by such adjuvants as surfactants WK, Renex 30 and Citowett Plus. The action of fosamine was found to be antagonised by the presence of alkaline earth metals in the spray water or high concentrations of phosphorus in the plant tissue (Weigel and Riggleman, 1978).

The absorption and translocation of ^{14}C-fosamine-ammonium has also been examined in susceptible and resistant seedling woody plants (Kitchen and Rieck, 1978). Higher levels of absorption and translocation were observed in susceptible species although translocation of ^{14}C was directed to the roots of all but one species in which accumulation occurred in the meristematic regions. They conclude that one mechanism of resistance of woody plants to this herbicide may be the physical exclusion of phytotoxic levels of the herbicide by the plant.

References

AKHAVEIN, A. A. and LINSCOTT, D. L. (1970). *Weed Sci.*, **18**, 378.

ALMODOVAR-VEJA, L. (1971). *Diss. Abstr. Int. B.*, **32**, 665.

AMELUNXEN, F. MORGENROTH, K. and PICKSAK, T. (1967). *Z. Pflanzenphysiol.*, **57**, 79.

ANDERSON, J. L. and THOMSON, W. W. (1973). *Residue Rev.*, **47**, 167.

ANDERSON, R. N., LINCK, A. J. and BEHRENS, R. (1962). *Weeds*, **10**, 1.

ASHTON, F. M. and CRAFTS, A. S. (1973). In *Mode of Action of Herbicides*, Wiley-Interscience, New York, London, 504 pp.

BABIKER, A. G. T. and DUNCAN, H. J. (1974). *Weed Res.*, **14**, 375.

BABIKER, A. G. T. and DUNCAN, H. J. (1975). *Pestic. Sci.*, **6**, 655.

BACHELARD, E. P. and AYLING, R. D. (1971). *Weed Res.*, **11**, 31.

BAIRD, D. D., UPCHURCH, R. P. and SELLECK, G. W. (1972). *Proc. 24th Ann. Calif. Weed Conf.*, 94.

BAKER, E. A. (1972). *The Effect of Environmental Factors on the Development of the Leaf Wax of* Brassica oleracea *var.* gemmifera, M.Sc. Thesis, University of Bristol.

BAKER, E. A. and BUKOVAC, M. J. (1971). *Ann. Appl. Biol.*, **67**, 243.

BAKER, E. A. and HOLLOWAY, P. J. (1970). *Phytochemistry*, **9**, 1557.

BAKER, E. A. and MARTIN, J. T. (1963). *Nature (London)*, **199**, 1268.

BALDWIN, B. C. (1963). *Nature (London)*, **198**, 872.

BARRENTINE, J. L. and FRANS, R. E., (1973). *Weed Sci.*, **21**, 445.

BARRENTINE, J. L. and WARREN, G. F. (1970). *Weed Sci.*, **18**, 373.

BASLER, E. (1973). *Pl. Physiol., Lancaster*, **51**, 63.

BASLER, E. (1978). *Proc. 31st Ann. Mtg. S. Weed Sci. Soc.*, 279.

BAUR, J. R., BOVEY, R. W. and RILEY, I. (1974). *Weed Sci.*, **22**, 481.

BAUR, J. R., BOWMAN, J. J. and BOVEY, R. W. (1973). *Abstr. Mtg. Weed Sci. Soc. Amer.*, 68.

BERGMANNOVA, K. and ZEMÄNEK, J. (1973). *Agrochemia*, **13**, 312.

BHAN, V. M., STOLLER, E. W. and SLIFE, F. W. (1970). *Weed Sci.*, **18**, 733.

BLACKMAN, G. E., BRUCE, R. S. and HOLLY, K. (1958). *J. Exp. Bot.*, **9**, 175.

BLANCHARD, F. A., MUELDER, U. V. and SMITH, G. N. (1960). *J. Agric. Fd. Chem.*, **8**, 124.

BLAND, P. D. and BRIAN, R. C. (1972). *Pestic Sci.*, **3**, 133.

BLAND, P. D. and BRIAN, R. C. (1975). *Pestic Sci.*, **6**, 419.

BLANTON, W. E. and BASLER, E. (1976). *Proc. 29th Ann. Mtg. S. Weed Sci. Soc.*, 403.

BOIZE, L., GUDIN, C. and PURDUE, G. (1976). *Ann. Appl. Biol.*, **84**, 205–11

BONDARENKO, D. D. (1957). *Proc. 14th North Central Weed Control Conf.*, 9–10.

BRADY, H. A. (1974). *Proc. 27th Ann. Meet. S. Weed Sci. Soc.*, 204.

BRIAN, R. C. (1966). *Weed Res.*, **6**, 292.

BRIAN, R. C. (1967). *Ann. Appl. Biol.*, **59**, 91.

BRIAN, R. C. (1969). *Ann. Appl. Biol.*, **63**, 117.

BRIAN, R. C. (1972). *Pestic Sci.*, **3**, 121.

BRIAN, R. C. and HEADFORD, D. W. R. (1968). *Proc. 9th Brit. Weed Control. Conf.*, 108.

BRIAN, R. C. and WARD, J. (1967). *Weed Res.*, **7**, 117.

BRUINSMA, J. (1967). *Acta Bot. Neerl*, **16**, 73.

BUKOVAC, M. J. (1965). *Proc. Amer. Soc. Hortic. Sci.*, **87**, 131.

BUKOVAC, M. J., SARGENT, J. A., POWELL, R. G. and BLACKMAN, G. E. (1971). *J. Exp. Bot.*, **22**, 598.

BURR, R. J. (1971). *Diss. Abstr. Int. B.*, **32**, 3111.

BURR, R. J. and WARREN, G. F. (1972). *Weed Sci.*, **20**, 324.

BURT, G. W. (1972). *Proc. N.E. Weed Sci. Soc., New York*, **26**, 71.

BYBEE, T. A., MESSERSMITH, C. G. and GIGAX, D. R. (1979). *Abstr. Mtg. Weed Sci. Soc. Amer.*, 2.

CAMPER, N. D. and MORELAND, D. E. (1971). *Weed Sci.*, **19**, 269.

CANNY, M. J. (1962). *Ann. Bot.*, **26**, 603.

CARSON, A. G. and BANDEEN, J. D. (1975). *Can. J. Pl. Sci.*, **55**, 795.

CASELEY, J. (1972). *Proc. 11th Brit. Weed Control Conf.*, 641.

CATCHPOLE, A. H. and HIBBITT, C. J. (1972). *Proc. 11th Brit. Weed Control Conf.*, 77.

CHALLEN, S. B. (1962). *J. Pharm. Pharmacol.*, **14**, 707.

CHAMBERS, T. L., RITCHIE, I. M. and BOOTH, M. A. (1976). *New Phytol.*, **77**, 43.

CLAUS, J. S. and BEHRENS, R. (1976). *Weed Sci.*, **24**, 149.

COATS, G. E. (1971). *Diss. Abstr. Int. B.*, **32**, 3102.

COATS, G. E. and FOY, C. L. (1974). *Weed Sci.*, **22**, 220.

COOK, G. T., BABIKER, A. G. T. and DUNCAN, H. J. (1977). *Pestic. Sci.*, **8**, 137.

CORBIN, F. T., SCHRADER, J. W. and COBLE, H. D. (1972). *Abstr. Mtg. Weed Sci. Soc. Amer.*, 47.

COUPLAND, D., TAYLOR, W. A. and CASELEY, J. C. (1978). *Weed Res.*, **18**, 123.

CRAFTS, A. S. (1956a). *Amer. J. Bot.*, **43**, 548.

CRAFTS, A. S. (1956b). *Hilgardia*, **26**, 335.

CRAFTS, A. S. (1959). *Pl. Physiol., Lancaster*, **34**, 613.

CRAFTS, A. S. (1960). *Weeds*, **8**, 19.

CRAFTS, A. S. (1964). In *The Physiology and Biochemistry of Herbicides* (Ed. L. J. Audus), Academic Press, London and New York, p. 75.

CRISP, C. E. (1965). *The Biopolymer Cutin*, Ph.D. Thesis, Univ. of Calif., Davis, 75.

CRISP, C. E. (1972). In *Pesticide Chemistry: Proc. 2nd Internat. IUPAC Cong. Pestic. Chem.* (Ed. A. S. Tahori), Gordon and Breach, London, Vol. I, Insecticides, p. 211.

CURRIER, H. B. and DYBING, C. D. (1959). *Weeds*, **7**, 195.

DAMONAKIS, M., DRENNAN, D. S. H., FRYER, J. D. and HOLLY, K. (1970). *Weed Res.*, **10**, 278.

DARWENT, A. L. and BEHRENS, R. (1972). *Weed Sci.*, **20**, 540.

DAVIES, P. J., DRENNAN, D. S. H., FRYER, J. D. and HOLLY, K. (1967). *Weed Res.*, **7**, 220.

DAVIES, P. J., DRENNAN, D. S. H., FRYER, J. D. and HOLLY, K. (1968). *Weed Res.*, **8**, 233.

Davies, P. J. and Seaman, D. E. (1968). *Weed Sci.*, **16**, 293.

Davis, D. E., Funderburk, H. H. and Sansing, N. G. (1959). *Weeds*, **7**, 300.

Davis, D. G. and Dusbabek, K. E. (1973). *Weed Sci.*, **21**, 16.

Davis, F. S., Meyer, R. E., Bauer, J. R. and Bovey, R. W. (1972). *Weed Sci.*, **20**, 264.

De Maria, M. E., Thaine, R. and Sarisalo, H. I. M. (1975). *J. Exp. Bot.*, **26**, 145.

Dexter, A. G., Burnside, O. C. and Lavy, T. L. (1966). *Weeds*, **14**, 222.

Dexter, A. G., Slife, F. W. and Butler, H. S. (1971). *Weed Sci.*, **19**, 721.

Diem, J. R. and Davis, D. E. (1974). *Weed Sci.*, **22**, 285.

Domir, S. C., Woolson, E. A., Kearney, P. C. and Isensee, A. R. (1976). *J. Agric. Fd. Chem.*, **24**, 1214.

Donnalley, W. F. and Ries, S. K. (1964). *Science N.Y.*, **145**, 497.

Donoho, C. W. Jr., Mitchell, A. E. and Bukovac, M. J. (1961). *Proc. Amer.Soc. Hortic. Sci.*, **78**, 96.

Dorozhkina, L. A., Kuz'minskaya, V. A., Kalinin, V. A. and Gruzdev, G. S. (1975). *Izv. Tim. Sel'skok. Akad.*, 156.

Duble, R. L., Holt, E. C. and McBee, G. G. (1968). *J. Agric. Fd. Chem.*, **17**, 1247.

Dybing, C. D. and Currier, H. B. (1961). *Pl. Physiol., Lancaster*, **36**, 169.

Edgerton, L. J. and Haesler, C. W. (1959). *Proc. Amer. Soc. Hortic. Sci.*, **74**, 54.

Eglington, G. and Hunneman, D. H. (1968). *Phytochemistry*, **7**, 313.

Eliasson, L. and Hallmén, U. (1973). *Physiol. Plant*, **28**, 182.

Eveling, D. W. and Elisa, M. Z. (1976). *Weed Res.*, **16**, 15.

Evert, R. F., Milton Tucker, C., Davis, J. D. and Deshpande, B. P. (1969). *Amer. J. Bot.*, **56**, 999.

Ezerzha, A. A., Kulikov, B. N., Mochalkin, A. I. and Popov, L. N. (1971). *Fiz. Biokhim. Kul'turn. Rast.*, **3**, 140.

Farnworth, J. and Davies, G. M. (1974). *Weed Res.*, **14**, 397.

Fellows, R. J. and Geiger, D. R. (1974). *Pl. Physiol., Lancaster*, **54**, 877.

Feng, K. A. (1976). *Pl. Physiol., Lancaster*, **57**, 64.

Field, R. J. and Phung-Hong-Thai (1979). *Proc. 32nd N.Z. Weed and Pest Control Conf.*, 309.

Fisher, D. B. (1975). *Pl. Physiol., Lancaster*, **56**, 555.

Fiveland, T. J. (1974). *Festskrift til Ugrasbiolog Torstein pa hans 70 ars dag, 16 Desenber, Orkanger, Norway*, 413.

Fiveland, T. J., Erickson, L. C. and Seely, C. V. (1972). *Weed Res.*, **12**, 155.

Flore, J. A. and Bukovac, M. J. (1974). *HortSci.*, **9**, 33.

Flore, J. A. and Bukovac, M. J. (1975). *HortSci.*, **10**, 333.

Forde, B. J. (1966). *Weeds*, **14**, 178.

Fox, H. W. and Zisman, W. A. (1950). *J. Colloid Sci.*, **5**, 514.

Foy, C. L. (1961). *Res. Progr. Rept., Western Weed Control Conf.*, 96.

Foy, C. L. (1962). *J. Agric. Fd. Chem.*, **12**, 473.

Foy, C. L. (1964). *Weeds*, **12**, 103.

Foy, C. L. and Smith, L. W. (1965). *Weeds*, **13**, 15.

Franke, W. (1961). *Amer. J. Bot.*, **48**, 683.

FRANKE, W. (1964a). *Nature (London)*, **202**, 1236.

FRANKE, W. (1964b). *Planta*, **63**, 279.

FRANKE, W. (1967). *Ann. Rev. Pl. Physiol.*, **18**, 281.

FRANKE, W. (1970). *Pestic. Sci.*, **1**, 164.

FRANKE, W. (1971). *Residue Rev.*, **38**, 81.

FURMIDGE, C. G. L. (1959). *J. Sci. Fd. Agric.*, **10**, 274.

FYKSE, H. (1975). *Weed Res.*, **15**, 165.

GIAQUINTA, R. (1976). *Pl. Physiol., Lancaster*, **57**, 872.

GOSS, J. R. (1978). *Diss. Abstr. Int. B.*, **39**, 23.

GOTTRUP, O., O'SULLIVAN, P. A., SCHRAA, R. J. and VANDEN BORN, W. H. (1976). *Weed Res.*, **16**, 197.

GREENE, D. W. and BUKOVAC, M. J. (1971). *J. Amer. Soc. Hortic. Sci.*, **96**, 240.

GREENE, D. W. and BUKOVAC, M. J. (1972). *Pl. Cell Physiol.* (Tokyo), **13**, 321.

GREENE, D. W. and BUKOVAC, M. J. (1974). *Amer. J. Bot.*, **61**, 100.

HADERLIE, L. C., SLIFE, F. W. and BUTLER, H. S. (1978). *Weed Res.*, **18**, 269.

HAI, T. V. and HUNG, T. M. (1975). *Plant and Soil*, **43**, 671.

HALL, W. C., JOHNSON, S. P. and LEINWEBER, C. L. (1954). *Texas Agr. Exp. Sta. Bull.*, **789**, 1.

HALLAM, N. D. (1970). *Planta*, **93**, 257.

HALLMÉN, U. (1974). *Physiol. Plant*, **32**, 78.

HALLMÉN, U. (1975). *Physiol. Plant*, **34**, 266.

HALLMÉN, U. and ELIASSON, L. (1972). *Physiol. Plant*, **27**, 143.

HAMILL, A. S., SMITH, L. W. and SWITZER, C. M. (1972). *Weed Sci.*, **20**, 226.

HAMILTON, R. J. (1977). In *Herbicides and Fungicides* (Ed. N. R. McFarlane), The Chemical Society, London, p. 81.

HARVEY, B. M. R., MULDOON, J. and HARPER, D. B. (1978). *Plant Cell and Env.*, **1**, 203.

HARVEY, R. G. and MUZIK, T. J. (1973). *Weed Sci.*, **21**, 135.

HAUSER, E. W. (1955). *Agron. J.*, **47**, 32.

HAUSER, F. W. and THOMPSON, J. (1954). *J. Agric. Fd. Chem.*, **2**, 680.

HEINEN, W. and BRAND, I. V. D. (1963). *Z. Naturforsch.*, **18 b**, 67.

HILTON, H. W., NOMURA, N. S., KAMEDA, S. S. and YAUGER, W. L. (1974). *Abstr. 3rd Int. Cong. Pestic. Chem., IUPAC, Helsinki*, 141.

HILTON, J. L. and CHRISTIANSEN, M. N. (1972). *Weed Sci.*, **20**, 290.

HOLLOWAY, P. J. (1969a). *J. Sci. Fd. Agric.*, **20**, 124.

HOLLOWAY, P. J. (1969b). *Ann. Appl. Biol.*, **63**, 145.

HOLLOWAY, P. J. (1970). *Pestic. Sci.*, **I**, 156.

HOLLOWAY, P. J. and DEES, A. H. B. (1971). *Phytochemistry*, **10**, 2781.

HOLLY, K. (1956). *Ann. Appl. Biol.*, **44**, 195.

HOLT, E. C., FAUBIAN, J. L., ALLEN, W. W. and MCBEE, G. G. (1967). *Weeds*, **15**, 13.

HULL, H. M. (1957). *Pl. Physiol., Lancaster*, **32**, (Suppl.), 43.

HULL, H. M. (1958). *Weeds*, **6**, 133.

HULL, H. M. (1970). *Residue Rev.*, **31**, 1.

HULL, H. M., MORTON, H. L. and MARTIN, R. D. (1973). *Abstr. Mtg. Weed Sci., Soc. Amer.*, 65.

HULL, H. M., MORTON, H. L. and WHARRIE, J. R. (1975). *Bot. Rev.*, **41**, 421.

HUNTER, J. H. and MCINTYRE, G. I. (1974). *Weed Sci.*, **22**, 167.

ILNICKI, R. D., THARRINGTON, W. H., ELLIS, J. F. and VISINSKI, E. I. (1965). *Proc. 19th N.E. Weed Control Conf.*, 295.

ISHIZUKA, K., KOBAYASHI, K. and MASHIMO, Y. (1975). *8th Int. Pl. Prot. Cong. Rept. and Info. Sect. 3. Chem. Control*, **1**, 358.

JANSEN, L. L. (1973). *Weed Sci.*, **21**, 130.

JEFFREE, C. E., BAKER, E. A. and HOLLOWAY, P. J. (1975). *New Phytol.*, **75**, 539.

JUNIPER, B. E. (1960). *J. Linn. Soc. (Bot.)*, **56**, 413.

KASSENBEER, H. (1969). *Schering A. G. Conf. (Berlin)*, 5.

KEELEY, P. E. and THULLEN, R. J. (1971). *Weed Sci.*, **19**, 601.

KELLS, J. J. and RIECK, C. E. (1978). *Proc. 31st Ann. Mtg. S. Weed Sci. Soc.*, 243.

KELLS, J. J. and RIECK, C. E. (1979). *Weed Sci.*, **27**, 235.

KEMPEN, H. M. (1970). *A Study of Monosodium Methane-arsonate in Plants*, Masters dissertation, University of Calif., Davis, pp. 129.

KENNEDY, C. D. (1974). *Pestic. Sci.*, **5**, 675.

KENNEDY, C. D. and HARVEY, J. M. (1972). *Pestic. Sci.*, **3**, 715.

KING, D. L. (1973). *Diss. Abstr. Int. B.*, **34**, 3028.

KING, D. L. and BAYER, D. E. (1972). *Proc. West Soc. Weed Sci.*, **25**, 37.

KING, M. G. and RADOSEVICH, S. R. (1978). *Proc. West Soc. Weed Sci.*, **31**, 40.

KIRKWOOD, R. C. (1972). *Proc. 11th Brit. Weed Control Conf.*, 1117.

KIRKWOOD, R. C. (1978). In *Herbicides and Fungicides, Factors Affecting Their Activity* (Ed. N. R. McFarlane), The Chem. Soc., pp. 67–80.

KIRKWOOD, R. C. (1979). In *Adv. in Pestic. Sci.* (Ed. H. Geissbuhler), Pergamon Press, p. 410.

KIRKWOOD, R. C., DALZIEL, J., MATLIB, A. and SOMERVILLE, L. (1968). *Proc. 9th Brit. Weed Control Conf.*, 650.

KIRKWOOD, R. C., DALZIEL, J., MATLIB, A. and SOMERVILLE, L. (1972). *Pestic. Sci.*, **3**, 307.

KITCHEN, L. M. and RIECK, C. E. (1978). *Proc. N.C. Weed Control Conf.*, **33**, 47.

KOBAYASHI, K. and ISHIZUKA, K. (1977). *J. Pestic. Sci.*, **2**, 59.

KOCH, W., MAJUMDAR, J., FYKSE, H. and RADEMACHER, B. (1969). *Schering A.G. Conf. (Berlin)*, 12.

KOGAN, M. and ARAYA. A. (1974). In *Assoc. Latinoamericana de Malezas 'ALAM' Soc. Colombiana de Control de Malezas Y Fisiologia Vegetal 'COMALFI' Resumenes de los Trabajos en el II Congreso ALAM Y VI Seminario, COMALFI, Calif.*, 76.

KOLATTUKUDY, P. E. (1970). *Lipidas*, **5**, 259.

KOLATTUKUDY, P. E. and WALTON, T. J. (1973). In *Progress in the Chemistry of Fats and Other Lipids* (Ed. R. T. Holman), Pergamon Press, Oxford, p. 121.

LEONARD, R. T. and SHANER, D. L. (1979). *Abstr. Mtg. Weed Sci. Soc. Amer.*, 98.

LINSCOTT, D. L. (1964). *J. Agric. Fd. Chem.*, **12**. 7.

LUCKWILL, L. C. and LLOYD-JONES, C. P. (1962). *J. Hort. Soc.*, **37**, 190.

LUMB, J. M. (1972). *Diss. Abstr. Int. B.*, **32**, 5062.

LUND-HØIE, K. (1972). *Meldinger fra Norges Landbrukshøgskole,* **51**, 1.

LUND-HØIE, K. (1973). *Meldinger fra Norges Landbrukshøgskole*, **52**, 15.

LUND-HØIE, K. (1976). *Meldinger fra Norges Landbrukshøgskole*, **55**, 26.

LUND-HØIE, K. (1979). *Meldinger fra Norges Landbrukshøgskole*, **58**, 24.

MCINTYRE, G. I. (1962). *Weed Res.*, **2**, 165.

MCINTYRE, G. I., FLEMING, W. W. and HUNTER, J. H. (1978). *Can. J. Bot.*, **56**, 715.

MCKINLAY, K. S., BRANDY, S. A., MORSE, P. and ASHFORD, R. (1972). *Weed Sci.*, **20**, 450.

MCWHORTER, C. G. (1966). *Weeds*, **14**, 191.

MCWHORTER, C. G. and JORDAN, T. N. (1976). *Physiol. Plant*, **38**, 166.

MARQUIS, L. Y., COMES, R. D. and YANG, C. P. (1979). *Weed Res.*, **19**, 335.

MARTIN, J. T. and JUNIPER, B. E. (1970). *The Cuticle of Plants*, Arnold, London, p. 347.

MATLIB, M .A., KIRKWOOD, R. C. and PATTERSON, J. D. E. (1971). *Weed Res.*, **11**, 190.

MAZLIAK, P. (1968). In *Progress in Phytochemistry* (Eds. L. Reinhold and Y. Liwschitz), Vol. 1, Interscience, New York, p. 49.

MÜLLER, F. and SANAD, A. (1975). *8th Int. Cong. Pl. Prot., Rept. and Info., Sect. 3*, 524.

NEUMANN, S. and JACOB, F. (1968). *Naturwissenschaften*, **53**, 89.

NORRIS, R. F. and BUKOVAC, M. J. (1968). *Amer. J. Bot.*, **55**, 975.

NORRIS, R. F. and BUKOVAC, M. J. (1971). *Pl. Physiol., Baltimore*, **49**, 615.

OLUNUGA, B. A., LOVELL, P. H. and SAGAR, G. R. (1977). *Weed Res.*, **17**, 213.

OSIPOW, L. I. (1964). In *Surface Chemistry*, Rheinhold, New York.

OVERBEEK, J. VAN., (1956). *Ann. Rev. Pl. Physiol.*, **7**, 355.

PALLAS, J. E. (1960). *Pl. Physiol., Lancaster*, **35**, 575.

PARR, J. F. and NORMAN, A. G. (1965). *Bot. Gaz.*, **126**, 86.

PETERSON, C. A. and EDGINGTON, L. V. (1976). *Pestic. Sci.*, **7**, 483.

PHUNG-HONG-THAI and FIELD, R. J. (1979). *Weed Res.*, **19**, 51.

PICKERING, E. R. (1965). *Foliar Penetration Pathways of 2,4-D, Monuron and Dalapon as Revealed by Historadioautography*, Ph. D. Diss., Univ. of Calif., Davis, 186 pp.

PIZZOLATO, T. D. and REGEHR, D. L. (1979). *Can. J. Pl. Sci.*, **57**, 1340.

PRASAD, R. and BLACKMAN, G. E. (1965). *J. Exp. Bot.*, **16**, 86.

PRASAD, R., FOY, C. L. and CRAFTS, A. S. (1967). *Weeds*, **15**, 149.

PREGO, I. A., MARODER, H. L. and SONVICO, V. (1976). In *Trabajos Y Resumenes, III Congreso Association Latinoamericana de Malezas 'AŁAM' Y VIII Reunion Argentina de Malegas Y su Control 'ASAM', Mar Del Plato, 1976, ASAM,* **1**, 201.

PULVER, E. L. and ROMERO, C. (1976). *Revis. Comalfi*, **3**, 94.

PUTNAM, A. R. and RIES, S. K. (1968). *Weed Sci.*, **16**, 80.

RADOSECVICH, S. R. and BAYER, D. E. (1979). *Weed Sci.*, **27**, 22.

REGEHR, D. L. and PIZZOLATO, T. D. (1979). *Proc. N.E. Weed Sci. Soc.*, **33**, 100.

REPP, G. (1958). *Z. Acker-u PflBau*, **107**, 49.

RICHARD, E. P. JR. (1978). *Diss. Abstr. Int. B.*, **39**, 26.

RICHARD, E. P. and SLIFE, F. W. (1979). *Weed Sci.*, **27**, 426.

RICHARDSON, R. G. (1976). *Weed Res.*, **16**, 375.

RIOUX, R., BANDEEN, J. D. and ANDERSON, G. W. (1974). *Can. J. Pl. Sci.*, **54**, 397.

ROBARDS, A. W. (1975). *Ann. Rev. Pl. Physiol.*, **26**, 13.

ROBERTSON, M. M. and KIRKWOOD, R. C. (1969). *Weed Res.*, **9**, 224.

ROBERTSON, M. M. and KIRKWOOD, R. C. (1970). *Weed Res.*, **10**, 94.

ROBERTSON, M. M., PARHAM, P. H. and BUKOVAC, M. J. (1971). *J. Agric. Fd. Chem.*, **19**, 754.

ROLSTON, M. P. and ROBERTSON, A. G. (1975). *Proc. 28th N.Z. Weed and Pest Control Conf.*, 54.

ROLSTON, M. P. and ROBERTSON, A. G. (1976). *Weed Res.*, **16**, 82.

SACHS, R. M. and MICHAEL, J. L. (1971). *Weed Sci.*, **19**, 558.

SAGAR, G. R. (1960). *Proc. 5th Brit. Weed Control Conf.*, 271.

SANAD, A. J. and MÜLLER, F. (1973). *Z. Pflanzenkrankheiten Pflanzenschutz*, **80**, 74.

SANDERBERG, C. L., MEGGITT, W. F. and PENNER, D. (1978). *Abstr. Mtg. Weed Sci. Soc. Amer.*, 73.

SANDS, R. and BACHELARD, E. P. (1973a). *New Phytol.*, **72**, 69.

SANDS, R. and BACHELARD, E. P. (1973b). *New Phytol.*, **72**, 87.

SARGENT, J. A. (1964). *Meded. LandbHogesch Opzoek-Stns, Gent*, 656.

SARGENT, J. A. (1965). *Ann. Rev. Pl. Physiol.*, **16**, 1.

SARGENT, J. A. (1966). *Proc. 8th Brit. Weed Control Conf.*, 804.

SARGENT, J. A. and BLACKMAN, G. E. (1962). *J. Exp. Bot.*, **13**, 348.

SARGENT, J. A. and BLACKMAN, G. E. (1965). *J. Exp. Bot.*, **16**, 24.

SARGENT, J. A. and BLACKMAN, G. E. (1969). *J. Exp. Bot.*, **20**, 542.

SARGENT, J. A. and BLACKMAN, G. E. (1972). *J. Exp. Bot.*, **23**, 830.

SARGENT, J. A., POWELL, R. G. and BLACKMAN, G. E. (1969). *J. Exp. Bot.*, **63**, 426.

SAVORY, B. M. (1968). *Proc. 9th Brit. Weed Control Conf.*, **1**, 102.

SCHAFER, D. E. and CHILCOTE, D. O. (1970). *Weed Sci.*, **18**, 729.

SCHNEPF, E. (1959). *Planta*, **52**, 644.

SCHÖNHERR, J. and BUKOVAC, M. J. (1970). *Planta*, **92**, 189–201.

SCHÖNHERR, J. and BUKOVAC, M. J. (1972). *Pl. Physiol., Baltimore*, **49**, 813.

SCHÖNHERR, J. and BUKOVAC, M. J. (1973). *Planta*, **109**, 73.

SCHULTZ, M. E. and BURNSIDE, D. C. (1979). *Abstr. Mtg. Weed Sci. Soc. Amer.*, 95.

SCHULTZ, M. E. and BURNSIDE, D. C. (1980). *Weed Sci.*, **28**, 149.

SCIFRES, C. J., BAUR, J. R. and BOVEY, R. W. (1973). *Weed Sci.*, **21**, 94.

SCKERL, M. M. and FRANS, R. E. (1969). *Weed Sci.*, **17**, 421.

SHARMA, M. P. and VANDEN BORN, W. H. (1970). *Weed Sci.*, **18**, 57.

SHARMA, M. P. and VANDEN BORN, W. H. (1973). *Weed Sci.*, **21**, 350.

SHARMA, M. P. and VANDEN BORN, W. H. (1978). *Abstr. Mtg. Weed Sci. Soc. Amer.*, 100.

SHARMA, M. P., VANDEN BORN, W. H., FRIESEN, H. A. and MCBEATH, D. K. (1976). *Weed Sci.*, **24**, 379.

SHARMA, M. P., VANDEN BORN, W. H. and MCBEATH, D. K. (1978). *Weed Res.*, **18**, 169.

SHIMABUKURO, R. H., WALSH, W. C. and HAERAUF, R. A. (1976). *Pestic. Biochem. Physiol.*, **6**, 115.

SHONE, M. G. T., LEE, R. B. and WOOD, A. V. (1972). *Ann. Rept. ARC Letcombe Lab. Wantage, U.K.*, 68.

SHONE, M. G. T. and WOOD, A. V. (1972). *Ann. Rept. ARC Letcombe Lab. Wantage, U.K.*, 32.

SIEBERT, R. and KÖCHER, H. (1972a). *Mitteilungen aus der Biologischen Bundesanstalt fur Land- und Forstwirtschaft Berlin-Dahlem*, **146**, 168.

SIEBERT, R. and KÖCHER, H. (1972b). *Z. Pflanzenkrankheiten Pflanzenschutz*, **79**, 463.

SILVA, FERNANDES, A. M. (1965). *Ann. Appl. Biol.*, **56**, 305.

SINGH, J. N. (1972). *Diss. Abstr. Int. B.*, **33**, 533.

SINGH, J. N., BASLER, E. and SANTELMANN, P. W. (1972). *Pestic. Biochem. Physiol.*, **2**, 143.

SINGH, S. P. and MÜLLER, F. (1979a). *Weed Res.*, **19**, 1.

SINGH, S. P. and MÜLLER, F. (1979b). *Weed Res.*, **19**, 171.

SKOSS, J. D. (1955). *Bot. Gaz.*, **117**, 55.

SLADE, P. and BELL, E. G. (1966). *Weed Res.*, **6**, 267.

SMITH, A. E. (1972a). *Physiol. Plant*, **27**, 338.

SMITH, A. E. (1972b). *Weed Sci.*, **20**, 46.

SMITH, C. N. and NALEWAJA, J. D. (1972). *Weed Sci.*, **20**, 36.

SMITH, G. N. and DYER, D. L. (1961). *J. Agric. Fd. Chem.*, **9**, 155.

SMITH, J. M. and SAGAR, G. R. (1966). *Weed Res.*, **6**, 314.

SMITH, L. W. and DAVIES, P. J. (1965). *Weed Res.*, **5**, 343.

SOMERVILLE, L. (1968). *Studies on the Mode of Action of Hydroxybenzonitrile Herbicides*, Ph.D. Diss., University of Strathclyde, Glasgow.

SOVONICK, S. A., GEIGER, D. R. and FELLOWS, R. J. (1974). *Pl. Physiol., Lancaster*, **54**, 886.

SPRANKLE, P., MEGGITT, W. F. and PENNER, D. (1975). *Weed Sci.*, **23**, 235.

SPRANKLE, P., PENNER, D. P. and MEGGITT, W. F. (1973). *Abstr. Mtg. Weed Sci. Soc. Amer., Atlanta, Georgia*, 75.

STILL, G. G., DAVIS, D. G. and ZANDER, G. L. (1970). *Pl. Physiol., Lancaster*, **46**, 307.

ST. JOHN, J. B. and HILTON, J. L. (1973). *Weed Sci.*, **21**, 477.

SUTTON, D. L. and FOY, C. L. (1971). *Bot. Gaz.*, **132**, 299.

TEMPLETON, A. R. and HURTT, W. (1972). *Proc. N. E. Weed Sci. Soc., New York*, **26**, 277.

THAINE, R. (1962). *J. Exp. Bot.*, **13**, 152.

THAINE, R., DE MARIA, M. E. and SARISALO, H. I. M. (1975). *J. Exp. Bot.*, **26**, 91.

THOMPSON, L. and SLIFE, F. W. (1969). *Weed Sci.*, **17**, 251.

THOMPSON, R. P. and SLIFE, F. W. (1974). *Abstr. Mtg. Weed Sci. Soc. Amer.*, 85.

THROWER, S. L., HALLAM, N. D. and THROWER, L. B. (1965). *Ann. Appl. Biol.*, **55**, 253.

TITOVA, A. A. (1978). *Gidrobiol. Zh.*, **14**, 110.

TURNER, D. J. and LOADER, P. C. (1972). *Proc. 11th Brit. Weed Control Conf.*, 654.

UPCHURCH, R. P. and BAIRD, D. D. (1972). *Proc. West Soc. Weed Sci.*, **25**, 41.

VEERASEKARAN, P., KIRKWOOD, R. C. and FLETCHER, W. W. (1976). *J. Linn. Soc.*, **73**, 247.

VEERASEKARAN, P., KIRKWOOD, R. C. and FLETCHER, W. W. (1977). *Weed Res.*, **17**, 33.

WATHANA, S., CORBIN, F. T. and WALDREP, T. W. (1972). *Weed Sci.*, **20**, 120.

WEBER, J. B. and SCOTT, D. C. (1966). *Science*, **152**, 1400.

WEIGEL, R. C. JR. and RIGGLEMAN, J. D. (1978). *Abstr. Mtg. Weed Sci. Soc. Amer.*, 87.

WESTWOOD, M. N. and BATJER, L. P. (1958). *Proc. Amer. Soc. Hortic. Sci.*, **72**, 35.

WHITE, J. A. and HEMPHILL, D. D. (1972). *Weed Sci.*, **20**, 478.

WHITECROSS, M. I. and ARMSTRONG, D. J. (1972). *Aust. J. Bot.*, **20**, 87.

WILKINSON, R. E. (1956). *The Physiological Activity of 2,2-Dichloropropionic Acid*, Ph.D. Diss., Univ. of Calif., Davis, p. 148.

WILKINSON, R. E. (1973). *Abstr. Mtg. Weed Sci. Soc. Amer.*, 71.

WILKINSON, R. E. (1974). *Pl. Physiol., Lancaster*, **53**, 269.

WILKINSON, R. E. and HARDCASTLE, W. S. (1969). *Weed Sci.*, **17**, 536.

WILKINSON, R. E. and SMITH, A. E. (1973). *Proc. 26th Ann. Mtg. South Weed Sci. Soc.*, 415.

WILLS, G. D. (1978). *Weed Sci.*, **26**, 509.

WILSON, B. J. (1974). *Diss. Abstr. Int. B*, **35**, 1997.

WILSON, B. J. and NISHIMOTO, R. K. (1974). *Weed Sci.*, **22**, 297.

WILSON, B. J. and NISHIMOTO, R. K. (1975). *Weed Sci.*, **23**, 289.

WILSON, L. A. and NORRIS, R. F. (1973). *Pl. Physiol., Lancaster*, **51**, (Suppl.), 47.

WOOD, G. H. and GOSNELL, J. M. (1965). *Proc. S. Afr. Sugar Technol. Assoc.*, 7.

WRIGLEY, G. (1973). *PANS*, **19**, 54.

YAMADA, Y., RASMUSSEN, H. P., BUKOVAC, M. J. and WITTWER, S. H. (1966). *Amer. J. Bot.*, **53**, 170.

YAMADA, Y., WITTWER, S. H. and BUKOVAC, M. J. (1965). *Pl. Physiol., Lancaster*, **40**, 170.

ZABKIEWICZ, J. A. and GASKIN, R. E. (1979). *Proc. 32nd N.Z. Weed and Pest Control Conf.*, 314.

ZANDSTRA, B. H. and NISHIMOTO, R. K. (1977). *Weed Sci.*, **25**, 268.

ZEMSKAYA, V. A. (1971). *Soviet Pl. Physiol.*, **18**, 626.

ZEMSKAYA, V. A., RAKITIN, YU. V., CHERNIKOVA, L. M. and KALIBERNAYA, Z. V. (1975). *Fiz. Rast.*, **22**, 1044.

ZETTERBERG, G., BUSK, L., ELOVSON, R., STARECDENHAMMAR, I. and RYTTMAN, H. (1977). *Mutation Res.*, **42**, 3.

ZWEEP, W. VAN DER (1965). *Z. Pflkrankh, Pflpatt, Pflschutz., (Sonderh 3)*, 123.

CHAPTER FOUR

The Absorption and Translocation of Soil-applied Herbicides

Root absorption and xylem transport

Since the discovery of the pre-emergence activity of 2,4-D (Anderson and Ahlgren, 1947; Anderson and Wolf, 1947) many root-absorbed compounds have been discovered which prevent weed seeds from germinating or which are absorbed by the roots or coleoptiles of young seedlings. While the mechanism of uptake and translocation of these soil-applied herbicides has probably received less attention from plant physiologists than have foliage-applied compounds, many interesting studies have been carried out on the route and mechanisms of uptake and translocation, and the influence of plant, herbicide and environmental factors on the efficiency of these processes. The selectivity of such herbicides has been found to depend on such subtle factors as differences in the rooting depth of crop and weed (fig. 4.1) and on differences in absorption, translocation and metabolism of the herbicides.

Unlike the foliage-applied compounds, soil-applied herbicides are not applied directly to the plant and the amount of active ingredient which reaches the absorbing surface of the roots depends upon a

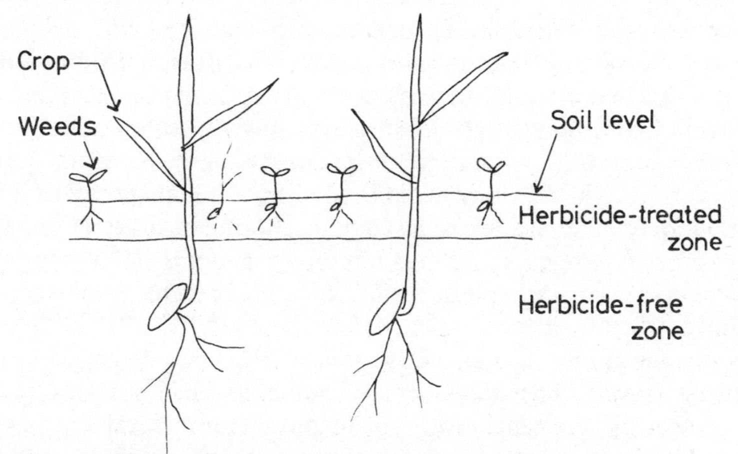

Fig. 4.1 A possible basis for the selectivity of a soil-applied herbicide.

complex of factors including the efficiency of transfer from the soil surface to the roots, root absorption, translocation to the shoots and accumulation at the active sites. Transfer of herbicide molecules from the soil surface to the rooting zone may involve water and air-phase diffusion, leaching and dynamic dispersion (Hartley, 1976); the level of soil absorption and the solubility of the applied compound may markedly influence the efficiency and rate of this movement (Hartley, 1976). In addition, in the soil, the herbicide is subject to microbial (Kearney and Kaufman, 1976) and non-biological degradation (Crosby, 1976), further reducing its potential activity (fig. 4.2).

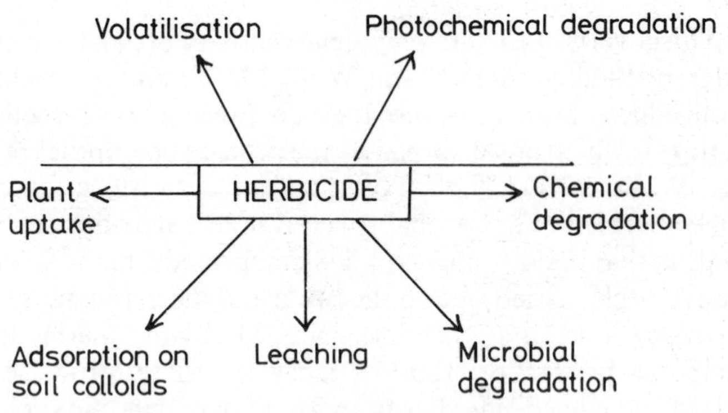

Fig. 4.2 The possible mechanisms of herbicide loss from soil.

Comparative studies have shown that plant roots absorb and generally translocate certain compounds very rapidly including simazine (Davis *et al.*, 1959), the uracils (Hilton *et al.*, 1964), diphenamid (Lemin, 1966; Golab *et al.*, 1966), dichlobenil (Verloop and Nimmo, 1969, 1970), linuron (Rieder *et al.*, 1970), diuron (Moyer, McKercher and Hance, 1972), benthiocarb (Eastin, 1975), propham (Burt and Corbin, 1978) and chlornitrofen (Shimotori and Kuwatsuka, 1978). Other compounds such as aminotriazole, dalapon or maleic hydrazide (MH) may be absorbed more slowly (Ashton and Crafts, 1973). While translocation in the xylem to the shoots is generally also rapid, variation in rate may occur according to plant species and the compound (e.g. diphenamid in tomato, Bermuda grass (*Cynodon dactylon*) and winged euonymus (*Euonymus alatus*) (Bingham and Shaver, 1971). For example, a variation in rate of transport was found for aminotriazole, dalapon and MH (Ashton and Crafts, 1973), and chlornitrofen (Shimotori and Kuwatsuka, 1978) may be translocated relatively slowly. The movement of some herbicides seems to be determined by concentration, for example, low levels of 2,4-D accumulate in the root while high concentrations, which cause rapid contact injury, may move throughout the plant (Ashton and Crafts, 1973).

Root absorption appears to take place largely through root hairs which are located just behind the root apical meristem (fig. 4.3). Several workers have described uptake as having an initial rapid phase followed by a continued and steady phase, for example in the case of atrazine (Vostral *et al.*, 1970) and napropamide (Barrett and Ashton, 1979). There is some controversy as to whether uptake is passive or governed by active metabolism. Several workers have reported that

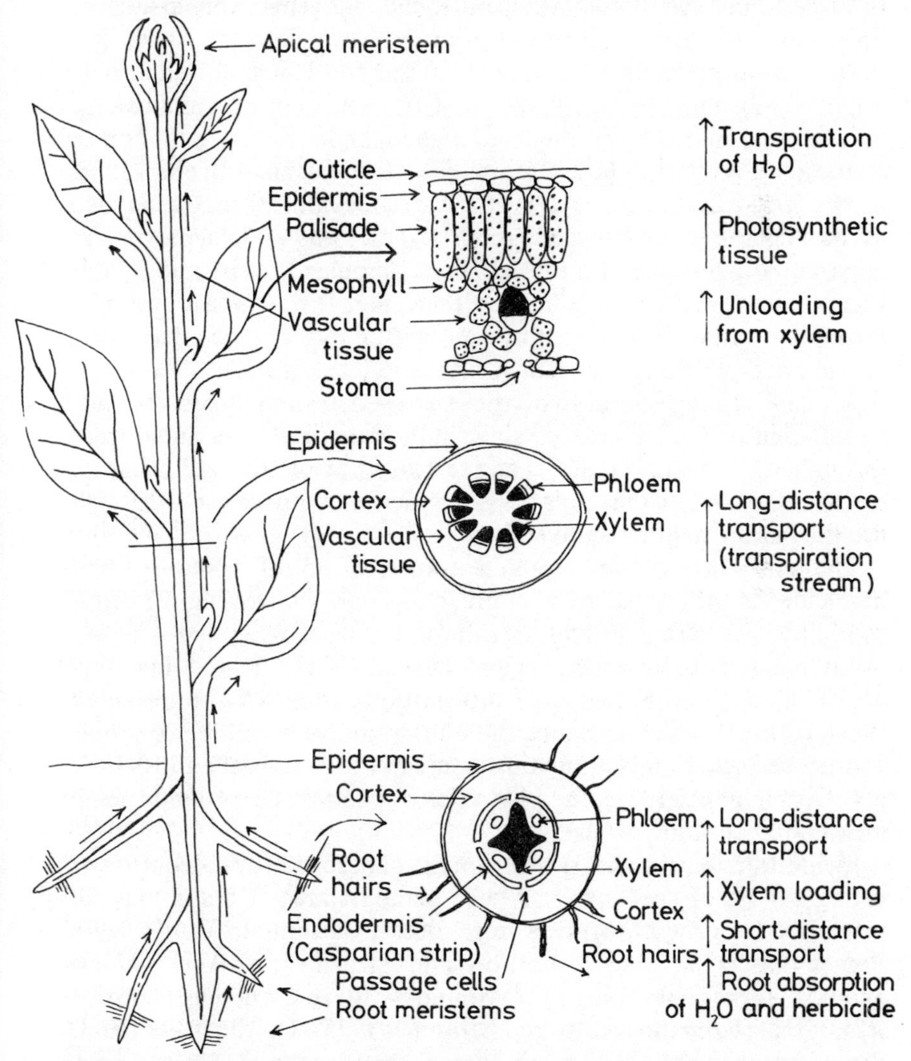

Fig. 4.3 A schematic view of the probable stages and route of uptake and translocation of a soil-applied systemic herbicide.

uptake is passive being governed in their test species by physical factors. Examples include monuron (Donaldson, 1967), chlorpropham (Ashton and Helfgott, 1966; Helfgott and Ashton, 1966; Helfgott, 1969; Rieder *et al.*, 1970), alachlor (Groenwold, 1971), monolinuron (Ostrowski, 1972). Other workers have suggested that the process involves active metabolism, for example in the case of linuron (Moody and Buchholtz, 1968) and picloram (Morrison and Vanden Born, 1975).

Transmission of water-soluble compounds across the cortex is believed to involve a symplastic route, such movement bypassing the impervious Casparian strip and allowing entry into the stele. The herbicide molecules are then believed to leak from the symplast into the apoplast ascending to the foliage in the transpiration stream. Leakage may be conditioned by gradients of high O_2 and low CO_2 in the cortex shifting to low O_2 and high CO_2 in the stele (Crafts and Broyer, 1938; Crafts, 1961). In the case of relatively water-insoluble compounds such as the substituted ureas, uracils and triazines, trans-cortical movement apparently takes place via the cell walls (apoplast) to the impervious Casparian strip. Here they must either enter the symplast and pass through, or alternatively continue apoplastic migration by dissolving in and passing through the Casparian strip into the stele. Ashton and Crafts (1973) suggest that to do this, such penetrating molecules must be sufficiently fat soluble to dissolve in the suberized layer and diffuse through it. They point out that the existence of the root pressure mechanism implies that polar water molecules cannot pass through the Casparian strip into the stele.

The efficiency of uptake and translocation of root-absorbed herbicides is influenced by a number of factors including (1) water solubility, (2) herbicide concentration, e.g. atrazine (Vostrol *et al.*, 1970), picloram (Isensee, Jones and Turner, 1971), EPTC (Marton *et al.*, 1978), (3) the presence of adventitious roots (Nishimoto and Warren, 1971) which enhance the absorption of a number of root-applied herbicides, including diuron, ametryne or terbacil, and (4) soil pH (Corbin *et al.*, 1971), the phytoxicity of certain compounds being maximal at certain pH levels.

In addition to root absorption, uptake of herbicide by coleoptiles or young shoots of seedlings has been demonstrated. Uptake may be primarily by root absorption, e.g. diuron (Nishimoto *et al.*, 1967; Prendeville *et al.*, 1967), linuron and atrazine (Walker, 1973a), pyrazon (now chloridazon) (5-amino-4-chloro-2-phenylpyridazin-3(2*H*)-one) (Romanovskaya and Straume, 1975). On the other hand, shoot uptake may be more rapid than root absorption, e.g. dichlobenil (Massini, 1961), trifluralin (Parker, 1966; Knake *et al.*,

1967), linuron and propachlor (Knake and Wax, 1968), EPTC (Oliver *et al.*, 1968; Gray and Joo, 1978), alachlor (Groenwold, 1971), monolinuron (Süss, Fuchsbichler and Siegmund, 1978). Volatility may be an advantage in the shoot uptake of certain herbicides such as the thiocarbamates (Parker, 1966), trifluralin (Negi and Funderburk, 1968; Swann and Behrens, 1969), EPTC (Oliver *et al.*, 1968); dinitroanilines (Ellis and Norton, 1976; Jacques and Harvey, 1979) and certain other herbicides (Knake and Wax, 1968). Depth of soil incorporation as well as seed placement may be critical factors (e.g. metolachlor (Dixon and Stoller, 1979)). Further, temperature and humidity may influence vapour emission from soil and thus indirectly affect uptake (Swann and Behrens, 1969).

Differences in the uptake of herbicide by seeds of different species have been reported, the rates of uptake being similar in living and dead seeds of the same species (Reider, Buchholtz and Kust 1970).

Some root-absorbed compounds are xylem and phloem mobile, e.g. chloridazon (Ashton and Crafts, 1973), picloram (O'Donovan and Vanden Born, 1979), while others are translocated only in the transpiration stream (e.g. ureas and s-triazines). The relative mobilities and distribution patterns of fairly closely related compounds may be quite different (e.g. Voss and Geissbühler, 1966).

Transpiration may be a major influence on the translocation of root-absorbed compounds, e.g. diuron (Walker, 1973b), monuron (Anon, 1976), norflurazon (Harden and Corbin, 1975), metflurazone (Motooka, Corbin and Worsham, 1977). In such cases, xylem transport of herbicide is influenced by the factors which affect the efficiency of the transpiration stream. Thus environmental conditions (e.g. temperature, humidity, light and wind) which increase the rate of water loss presumably affect the transportation of herbicides to the leaves. There is evidence that certain herbicides may inhibit transpiration and thus inhibit their own transportation, e.g. diuron (Ahmed and Fletcher, 1980). Uptake and movement is enhanced by low humidity, e.g. certain triazines (Sheets, 1961; Uhlig, 1968) and diphenamid (Schultz and Tweedy, 1972), increase in temperature, e.g. triazines (Wax and Behrens, 1965; Minshall, 1969), chlorpropham (Rieder *et al.*, 1970) and increase in soil moisture, e.g. bromacil (Schreiber, 1972). An interesting variation in effect was reported in the case of trifluralin by Hawxby, Basler and Santelmann (1972), who found that absorption by excised lateral roots of peanut was greatest at low temperature while translocation was increased by a rise in temperature; the implications of these findings are mentioned later.

Some workers have failed to find any relationship between

selectivity and uptake and/or translocation, e.g. chlorpropham (Prendeville *et al.*, 1968), EPTC (Marton *et al.*, 1978), metobromuron (Hogue, 1978) and barban (Kobayashi and Ishizuka, 1979), while others have found that selective activity could be related to differences in uptake, translocation or metabolism, e.g. chloroxuron (Abei and Ebner, 1962; Geissbühler *et al.*, 1963), simazine (Freeman *et al.*, 1964), di-allate, tri-allate (Appleby *et al.*, 1965), fluorodifen (Eastin, 1972), pyrnachlor (Silvia and Putnam, 1973), metribuzin (Hargroder and Rogers, 1974), metflurazone, norflurazon (Strang and Rogers, 1974), picloram (Morrison and Vanden Born, 1975), terbutryne and prometryne (Eshel, Kovacs and Rubin, 1975a), bifenox (Leather and Foy, 1976), dipropetryn and prometryne (Basler, Murray and Santelmann, 1978), ethalfluralin and trifluralin (Vandeventer, Meggitt and Penner, 1979).

In considering the root absorption and translocation of individual herbicides the groups haloalkanoic acids, phenoxyalkanoic acids, aromatic acids and nitrophenols have been omitted since they are normally foliage applied. The uptake/translocation characteristics of these groups are described in chapter 3.

Uptake and translocation of soil-applied herbicides

AMIDES

Compounds belonging to this group are generally taken up readily by the roots and transported to the foliage. For instance root absorption of **diphenamid** appears to occur rapidly, most of the herbicide accumulating in the leaves (Lemin, 1966; Golab *et al.*, 1966). The rate of uptake and translocation of ^{14}C-diphenamid has been found to vary according to the species tested (Bingham and Shaver, 1971). For example, apoplastic translocation of diphenamid was rapid in tomato seedlings, intermediate in Bermuda grass (*Cynodon dactylon*) and slow in winged euonymus (*Euonymus alatus*). Light and humidity regimes also have an effect on absorption (and degradation) of ^{14}C-diphenamid. For example, tomato plants grown under low light, low humidity conditions accumulated higher levels of diphenamid in the shoots than did those grown under high light, high humidity conditions (Schultz and Tweedy, 1972). Thus phytotoxicity of diphenamid to tomato (normally resistant) could be greatly enhanced by low light and low humidity conditions. Meissner and Oosthuizen (1976) also found that the inhibitory effect of diphenamid on root growth of tomato increased with temperature (35–45°C), with complete inhibition at the highest temperature. These results are consistent with the results of Rice and Putnam (1980) who found that while uptake and metabolism of ^{14}C-diphenamid by tomato plants in the cotyledon stage were not greatly affected by temperature or pH, translocation to the shoot was reduced under low temperature and high pH.

Growth regulator treatment appears to enhance the root absorption and translocation of certain of these herbicides. For example the root uptake of **naptalam** from culture solution by bean plants was greatly enhanced by GA (10^{-4}–10^{-3} M) (Devlin and Yaklich, 1971), while uptake of naptalam by *Potamogeton pectinatus* was also increased by 2,4-D (10^{-7}–10^{-4} M) and 4-CPA (4-chlorophenoxyacetic acid) (10^{-5}–10^{-4} M) (Devlin, 1974). Uptake of naptalam by *Potamogeton nodusus* increased in the presence of GA (50–100 p.p.m.) and fenoprop (50 and 100 p.p.m.), the latter indicating adverse effects on membrane permeability (Devlin, 1974). Subsequently it was reported that uptake of naptalam by roots and shoots of wheat (cv. Merscopa) and soybean (cv. York) was enhanced when naptalam was applied simultaneously with SADH(succinic acid-2,2-dimethylhydrazide) but not with GA (Devlin and Karczmarczyk, 1977).

The uptake, translocation and adsorption of **napropamide** has been examined in oats and in *Agropyron repens* by Carlson, Lignowski and Hopen (1975). The compound was most phytotoxic to oat when placed in the seed zone and to *A. repens* when placed in the rhizome zone and this coincided with rapid absorption. Foliage applications were not effective. Bioassay and autoradiographic studies also showed napropamide to be more effective when applied to the root zone than to the foliage (Wu, Buehring and Santelmann, 1974) and this herbicide was the most active of 122 compounds tested as growth inhibitors of buds of couch grass (*A. repens*) (Harvey and Baker, 1974). The napropamide fluxes in isolated roots of maize (*Zea mays*) were investigated by Barrett and Ashton (1979). They found that influx into maize root segments was initially rapid and then continued at a steady but slower rate over a period of 8 h. The total influx was divided into two major fractions: (1) a fraction which could be removed by washing with unlabelled napropamide and (2) a residual fraction. Both the total influx and the residual fraction were reduced by anaerobic conditions having Q10s of 1.5 and 2 respectively between 15 and 25°C. A linear increase in both fractions was observed with increasing napropamide concentration from 10^{-7} to 10^{-5} M. Barrett and Ashton concluded that influx of napropamide appears to be governed by the tightly bound residual fraction.

NITRILES

In contrast to the hydroxybenzonitriles (HBNs), **dichlobenil** and **chlorthiamid** are essentially pre-emergence herbicides which are absorbed by the roots and transported to a limited degree in the xylem in the transpiration stream. However, shoot absorption of dichlobenil by beans (*Phaseolus vulgaris*) exposed to a saturated atmosphere of the chemical has been reported (Massini, 1961).

Massini (1961) first reported that dichlobenil was readily absorbed by the roots of bean (*P. vulgaris*) and transported to the leaves in the xylem, movement being slower than water due to the affinity of dichlobenil for plant tissue. Later studies using *Alternathera philoxeroides* and bean (*P. vulgaris*) confirmed that dichlobenil was readily root absorbed. Leaf absorption in these species was however, slight (Pate and Funderburk, 1966; Verloop and Nimmo, 1969). Verloop and Nimmo (1969, 1970) reported that dichlobenil was absorbed through the roots of bean (*P. vulgaris*), wheat (*Triticum vulgare*) and rice (*Oryza sativa*) and rapidly translocated to the leaves; the quantity of herbicide lost as vapour was species dependent. Price and Putnam (1969) found that dichlobenil was readily absorbed by the roots of maize (*Zea mays*) but not retained. The highest concentration of dichlobenil in the shoot was obtained at 4 h, but subsequently it declined over a period of 72 h. Sikka, Lynch and Lindenberger (1974) examined the uptake and fate of dichlobenil by the aquatic species alligator weed (*Alternathera philoxeroides*) (susceptible) and *Myriophyllum brasiliense* (resistant). Dichlobenil was readily absorbed and translocated by both species, leakage from the roots being noted in the case of *A. philoxeroides* but not in *M. brasiliense*.

The uptake, translocation and metabolism of ^{14}C-dichlobenil has also been examined in selected aquatic species (*Berula erecta*, resistant; *Rorippa nasturtium-aquaticum*, moderately susceptible; *Phragmites communis*, susceptible) (Mottley and Kirkwood, 1978). Ten days after root treatment with 5 p.p.m. ^{14}C-dichlobenil in a vapour-trapping device, the amount of ^{14}C absorbed and translocated by *P. communis* was four to six times greater than that for the other two species; the loss of ^{14}C-dichlobenil vapour from the leaves of *R. nasturtium-aquaticum*, was six to eleven times greater than that of the other two species. The distribution of ^{14}C in the roots, stems and vapour of *R. nasturtium-aquaticum* was uniform, contrasting with *B. erecta* where ^{14}C appeared to be restricted in the roots, and *P. communis* where ^{14}C accumulated in the rhizomes and leaves. Differences in metabolism of dichlobenil in these aquatic species were also involved in selectivity.

ANILIDES

Compounds of the anilide group are generally readily absorbed by the roots and transported in the xylem to the shoots. For example, **propachlor** is absorbed by the roots of maize and soybean (Jaworski, 1964; Jaworski and Porter, 1965), though it may be more readily taken up from the soil by the shoot than by the roots (Knake and Wax, 1968). Foliage absorption of certain anilides such as **dicryl** (3′,4′-dichloro-2-methylacrylanilide) (Porter *et al.*, 1960), **propanil** (Takematsu and Yanagishima, 1963; Adachi *et al.*, 1966; Yih *et al.*, 1968) and **solan** (pentanochlor) (Colby and Warren, 1965) has also been reported but translocation appears to be very limited.

Studies carried out with ^{14}C-**alachlor** in maize and oats suggest that it is absorbed primarily in a passive fashion, uptake being only slightly reduced by metabolic inhibitors and only slightly increased by rise in temperature (Groenwold, 1971). More alachlor is taken up by the shoot region of these species than via the roots, whereas the reverse is true with soybeans and cucumber. Root absorption and xylem transport of this herbicide (or its metabolites) have been reported in soybean (resistant) and especially in wheat (susceptible) (Chandler, 1972; Chandler, Basler and Santelmann, 1974). Rice and Putnam (1980) reported that uptake of ^{14}C-labelled alachlor by snapbeans (*Phaseolus vulgaris*) was greater at higher temperatures, accumulation occurring in the roots where it was metabolised. However, under low temperature and high pH conditions, less herbicide was metabolised and the ^{14}C was more evenly distributed throughout the plant (also Rice, 1978). While application to the primary leaf tissue of both species resulted in uptake, translocation in soybean was limited (Chandler, 1972).

The basis of selectivity of ^{14}C-**pyrnachlor** in onion seedlings (resistant) and proso millet (*Panicum miliaceum*) (susceptible) has been investigated by Silvia and Putnam (1973). After root uptake from nutrient solution, onion contained more ^{14}C than did *P. miliaceum* though translocation was greater in the latter; relatively polar metabolites were formed in both species after 5 days.

The action of naphthalic anyhdride as a seed protectant against **metolachlor** was investigated using sorghum, maize and soybean seedlings (Ahrens and Davis, 1978). In sorghum, metolachlor moved upward from root to shoot within 3 h and continued to accumulate up to the final harvest at 96 h; in maize and soybean accumulation occurred in the shoot for up to 48 h. Naphthalic anhydride apparently decreased the amount of metolachlor translocated to the shoot in sorghum and maize but not in soybean. Another 'herbicide safener'

CGA-43089 (α-[cyano-methoximino]benzacetonitrile) has been discovered which will also allow the safe use of metolachlor in grain sorghum (*Sorghum bicolor*) (Ellis *et al.*, 1980) and other crop species (Nyffeler, Gerber and Hensley, 1980).

In another study involving metolachlor, Dixon and Stoller (1979) investigated the sites of uptake by maize and yellow nutsedge (*Cyperus esculentus*), the herbicide being incorporated into the soil layer at 1 or 4 p.p.m. above, within or below the seed zone. In maize the site of uptake was in the seed and shoot zones but no toxic effects were seen in *C. esculentus* irrespective of the site of uptake.

NITROPHENYL ETHERS

In this group of compounds **nitrofen, fluorodifen, chlornitrofen** and **acifluorofen** may be applied pre- or post-emergence, though on balance they are probably more often used as soil-applied pre-emergence herbicides. While root absorption generally appears to be rapid, xylem transport to the shoot may be restricted and differentials in the efficiency of root absorption and translocation may influence selectivity. Similarly, while foliage absorption may occur relatively readily in some species, symplastic transport appears to be restricted, movement being acropetal in nature.

Root absorption of fluorodifen by soybean resulted in some translocation from the roots to the foliage (Ebner *et al.*, 1968), while Fadayomi (1976) reported some movement of root-applied ^{14}C-oxyfluorfen ([2-chloro-4-trifluoromethylphenyl]-3-ethoxy-4-nitrophenyl ether), fluorodifen or nitrofen to the shoots of sorghum and pea. Similarly, Shimotori and Kuwatsuka (1978) found that ^{14}C-chlornitrofen and its amino derivatives were rapidly absorbed by the roots of rice plants, especially from water culture rather than soil; however, translocation to the foliage was again slow, especially in the case of the amino derivative. Similar results were obtained for ^{14}C-**nitrofluoren** and ^{14}C-**oxyfluorfen** in *Vica faba* and green foxtail (*Setaria viridis*) (Vanstone and Stobbe, 1978). Eastin (1972) found evidence of greater absorption and translocation of fluorodifen (1 p.p.m.) in seedlings of cucumber (sensitive) than groundnut (resistant). Similarly, ^{14}C-**bifenox** was absorbed more readily by seedlings of *Abutilon theophrasti* (susceptible) than by those of maize or soybean (resistant) (Leather and Foy, 1976). Translocation of ^{14}C-bifenox to the shoot was extensive in the case of *A. theophrasti* but absent in the crop species. When applied to these species in nutrient solution, however, root-absorbed ^{14}C-bifenox was translocated to the shoots of all the test species, though differences were evident in the distribution of the label in the leaf. In maize and soybean, ^{14}C was confined to the primary and secondary leaf veins, while in *A. theophrasti*, distribution was general throughout the leaf tissues (Leather and Foy, 1976). Ohyama and Kuwatsuka (1976) and Kuwatsuka *et al.* (1976) found that bifenox absorbed by the roots of rice (in soil) was hardly translocated but that the free acid was translocated.

It seems likely that the site of action of these root-applied nitrophenyl ethers lies in the chloroplasts. Eastin (1972) found that higher concentrations of fluorodifen reached the chloroplasts of cucumber (S) than of groundnut (R). Accumulation of fluorodifen within the chloroplasts has also been reported by Boulware and Camper (1973) who observed absorption of this compound (and trifluralin) by isolated leaf cells of *Zinnea elegans* and by protoplasts of immature tomato fruit over a 2–6 h incubation period. Within the protoplast, trifluralin was distributed evenly whereas fluorodifen was concentrated in the chloroplast nuclei fraction only.

When used post-emergence, the basis of selectivity of these compounds may involve differences in foliage retention, absorption and translocation. For example, when nitrofen was applied to susceptible and resistant cultivars of cabbage, selective activity was found to be due to differences in herbicide retention and penetration, contact action resulting from loss of membrane integrity (Pereira, 1970). Similar relationships have been reported by Hawton and Stobbe (1971) and effects on membrane permeability of beetroot sections noted.

Symplastic translocation of foliage-applied nitrophenyl ethers appears to depend on species. Foliage treatment of soybean with fluorodifen did not appear to result in any appreciable symplastic translocation (Ebner *et al.*, 1968) while differential translocation of ^{14}C-bifenox was observed in maize, soybean and *A. theophrasti* (Leather and Foy, 1976). The latter found that in maize and soybean only slight acropetal movement occurred, while in *A. theophrasti* 3% of the applied label was translocated from the treated leaf to the stem.

NITROANILINES

The major route of absorption and translocation of the nitroaniline herbicides is a matter of some controversy but the predominant evidence suggests that they are readily absorbed by roots and shoots though generally translocation is minimal (Ashton and Crafts, 1973).

Absorption of **trifluralin** has been reported by both root and emerging shoots of sorghum seedlings (Parker, 1966) or green foxtail

(*Setaria viridis*) (Knake *et al.*, 1967). Shoot uptake of trifluralin by foxtail millet (*Setaria italica*) and proso millet (*Panicum miliaceum*) was more efficient than was root uptake. Microautoradiography of cotton and soybean seedlings treated with ^{14}C-trifluralin showed that the herbicide was retained on the root surfaces, entry only occurring through breaks in the epidermis (Strang and Rogers, 1971a, b). Some translocation of ^{14}C occurred into the leaves of cotton but not into the leaves of soybean. The uptake and movement of trifluralin (and other substituted aniline herbicides) in peanut seedlings was studied using g.l.c. by Kerchersid *et al.* (1969). Translocation was extensive though it declined with age of the plant and they suggested that trifluralin may circulate from apoplastic to symplastic systems and accumulate in regions of high lipid content such as the cotyledons. Hawxby *et al.* (1972) found that absorption of trifluralin by excised lateral roots of peanut was greatest at low (21°C) as opposed to high (38°C) temperatures; although translocation into the hypocotyls, tops and cotyledons generally increased with temperature, the amounts were small. They suggest that injury to trifluralin-treated seedlings in the field may be due to excessive accumulation of herbicide in the roots during cool conditions. Similar findings were obtained for a range of test species – soybean, pigweed (*Amaranthus* sp.), sorghum and barnyard grass (*Echinochloa crus-galli*) (Hawxby *et al.*, 1973).

The absorption, translocation and metabolism of ^{14}C-**ethalfluralin** and ^{14}C-trifluralin in *Solanum nigrum* from two seed sources was studied by Vandeventer, Meggitt and Penner (1979). Differences in the response by plants from these different seed sources was partly attributable to differences in the rate of absorption of both herbicides and differences in susceptibility to the two herbicides were due to differences in absorption and metabolism.

Temperature effects may influence the toxicity of other nitroaniline herbicides. For example, Hawxby and Basler (1976) investigated the absorption, translocation and metabolism of ^{14}C-**profluralin** (and ^{14}C-**dinitramine**) in sorghum (RS6-12), *E. crus-galli*, soybean (cv. 'Dare') and *Amaranthus palmeri*, after treatment with the herbicide in nutrient solution for 24 h at 16° C and 38° C. Little apoplastic translocation of ^{14}C-profluralin occurred, though ^{14}C-dinitramine accumulated in the shoots to a greater extent at 38° C than at 16° C; metabolism of the herbicide was unaffected by temperature. They suggest that these findings may explain the excess toxicity of profluralin at 16°C and dinitramine at 38°C in normally resistant species. Light and temperature (inhibitory) affected absorption of trifluralin (and fluorodifen) by cells isolated from *Zinnia* sp. (Camper, 1972).

Perhaps the most important factor affecting the activity of dinitroaniline herbicides is volatility (Ellis and Norton, 1976). They are mobile in a gaseous state which is herbicidally active, diffusion being greatest under dry soil conditions. Due to their non-ionic nature, their activity is relatively unaffected by minerals and soil pH, though it may be reduced by lipophilic adsorption to organic matter. Jacques and Harvey (1979) found that, in general, vapour absorption of dinitroaniline herbicides by oats (*Avena sativa*) and peas (*Pisum sativum*) was correlated with rates of herbicide volatilisation. All the dinitroaniline herbicides examined (^{14}C-labelled **benfluralin**, dinitramine, **oryzalin**, profluralin and trifluralin) were absorbed by the roots and shoots of germinating oats and peas. Some translocation of herbicide from root to shoot was observed in peas, but no shoot to root transport was detected in either species.

CARBAMATES

Some carbamate herbicides are characterised by low water solubilities. They are normally soil-applied, root-absorbed compounds having little phytotoxicity when applied to foliage (e.g. **propham** and **chlorpropham**). Others such as **barban**, **phenmediphan** and **asulam** are relatively water soluble and readily foliar absorbed; their absorption/translocation characteristics are dealt with in chapter 3.

Comparison of the absorption and movement of ^{14}C-ring- or side-chain-labelled chlorpropham by foliage or roots of redroot pigweed (*Amaranthus retroflexus*), pale smartweed (*Polygonum lapathifolium*) and parsnip revealed that absorption occurred by both routes though only apoplastic transport was evident (Prendeville *et al.*, 1968); selectivity could not be attributed to interspecies variation in absorption or translocation. Studies of the control of dodder (*Cuscuta* sp.) in alfalfa suggested that the foliage of alfalfa restricted air movement of the herbicide vapour sufficiently to allow vapour toxicity to the dodder (Slater *et al.*, 1969). Control of this weed is enhanced by incorporation with chlorpropham of *p*-chlorophenyl-*N*-methyl-carbamate (PCMC) (1.65 kg/ha) which is a microbial inhibitor and presumably delays microbial breakdown of chlorpropham in the soil.

The absorption and translocation of chlorpropham by germinating seedlings of soybean, maize, peanut and castor beans has been studied by Ashton and Helfgott (1966); Helfgott and Ashton (1966) and Helfgott (1969). They found that during the initial stages of germination, absorption of chlorpropham is essentially a physical process which is independent of and more rapid than water uptake, the seed coat appearing to act as a barrier for penetration. Very little of the absorbed radioactivity is translocated and that which is appears to move in the apoplast. The uptake of chorpropham by soybean seeds is directly related to concentration and increases with rise in temperature (10–30°C) (Rieder *et al.*, 1970). They concluded that absorption of chlorpropham was largely a physical process which required seed hydration, though absorption continued after cessation of water uptake. However this view has been questioned by Helfgott (1969) who reported that dry seeds could absorb chlorpropham vapour.

The fate of root-applied chlorpropham within the treated plant has been further clarified by the work of Still and Mansager (1973a) who found that root-treated cucumber could absorb, translocate and metabolise ^{14}C-chlorpropham. Even after 3 days, polar products and solid residues were found in roots, stems and leaves. However, these metabolites were not translocated once they were formed in either roots or shoots. It appears that chlorpropham is converted to the 4-hydroxy derivative which is then conjugated with certain plant components.

These authors also found evidence of rapid uptake, translocation and metabolism of root-applied ^{14}C-propham, the parent compound being completely metabolised after 3 days. Polar products and non-extractable residues were found in roots, stems and leaves (Still and Mansager, 1973b). Similarly in lucerne (cv. Moapa), propham was root absorbed, translocated and metabolised to form a number of polar metabolites, including the 2-hydroxy and 4-hydroxy derivatives (Zurquiyah, Jordan and Jolliffe, 1976). The uptake, translocation and metabolism of propham by wheat (*Triticum aestivum*), sugar-beet (*Beta vulgaris*) and alfalfa (*Medicago sativa*) were studied by Burt and Corbin (1978). They found that absorption and translocation in all species was rapid, but the greatest differences were observed in the rates of metabolism; after a 4-day liquid culture treatment most rapid breakdown occurred in leaves of sugar-beet (95%) and wheat (91%) and least in lucerne (36%).

Normally barban is foliage applied, but the selective herbicidal action of root-applied barban on oat and wheat plants has been reported by Kobayashi and Ishizuka (1979). The concentration of barban was greater in oats (susceptible) than wheat (tolerant) and these differences are attributed to different rates of translocation from roots to shoots.

THIOCARBAMATES

The thiocarbamates are normally soil applied, absorbed through the root, coleoptile or stem base and translocated in the apoplast. Many are, however, capable of foliage absorption and phloem translocation

(Ashton and Crafts, 1973) and are dealt with in chapter 3.

Di-allate is absorbed by the coleoptiles of emerging grass seedlings (Appleby *et al.*, 1965) and selective activity may depend on differential absorption resulting from differences in depth of seeding and depth of herbicide incorporation. For example, the selectivity of di-allate (or **tri-allate**) between wild oat and barley or wheat depends on such factors (Parker, 1963), though differential uptake of tri-allate by seedlings of barley (cv. Moravian) and wild oat (*Avena fatua*) appears not to be the basis of selectivity (Thiele and Zimdahl, 1976). In the latter case barley (resistant) accumulated greater amounts of tri-allate than did wild oats (susceptible) without injury.

Like di-allate the herbicidal activity of **EPTC** appears to depend on dosage, depth of incorporation and seed placement as well as species of weed involved. It is apparently xylem and phloem mobile. For example, application of ^{35}S-EPTC as vapour to the leaves of a number of crop and weed species resulted in absorption and symplastic translocation while root application resulted in rapid absorption and apoplastic distribution within the shoot (Yamaguchi, 1961). Certain other thiocarbamates show similar xylem and phloem mobility, e.g. **vernolate, pebulate, butylate** and **molinate** (Ashton and Crafts, 1973).

Differences in susceptibility to EPTC may be associated with the sites of absorption. For example, Oliver *et al.* (1968) studied root/shoot uptake of ^{14}C-EPTC by seedlings of barley, wheat, oats, sorghum and giant foxtail (*Setaria faberi*) and found that tolerance was greatest in barley in which the roots were the major site of absorption, increasing susceptibility being evident in the remaining species in which shoot uptake was equal to, or greater than, root absorption. On the other hand, selectivity of EPTC in these species appears not to be due to differences in root uptake and translocation. Uptake of ^{14}C-EPTC applied in nutrient culture to brome grass (*Bromus inermis*) increased with time and concentration (Marton *et al.*, 1978); distribution of ^{14}C within the roots and shoots being uniform and similar to that obtained for EPTC in lucerne and maize.

The mode of action of *N,N*-diallyl-2,2-dichloroacetamide (R-25 788) as an EPTC antidote has been examined in maize seedlings (Chang, Stephenson and Bandeen, 1974). Application of R-25 788 2 days after pretreatment with EPTC in the nutrient solution, significantly reduced EPTC injury. This is attributed to the enhanced release of $^{14}CO_2$ and ^{14}C-EPTC vapour, though the magnitude of these effects does not seem adequate to explain the full mode of action of R-25 788. Gray and Joo (1978) exposed different parts of maize seedlings to vapours of EPTC and the same antidote (R-25 788) and found that more severe injury occurred when the herbicide was applied to the soil in the shoot zone, R-25 788 giving best protection when applied to that zone. On the other hand, root uptake and translocation of EPTC (10^{-5}–10^{-3} M) were unaffected by the EPTC antidote (10^{-5} M). **Thiobencarb** is readily absorbed by both roots and leaves, systemic translocation occurring in the xylem and phloem respectively; selectivity appears to be due to differentials in absorption, translocation or metabolism. Nakamura, Ishikawa and Kuwatsuka (1974) found that root absorption and translocation of thiobencarb occurred more readily in *Echinochloa crus-galli* (S) than in rice (R) plants, rapid metabolism of thiobencarb occurring in both species. Foliage application resulted in translocation into other leaves and into the roots of some species. The mode of action of this herbicide in rice has also been studied by Eastin (1975). He treated seedlings at the 3-leaf stage with ^{14}C-thiobencarb (1 p.p.m.) applied to the roots in nutrient solution or as 5 μl droplets (2.75 μg) to the centre of the adaxial surface of the 1st, 2nd or 3rd leaf. Thiobencarb was readily absorbed by the roots and ^{14}C translocated throughout the plant while foliage application resulted in translocation of ^{14}C into the roots. The level of non-extractable ^{14}C increased with time indicating a possible complexing of thiobencarb and/or its metabolites with certain natural plant components.

UREAS

Monuron, like most other urea herbicides, is rapidly absorbed by the roots and transported in the xylem; foliage absorption on the other hand, is relatively slow and symplastic transport nil. Indeed the substituted ureas have long been considered to be classical examples of solely xylem-translocated herbicides. Early studies with monuron prior to the availability of radiolabelled compounds suggested that root absorption was followed by translocation to the foliage in the transpiration stream (e.g. Bucha and Todd, 1951; McCall, 1952; Muzik *et al.*, 1954). These findings were confirmed by studies using ^{14}C-labelled monuron (Haun and Peterson, 1954; Fang *et al.*, 1955; Crafts and Yamaguchi, 1958, 1960; Crafts, 1959, 1962, 1967; Leonard *et al.*, 1966; Pereira *et al*, 1963; Smith and Sheets, 1966).

Controversy exists as to whether uptake of these herbicides is passive or active. Root uptake of **linuron** by isolated roots of soybean was found to be temperature dependent (5–30°C), indicating the involvement of active metabolism (Moody and Buchholtz, 1968), whereas uptake of monuron by barley roots was not temperature dependent suggesting passive uptake (Donaldson, 1967). After penetrating the root, movement occurs across the cortex through the Casparian strip of the endodermis and into the stele. The route of transcortical movement of monuron and **diuron** has been examined by Donaldson (1967) who found evidence to suggest that monuron, which is relatively non-polar, passively penetrated the membranes of the cortical cells of barley, by-passed the Casparian strip and moved into the stele. Histoautoradiography of the roots of diuron-treated plants indicated that this compound was located almost entirely in the cell walls, particularly in the stele.

Studies of the importance of metabolic root pressure and transpiration on the translocation of monuron from the growth medium to the leaves have been examined under different environmental conditions. There was evidence to suggest that transpiration exerts a major influence on the translocation of monuron, while the solute transfer mechanism involved in root pressure appears to have little or no effect (Anon, 1976).

The absorption and translocation of diuron is essentially similar to that of monuron (Bayer and Yamaguchi, 1965; Leonard *et al.*, 1966; Leonard and Glenn, 1968). Absorption of both herbicides appears to occur primarily by the roots rather than via the emerging shoot (Nishimoto *et al.*, 1967; Prendeville *et al.*, 1967). This contrasts with the findings of Knake and Wax (1968) whose evidence suggests that linuron may be absorbed by the emerging shoot rather than by the root zone. However Walker (1973a) found that uptake of ^{14}C-linuron (and ^{14}C-ring-labelled atrazine) by wheat seedlings growing in sandy-loam soil was proportional to the fraction of the root system exposed to the treated soil. He estimated that this factor would offset the reduction in herbicide concentration in the soil following incorporation of a standard dose to different depths. Other factors which influence uptake of diuron and presumably other urea herbicides include solution concentration and the volume of water transpired.

Transport of diuron appears to be rapid and efficient, as much as 50–60% of the diuron applied to the root being accumulated in the shoots of barley after 7 days (Moyer, McKercher and Hance, 1972). In cotton, ^{14}C-diuron was found to accumulate in the lysigenous glands and trichomes on the leaves and this may be an important factor explaining tolerance of this species to diuron (Strang and Rogers, 1971a) since this material is presumably immobilised and unavailable at the sites of action.

Linuron has similar characteristics to monuron and diuron. Studies of the release of this compound from excised roots of soybean suggest that the herbicide molecules released were those which diffused from free space rather than from adsorptive uptake sites (Rieder *et al.*, 1970). Uptake was again rapid, being concentration and temperature dependent, as much as 83% of the applied dose being absorbed after

48 h at 30°C. However, in living and dead seeds the uptake rate was found to be similar suggesting that in this case the uptake process was essentially a physical one requiring seed hydration. Ostrowski (1972) believes that absorption of **monolinuron** by *S. alba* is mainly a physical process, though he found that herbicide uptake continued after water uptake had ceased. Certainly absorption and translocation to the shoots of oat plants is rapid and sustained, Süss, Fuchsbichler and Siegmund (1978) finding that one half to two thirds of the initial activity is taken up by the plants within 3 weeks and 35–57% of the initial activity is translocated into the shoots after 9 weeks.

Movement of linuron is also essentially apoplastic though some interspecies variation in the rate has been reported. For example, Börner (1964, 1965) found that linuron was rapidly translocated to the shoots of *Sinapis arvensis* (susceptible) following root application; but transport within the bean (*P. vulgaris*) (resistant) was relatively slow. Again, movement of linuron through vertical 6 mm long sections of bean petiole was initially found to be relatively slow compared with aminotriazole, chlorpropham, or 2,4-D; later however, the rate of transport of linuron increased markedly (Taylor and Warren, 1968). Pretreatment with certain metabolic inhibitors stimulated the movement of linuron suggesting that membrane permeability was increased or active retention reduced.

The efficiency of uptake of ^{14}C-linuron by soybean seedlings from nutrient solution was influenced by the levels of calcium and nitrogen in the nutrient solution (Rivera and Penner, 1978) while in peas, linuron was absorbed and translocated for a longer duration (192 h) compared with monolinuron (48 h) (Budimir, 1979). The selective activity of linuron in carrot and parsnip (resistant) and turnip and lettuce (susceptible) has been attributed to relatively low levels of translocation to the shoots in the resistant as compared to susceptible species (30% and 60% of the recovered material respectively) (Walker and Featherstone, 1972). Similarly, the greater resistance of parsnip (*Pastinaca sativa*) to linuron (and **chlorbromuron**) has been attributed to lower absorption and translocation than was found in maize or *Amaranthus retroflexus* (Palm, 1971).

Differential absorption and translocation have also been found to determine selectivity in the case of **chloroxuron**. Studies with smallflower galinsoga (*Galinsoga paraviflora*) and wild buckwheat (*Polygonum convolvulus*) showed that this compound was rapidly absorbed by the roots and translocated into the shoot, about twice as much being taken up and transported in galinsoga than in buckwheat (Aebi and Ebner, 1962; Geissbühler *et al.*, 1963). Translocation is reduced by increased relative humidity, and diminished light and temperature, all factors which influence the rate of transpiration. Like other urea herbicides foliar treatment with chloroxuron results in apoplastic transport within the leaf but no basipetal movement out of the treated leaf.

The basis of selectivity of **metobromuron** and chlorbromuron to tomato (sensitive to both), carrot (moderately susceptible to both) and coriander (*Coriandrum sativum*) (resistant to both), has been examined by Hogue (1978). The physiological tolerance of coriander to chlorbromuron appears to be due primarily to lack of translocation of this photosynthetic inhibitor to the shoots. However, the greater tolerance of carrot and coriander compared with tomato towards metobromuron, could not be explained on the basis of differential absorption or metabolism. Similar studies have been carried out on the uptake, translocation and metabolism of ^{14}C-**tetrafluron** by cotton, jimson weed (*Datura stramonium*), peanuts and prickly sida (*Sida spinosa*) (Pinto and Corbin, 1979). Absorption and translocation of radioactivity was more rapid in the weed species than in cotton or groundnuts, with the accumulation of ^{14}C round the leaf margins; conversely metabolism of the tetrafluron was more rapid in the crop species.

TRIAZINES

As in the case of the urea herbicides, the triazines are mainly root absorbed and xylem transported, and again like the ureas, foliage absorption does not result in symplastic transport.

There is a considerable literature associated with the absorption and translocation of triazines and only a representative selection is presented here. Davis *et al.*, (1959) reported very rapid absorption of ^{14}C-**simazine** by the roots of maize, cotton and cucumber with translocation of ^{14}C to the leaves of cucumber within $^1/_2$ h. The distribution pattern within the leaves differed in monocotyledons and dicotyledons. In maize leaves the distribution of ^{14}C was uniform, while in cucumber ^{14}C-simazine and its metabolites accumulated round the leaf edges; in cotton accumulation occurred in the lysigenous glands. The latter distribution has also been reported for simazine (Foy, 1962), for **ipazine** (2-chloro-4-[diethylamino]-6-[isopropylamino]-s-triazine) and other triazines (Foy, 1964), prometryne (Sikka and Davis, 1968), and terbutryne (Rubin, 1976).

The root absorption of these soil-applied herbicides appears not to be uniform but consists of two phases. For example, Vostral *et al.*, (1970) studying the uptake of atrazine found it to consist of a rapid initial phase (30 min) followed by a relatively slow continuous phase during which the rate of absorption is less than 10% of that occurring during the initial phase. The rate of uptake was enhanced by temperature and herbicide concentration. They suggest that the initial rapid phase of uptake may reflect movement of the herbicide into the root apoplast.

There is considerable evidence to demonstrate that the rate of root absorption and translocation of the triazines is proportional to the amount of water absorbed and/or rate of transpiration (Sheets, 1961; Foy and Castelfranco, 1961; Grover, 1962; Davis *et al.*, 1965; Wax and Behrens, 1965; Lund-Høie, 1969; Vostral *et al.*, 1970; Minshall, 1972; Walker and Featherstone, 1973). However, these triazine herbicides may themselves reduce the transpiration rate and thus reduce their own absorption and translocation. This has been reported for atrazine (Smith and Buchholtz, 1962, 1964; Wills *et al.*, 1963; Sikka *et al.*, 1964; Graham and Buchholtz, 1968) which is believed to inhibit transpiration by inhibiting photosynthesis, thus causing an increase in the CO_2 content of the sub-stomatal cavity and resulting stomatal closure. This inhibition of transpiration is believed to reduce the absorption and xylem translocation of the herbicide. While agreeing that such a mechanism may operate in the first few days, Shimabukuro and Linck (1967) believe that in the longer term, inhibition may be due to reduced root absorption due to a reduction in carbohydrate concentration resulting in injury to the root tissues. More recently Ahmed and Fletcher (1980) found diuron to be the most effective of several herbicides in reducing transpiration in corn (*Zea mays*) plants. They concluded that the increase in water efficiency indicates that transpiration was reduced more than photosynthesis.

The effect of certain environmental conditions on the rate of absorption and translocation of triazine herbicides has been studied by several workers. Uptake of various triazines has been reported to be influenced by light (Plaisted and Ryskoewich, 1962), temperature (Wax and Behrens, 1965; Minshall, 1969), humidity (Sheets, 1961; Uhlig, 1968), addition of potassium nitrate (KNO_2) or urea to the soil (Minshall, 1969), soil pH (Mercer and Shone, 1971) and the rate of desorption from certain soils (Walker, 1972a, b).

Many studies on the basis of selectivity of the triazine herbicides have been carried out and differentials found in the absorption, translocation and rate of degradation of the herbicide. For example, absorption and translocation of ^{14}C-simazine by white pine (*Pinus strobus*) (resistant) is very much less than in red pine (*Pinus resinosa*) (susceptible) (Freeman *et al.*, 1964; Roeth and Lavy, 1971). Again, differences in uptake, distribution and metabolism appear to account for selectivity of simazine in black walnut and yellow poplar seedlings

(Wickman and Byrnes, 1975), crabgrass (*Digitaria sanguinalis*) and witchgrass (*Panicum capillare*) (Robinson and Greene, 1976). In the latter case, the greater sensitivity of *P. capillare* to simazine and atrazine appears to be due to the increased translocation to the shoots and to reduced metabolism of the herbicides by this plant. Similarly, susceptibility of certain species to atrazine and cyanazine was found to coincide with high levels of absorption and low metabolism (Thompson and Slife, 1973). Again, tolerance of cucumber (*Cucumis sativus*) to atrazine coincided with less root biomass and greater leaf area, lower translocation and greater localisation of ^{14}C in the roots and xylem (Werner and Putnam, 1980).

Susceptibility of wheat to **terbutryne** appears to be correlated with accumulation within the foliage of the plant and inability to degrade the terbutryne molecule (Dudek, Basler and Santelmann, 1973). The main factors responsible for the selectivity of terbutryne and **prometryne** in peas is availability from the soil, efficiency of translocation, and detoxification (Eshel, Kovacs and Rubin, 1975a). These authors also examined the effect of soil replacement and shoot uptake of terbutryne and prometryne on peas and found that the absorbed amount of each herbicide decreased in the order root system, lower epicotyl, upper epicotyl. Translocation of these herbicides to the foliage was higher from the upper than the lower part of the epicotyl (Eshel, Kovacs and Rubin, 1975b). A comparison of the absorption and distribution of ^{14}C-terbutryne and ^{14}C-fluometuron by germinating seeds of cotton and snapbean (*Phaseolus vulgaris*) was carried out by Rubin and Eshel (1978). They found that seeds of the susceptible snapbean absorbed larger amounts of both herbicides from soil than did the resistant cotton seedlings. The explanation lies in the fact that the herbicides accumulate mainly in the seed coat of cotton which is shed following emergence with the result that no translocation occurs to the seedling leaves. On the other hand, in snapbeans the herbicides accumulate in the cotyledons which then serve as a source for acropetal flow to the developing foliage.

The different rates of translocation of ^{14}C-**dipropetryne** and ^{14}C-prometryne which occur in seedlings of oats and maize have been investigated by Basler, Murray and Santelmann (1978). Despite similar rates of root absorption the ^{14}C of prometryne was more readily translocated than the ^{14}C of dipropetryne and translocation was greater in oats than in maize. The ^{14}C label of dipropetryne was immobilised in excised leaves after about 4 h and the ^{14}C label of prometryne was initially more mobile in excised leaves of oats than maize.

The relative contribution of foliage and root absorption,

translocation and metabolism to **cyanazine**, toxicity and selectivity in *Panicum dichotomiflorum, Setaria viridis* and maize have been studied by Kern, Meggitt and Penner (1975). They concluded that the selectivity of this compound was based not only on differences in foliage uptake but also on the differential uptake between foliage and roots. Under conditions favouring uptake by roots the degree of selectivity may be diminished.

The effect of certain triazine herbicides may be influenced by combination with certain other herbicides. For example, interactions between ^{14}C-simazine (or atrazine, prometryne and linuron) and trifluralin or nitralin were observed when these compounds were applied in combination to the upper 5 cm root region of pea and soybean (O'Donovan and Prendeville, 1976). A reduction in foliar injury was associated with reduced uptake and subsequent transport to the foliage of ^{14}C-labelled simazine in vetch and pea and ^{14}C-labelled atrazine in soybean. The reduced uptake was attributed to lack of root development on shoots treated with trifluralin. Again, reduced translocation of ^{14}C-atrazine was believed to be the reason for the antagonistic effect of foliage-applied 2,4-D in *Phaseolus vulgaris* (Dorozhkina *et al.*, 1974). EPTC (0.5 p.p.m. w) however, greatly enhanced the absorption of ^{14}C-atrazine by wheat (cv. Nisu) grown in half-strength Hoaglands solution (Wilkinson and Karunen, 1976).

The addition of other pesticides may have adverse or beneficial influences on the absorption or translocation of triazine herbicides. For example Penner (1974) found that pre-emergence application of the insecticides disulfoton (*O,O*-diethyl *S*-2-ethylthioethylphosphorodithioate), diazinon (*O,O*-diethyl *O*-2-isopropyl-6-methylpyrimidin-4-yl) or fensulfothion (*O,O*-diethyl *O*-4-methylsulphinylphenylphosphorothioate) (1.12 kg/ha), enhanced the distribution of ^{14}C-atrazine into the primary leaves of 7-day-old soybean seedlings.

The unsymmetrical triazine **metribuzin** may be applied either pre-emergence or post-emergence. The behaviour and fate of this compound in soybean (resistant) and hemp sesbania (*Sesbania exaltata*) (susceptible) was compared by Hargroder and Rogers (1974). The uptake and translocation of metribuzin was greater in plants grown in soil than in nutrient solution. In each species, the herbicide or its ^{14}C metabolites were translocated from the root to shoot, but, in *S. exaltata* more label accumulated in all the leaves while in soybean it tended to accumulate in the roots and lower leaves. Metabolism was more rapid in soybean, the metabolite (6-t-butyl-1,2,4-triazine-3,5-[24,4*H*]dione) being relatively non-toxic. These differences in absorption, translocation and metabolism are believed to account for selective activity between these species.

PYRIDINES

While **picloram** appears to be largely foliage absorbed, root absorption and xylem transport have been reported. Its root uptake from nutrient solution by oats and soybeans was found to increase with herbicide concentration and temperature, but decrease with increase in pH from 3.5 to 4.5, and increasing concentrations of the metabolic inhibitors DNP, sodium azide and sodium arsenite (Isensee, Jones and Turner, 1971). Low concentrations of picloram (10^{-6}–10^{-5} M) stimulated xylem transport and high concentrations inhibited it. Morrison and Vanden Born (1975) also found evidence that root absorption was governed at least in part by active mechanisms. They examined the uptake of picloram by roots of lucerne (sensitive) and barley (tolerant) and reported that the initial phase of uptake is governed by passive processes while continuing absorption is controlled in part by active mechanisms. Generally total uptake of picloram after 4 h was much greater in lucerne than barley. Subsequently O'Donovan and Vanden Born (1979) carried out a microautoradiographic study of ^{14}C-picloram distribution in soybean (*Glycine max*) following root absorption. In the roots, radioactivity appeared to be associated mainly with the central protoplast and vascular regions of the root cortical cells rather than with the cell walls, indicating the route of transcortical movement to the xylem. In the stem, ^{14}C was evident in both the xylem and phloem with little movement into the other tissues. Considerably more radioactivity was found in the younger apical leaves than in the older primary leaves. They concluded that picloram is freely mobile in both xylem and phloem.

PYRIDAZINES

Chloridazon is essentially a root-absorbed, xylem-transported herbicide though it is also foliage absorbed and translocated to some extent in the phloem (Ashton and Crafts, 1973). Selectivity appears to depend upon differential root absorption and upon differences in the rate of inactivation by crop and weed.

The ability of weed plants to inactivate chloridazon has been examined by Romanovskaya and Straume (1975). They sprayed 'Fenazon' (chloridazon, 80%) at 2 and 6 kg product/ha onto seedlings of fourteen species of dicotyledonous and monocotyledonous weeds growing in pots and onto the surface of soil in which rhizomes of *Agropyron repens* were planted. Most of the weed species inactivated the herbicide, depending on its rate of accumulation in cells, the phytotoxicity of chloridazon being directly proportional to the accumulation of unbound herbicide in the plant cells. In the case of *A. repens*, 6 kg chloridazon/ha placed just beneath the rhizomes completely killed them, whereas this dose applied to the soil surface merely checked their growth; clearly root absorption was the major route of entry.

Similarly **metflurazone** (4-chloro-5-[dimethylamino]-2-[(3-trifluoromethyl)phenyl]-3(2*H*)-pyridazinone) is readily root absorbed. Devlin and Yaklich (1972) studied the mode of action of this herbicide in cranberry (*Vaccinium macrocarpon* cv. Early black) and found it to be root absorbed and translocated to the leaves, though most of the label remained in the roots. After 8 days in the roots, metflurazone was metabolised in the roots to **norflurazon** and other metabolites. It was later demonstrated that both metflurazone and norflurazon were translocated into the shoot of cranberry grown in water solutions containing these herbicides, shoot accumulation increasing with time (Yaklich, Karczmarczyk and Devlin, 1974a, b).

The basis of selectivity of metflurazone and norflurazon in cotton (cv. Coker 203), maize (WF 9) and soybean (cv. Lee) was found to be due to differences in absorption and translocation (Strang and Rogers, 1974). Most of the ^{14}C absorbed by cotton (resistant) was retained in the roots while significantly more ^{14}C was translocated to the shoots of soybean (susceptible) and maize (very susceptible). Interspecific differences in metabolism were also reported (see chaper 6).

The uptake and translocation of norflurazon in groundnuts provided further evidence of its xylem-mobile characteristics (Harden and Corbin, 1975). Uptake was passive being proportional to the concentration of herbicide in the culture solution and rapid, 98% of the applied ^{14}C being absorbed in 192 h. Translocation to the shoots was also rapid, ^{14}C being detected in the upper stems and foliage within 6 h of treatment. They found that mature leaves were uninjured while those in the developmental stage were bleached, yet the latter contained less ^{14}C than did the former. This apparent anomaly is explained by the low level of transpiration in the bleached tissue, uptake of norflurazon having been shown to be related to water uptake and therefore transpiration. Similar findings were reported for the uptake and translocation of ^{14}C-metflurazone in soybean (cv. Kent) and sicklepod (*Cassia obtusifolia*) (Motooka, Corbin and Worsham, 1977). Uptake of this compound was also proportional to its concentration in the culture solution and to transpiration, provided transpiration and the uptake mechanisms were not limited. Translocation followed the apoplastic route in both species, higher concentrations of herbicide accumulating in the upper leaves and apical meristems of *C. obtusifolia* than in soybean.

PYRIMIDINES (URACILS)

The uracils (terbacil, bromacil and lenacil) are normally soil-applied herbicides, readily root absorbed and xylem translocated to the leaves thus resembling the substituted-urea herbicides in their site and mode of action (Hilton *et al.*, 1964). When foliage applied they appear to have limited phytotoxicity. For example, Hurtt (1971) examined the effect of certain adjuvants (0.2% Tween 20, 0.2% WK and 10% AL 411-F, singly or blended) on the foliar absorption of **bromacil** by Black Valentine beans. The low phytotoxicity of foliar-applied bromacil was attributed to limited foliar absorption. The uptake and distribution of soil-applied bromacil was enhanced by increased soil moisture (Schreiber, 1972), this being partly due to the more rapid uptake and translocation of bromacil from root to foliage.

The physiology of transport of ^{14}C-**lenacil** after root application to sugar-beet was examined by Sieber, Siegel and Göhler (1973). They found that transport occurred acropetally in the xylem, and that the compound broke down rapidly. After 28 days, the ratio of unchanged chemical to metabolites was 1:97 in the shoots and 1:61 in the roots; neither lenacil nor its metabolites were found in the soluble protein fraction.

Addition of the insecticide fonofos to a nutrient solution containing ^{14}C-**terbacil** was investigated in peppermint (*Mentha piperita* cv. Todd's Mitchum) (Weirich, Nelson and Appleby, 1977). During a 3-day period, movement into the leaves almost doubled, a high proportion of the total radioactivity being intact herbicide. They suggest that results could account for damage to peppermint resulting from closely timed applications of fonofos and terbacil.

HETEROCYCLIC NITROGEN COMPOUNDS (UNCLASSIFIED)

Aminotriazole, the best known of this group of herbicides is normally used as a foliage-applied systemic herbicide while certain others (e.g. **difenzoquat** and **metamitron**) are foliage-applied contact compounds (see chapter 3).

References

ADACHI, M., TONEGAWA, K. and UEJIMA, T. (1966). *Noyaku Seisan Gijyutsu*, **14**, 19.

AEBI, H. and EBNER, L. (1962). *Proc. 6th Brit. Weed Control Conf.*, 479.

AHMED SHABIR and FLETCHER, R. A. (1980). *Weed Sci.*, **28**, 180.

AHRENS, W. H. and DAVIS, D. E. (1978). *Proc. 31st Mtg. S. Weed Sci. Soc.*, 249.

ANDERSON, J. .C. and AHLGREN, G. (1947). *Down to Earth*, **3**, 16.

ANDERSON, J. C. and WOLF, D. E. (1947). *Amer. Soc. Agron. J.*, **39**, 341.

ANON (1976). *Res. Branch Rept., 1975, Agric., Canada*, 199.

APPLEBY, A. P., FURTICK, W. R. and FANG, S. C. (1965). *Weed Res.*, **5**, 115.

ASHTON, F. M. and CRAFTS, A. S. (1973). *Mode of Action of Herbicides*, John Wiley & Sons, New York, London.

ASHTON, F. M. and HELFGOTT, S. (1966). *Proc. 18th Calif. Weed Conf.*, 8.

BARRETT, M. and ASHTON, F. M. (1979). *Abstr. Mtg. Weed Sci. Soc. Amer.*, 102.

BASLER, E., MURRAY, D. S. and SANTELMANN, P. W. (1978). *Weed Sci.*, **26**, 358.

BAYER, D. E. and YAMAGUCHI, S. (1965). *Weeds*, **13**, 232.

BINGHAM, S. W. and SHAVER, R. (1971). *Weed Sci.*, **19**, 639.

BÖRNER, H. (1964). *Z. Pflanzenkrankh. Sonderheft*, **2**, 41.

BÖRNER, H. (1965). *Z. Pflanzenkrankh.*, **72**, 449.

BOULWARE, M. A. and CAMPER, N. D. (1973). *Weed Sci.*, **21**, 145.

BUCHA, H. C. and TODD, C. W. (1951). *Science*, **114**, 493.

BUDIMIR, M. (1979). *Zastita Bilja*, **30**, 211.

BURT, M. E. and CORBIN, F. T. (1978). *Weed Sci.*, **26**, 296.

CAMPER, N. D. (1972). *Proc. 25th Ann. Mtg. S. Weed Sci. Soc.*, 445.

CARLSON, W. C., LIGNOWSKI, E. M. and HOPEN, H. J. (1975). *Weed Sci.*, **23**, 148.

CHANDLER, J. M. (1972). *Diss. Abstr. Int. B.*, **33**, 530.

CHANDLER, J. M., BASLER, E. and SANTELMANN, P. (1974). *Weed Sci.*, **22**, 253.

CHANG, F. T., STEPHENSON, G. R. and BANDEEN, J. D. (1974). *Abstr. Mtg. Weed Sci. Soc. Amer.*, 1.

COLBY, S. R. and WARREN, G. F. (1965). *Weeds*, **13**, 257.

CORBIN, F. T., UPCHURCH, R. P. and SELMAN, F. L. (1971). *Weed Sci.*, **19**, 233.

CRAFTS, A. S. (1959). *Pl. Physiol., Lancaster*, **34**, 613.

CRAFTS, A. S. (1961). *The Chemistry and Mode of Action of Herbicides*, Interscience Publishers, New York, 269 pp.

CRAFTS, A. S. (1962). *Intern. J. Appl. Radiation Isotopes*, **13**, 407.

CRAFTS, A. S. (1967). *Hilgardia*, **37**, 625.

CRAFTS, A. S. and BROYER, T. C. (1938). *Amer. J. Bot.*, **25**, 525.

CRAFTS, A. S. and YAMAGUCHI, S. (1958). *Hilgardia*, **27**, 421.

CRAFTS, A. S. and YAMAGUCHI, S. (1960). *Amer. J. Bot.*, **47**, 248.

CROSBY, D. G. (1976). In *Herbicides, Physiology, Biochemistry, Ecology* (Ed. L. J. Audus), Vol. II, Academic Press, London and New York, p. 65.

DAVIS, D. E., FUNDERBURK, H. H. and SANSING, N. G. (1959). *Weeds*, **7**, 300.

DAVIS, D. E., GRAMLICH, J. V. and FUNDERBURK, H. H. (1965). *Weeds*, **13**, 252.

DEVLIN, R. M. (1974). *Proc. N.E. Weed Sci. Soc., Philadelphia*, **28**, 99.

DEVLIN, R. M. and KARCZMARCZYK, S. J. (1977). *Weed Sci.*, **25**, 142.

DEVLIN, R. M. and YAKLICH, R. W. (1971). *Proc. N.E. Weed Sci. Soc.*, **25**, 88.

DEVLIN, R. M. and YAKLICH, R. W. (1972). *Proc. N.E. Weed Sci. Soc., New York*, **26**, 72.

DIXON, G. A. and STOLLER, E. W. (1979). *Abstr. Mtg. Weed Sci. Soc. Amer.*, 98.

DONALDSON, T. W. (1967). *Absorption of the Herbicides 2,4-D and Monuron by Barley Roots*, Ph.D. Thesis, University of Calif., Davis, 124 pp.

DOROZHKINA, L. A., KALININ, V. A., KUZ'MINSKAYA, V. A. and GBIANZA, T. A. (1974). *Khim. Sel'sk. Khoz.*, **12**, 127.

DUDEK, C., BASLER, E. and SANTELMANN, P. W. (1973). *Weed Sci.*, **21**, 440.

EASTIN, E. F. (1972). *Abstr. Mtg. Weed Sci. Soc. Amer. St. Louis*, 93.

EASTIN, E. F. (1975). *Proc. 28th Ann. Mtg. S. Weed Sci. Soc.*, 306.

EBNER, L., GREEN, D. H. and PANDE, P. (1968). *Proc. 9th Brit. Weed Control Conf., 1026.*

ELLIS, J. F. and NORTON, J. A. (1976). *Proc. N.E. Weed Sci. Soc., Boston*, 52 (Suppl.).

ELLIS, J. F., PEEK, J. W., BOEHLE, J. JR. and MÜLLER, G. (1980). *Weed Sci.*, **28**. 1.

ESHEL, Y., KOVACS, M. and RUBIN, B. (1975a). *Pestic. Biochem. Physiol.*, **5**, 295.

ESHEL, Y., KOVACS, M. and RUBIN, B. (1975b). *Weed Res.*, **15**, 369.

FADAYOMI, R. O. (1976). *Diss. Abstr. Int. B.*, **37**, 1053.

FANG, S. C., FREED, V. H., JOHNSON, N. H. and COFFEE, D. R. (1955). *J. Agric. Fd. Chem.*, **3**, 400.

FOY, C. L. (1962). *Proc. 14th Calif. Weed Conf.*, 82.

FOY, C. L. (1964). *Weeds*, **12**, 103.

FOY, C. L. and CASTELFRANCO, P. (1961). *Pl. Physiol., Lancaster, Abstr.*, **35**, XXVIII.

FREEMAN, F. W., WHITE, D. P. and BUKOVAC, M. J. (1964). *Forest Sci.*, **10**, 330.

GEISSBÜHLER, H., HASELBACH, C., AEBI, H. and EBNER, L. (1963). *Weed Res.*, **3**, 181.

GOLAB, T., HERBERG, R. J., PARKA, S. J. and TEPE, J. B. (1966). *J. Agric. Fd. Chem.*, **14**, 592.

GRAHAM, J. C. and BUCHHOLTZ, K. (1968). *Weed Sci.*, **16**, 389.

GRAY, R. A. and JOO, G. K. (1978). In *Chemistry and Action of Herbicide Antidotes* (Eds. F. M. Pallas and J. E. Casida), Academic Press, New York, p. 67.

GROENWOLD, B. E. (1971). *Diss. Abstr. Int. B.*, **32**, 1318.

GROVER, R. (1962). *Pl. Physiol., Lancaster*, **37**, 12.

HARDEN, J. S. and CORBIN, F. T. (1975). *Proc. 28th Ann. Mtg. S. Weed Sci. Soc.*, 307.

HARGRODER, T. G. and ROGERS, R. L. (1974). *Weed Sci.*, **22**, 238.

HARTLEY, G. S. (1976). In *Herbicides, Physiology, Biochemistry, Ecology* (Ed. L. J. Audus), Vol. II, Academic Press, London and New York, p. 1.

HARVEY, R. G. and BAKER, C. R. (1974). *Weed Res.*, **14**, 57.

HAUN, J. R. and PETERSON, J. H. (1954). *Weeds*, **3**, 177.

HAWTON, D. and STOBBE, E. H. (1971). *Weed Sci.*, **19**, 42.

HAWXBY, K. and BASLER, E. (1976). *Weed Sci.*, **24**, 545.

HAWXBY, K., BASLER, E. and SANTELMANN, P. W. (1972). *Weed Sci.*, **20**, 285.

HAWXBY, K., BASLER, E. and SANTELMANN, P. W. (1973). *Proc. 26th Ann. Mtg. S. Weed Sci. Soc.*, 415.

HELFGOTT, S. (1969). *Absorption, Translocation and Metabolism of Chlorpropham (Isopropyl-*m*-carbanilate) by Germinating Seeds*, Ph.D. Diss., University of Calif., Davis, 207 pp.

HELFGOTT, S. and ASHTON, F. M. (1966). *Abstr. Weed Sci. Soc. Amer.*, 47.

HILTON, J. L., MONACO, T. J., MORELAND, D. E. and GENTNER, W. A. (1964). *Weeds*, **12**, 129.

HOGUE, E. J. (1978). *J. Env. Sci. Health B.*, **13**, 323.

HURTT, W. (1971). *Proc. N.E. Weed Sci. Soc.*, **25**, 153.

ISENSEE, A. R., JONES, G. E. and TURNER, B. C. (1971). *Weed Sci.*, **19**, 727.

JACQUES, G. L. and HARVEY, R. G. (1979). *Weed Sci.*, **27**, 371.

JAWORSKI, E. G. (1964). *J. Agric. Fd. Chem.*, **12**, 33.

JAWORSKI, E. G. and PORTER, C. A. (1965). *Abstr. 149th Mtg. Amer. Chem. Soc.*, 21A.

KEARNEY, P. C. and KAUFMAN, D. D. (1976). *Herbicides Chemistry, Degradation and Mode of Action*, Vols. I and II. Marcel Dekker, Inc., New York.

KERCHERSID, M. L., BOSELL, T. E. and MERKLE, M. G. (1969). *Agron. J.*, **61**, 185.

KERN, A. D., MEGGITT, W. F. and PENNER, D. (1975). *Weed Sci.*, **23**, 277.

KNAKE, E. L., APPLEBY, A. P. and FURTICK, W. R. (1967). *Weeds*, **15**, 228.

KNAKE, E. L. and WAX, L. M. (1968). *Weed Sci.*, **16**, 393.

KOBAYASHI, K. and ISHIZUKA, K. (1979). *Weed Res., Japan*, **24**, 23.

KUWATSUKA, S., NIKI, Y., OYAMADA, M., SHIMOTORI, H. and OHYAMA, H. (1976). *Proc. 5th Asian-Pacific Weed Sci. Soc. Conf., Tokyo, Japan, 1975*, 223.

LEATHER, G. R. and FOY, C. L. (1976). *Proc. N.E. Weed Sci. Soc., Boston*, **30**, 123.

LEMIN, A. J. (1966). *J. Agric. Fd. Chem.*, **14**, 109.

LEONARD, O. A. and GLENN, R. K. (1968). *Weed Sci.*, **16**, 352.

LEONARD, O. A., LIDER, L. A. and GLENN, R. R. (1966). *Weed Res.*, **6**, 37.

LUND-HØIE, K. (1969). *Weed Res.*, **9**, 142.

MCCALL, G. L. (1952). *Agric. Chem.*, **7**, 40.

MARTON, A. F., APRÓKOVÁCS, A. V., KÓMIVES, T., FODOR-CSORBA, K. and DUTKA, F. (1978). *Proc. 18th Hungarian Ann. Mtg. Biochem., Salgótarján, 1978*, 111.

MASSINI, P. (1961). *Weed Res.*, **1**, 142.

MEISSNER, R. and OOSTHUIZEN, A. (1976). *Crop Prod.*, **5**, 99.

MERCER, E. R. and SHONE, M. G. T. (1971). In *Ann, Rept. ARC Letcombe Lab. Wantage, Berks. U.K.*, 29.

MINSHALL, W. H. (1969). *Weed Sci.*, **17**, 197.

MINSHALL, W. H. (1972). *Abstr. 1972 Mtg. Weed Sci. Soc. Amer., St. Louis*, 44.

MOODY, K. and BUCHHOLTZ, K. P. (1968). *WSSA Abstr.*, 18.

MORRISON, I. N. and VANDEN BORN, W. H. (1975). *Can. J. Bot.*, **53**, 1774.

MOTOOKA, P. S., CORBIN, F. T. and WORSHAM, A. D. (1977). *Weed Sci.*, **25**, 30.

MOTTLEY, J. and KIRKWOOD, R. C. (1978). *Weed Res.*, **18**, 187.

MOYER, J. R., MCKERCHER, R. B. and HANCE, R. J. (1972). *Can. J. Pl. Sci.*, **52**, 668.

MUZIK, T. J., CRUZADO, H. J. and LOUSTALOT, A. J. (1954). *Bot. Gaz.*, **116**, 65.

NAKAMURA, Y., ISHIKAWA, K. and KUWATSUKA, S. (1974). *Agric. Biol. Chem.*, **38**, 1129.

NEGI, N. S. and FUNDERBURK, H. H. (1968). *WSSA Abstr.*, 37.

NISHIMOTO, R. K., APPLEBY, A. P. and FURTICK, W. R. (1967). *WSSA Abstr.*, 46.

NISHIMOTO, R. K. and WARREN, G. F. (1971). *Weed Sci.*, **19**, 156.

NYFFELER, A., GERBER, H. R. and HENSLEY, J. R. (1980). *Weed Sci.*, **28**, 6.

O'DONOVAN, J. T. and PRENDEVILLE, G. N. (1976). *Weed Res.*, **16**, 331.

O'DONOVAN, J. T. and VANDEN BORN, W. H. (1979). *Abstr. 1979 Mtg. Weed Sci. Soc. Amer.*, 92.

OHYAMA, H. and KUWATSUKA, S. (1976). *Proc. 5th Asian-Pacific Weed. Sci. Soc. Conf., Tokyo, Japan, 1975*, 227.

OLIVER, L. R., PRENDEVILLE, G. N. and SCHREIBER, M. M. (1968). *Weed Sci.*, **16**, 534.

OSTROWSKI, J. (1972). *Roczniki Gleboznawcze*, **23**, 51.

PALM, H. L. (1971). *Diss. Abstr. Int. B.*, **32**, 3113.

PARKER, C. (1963). *Weed Res.*, **3**, 259.

PARKER, C. (1966). *Weeds*, **14**, 117.

PATE, D. A. and FUNDERBURK, H. H. (1966). In *Use of Isotopes in Weed Research*, IAEA Symposium, Vienna, p. 17.

PENNER, D. (1974). *Agron. J.*, **66**, 107.

PEREIRA, J. F. (1970). *Some Plast Responses and the Mechanism of Selectivity of Cabbage Plants to Nitrogen*, Ph.D. Diss., Univ. of Illinois, Urbana.

PEREIRA, J. F., CRAFTS, A. S. and YAMAGUCHI, S. (1963). *Turrialba*, **13**, 64.

PINTO, H. and CORBIN, F. T. (1979). *Abstr. 1979 Mtg. Weed Sci. Soc. Amer.*, 91.

PLAISTED, P. H. and RYSKOEWICH, D. P. (1962). *Pl. Physiol., Lancaster, Abstr.*, **37**, XXV.

PORTER, W. K. JR., THOMAS, C. H., SLOANE, W. L. and MELVILLE, D. R. (1960). *Proc. 13th S. Weed Conf.*, 30.

PRENDEVILLE, G. N., ESHEL, Y., JAMES, C. S., WARREN, G. F. and SCHREIBER, M. M. (1968). *Weed Sci.*, **16**, 432.

PRENDEVILLE, G. N., ESHEL, Y., SCHREIBER, M. M. and WARREN, G. F. (1967). *Weed Res.*, **7**, 316.

PRICE, H. C. and PUTNAM, A. R. (1969). *J. Agric. Fd. Chem.*, **17**, 135.

RICE, R. P. JR. (1978). *Diss. Abstr. Int. B.*, **39**, 14.

RICE, R. P. JR. and PUTNAM, A. R. (1980). *Weed Sci.*, **28**, 131.

RIEDER, G., BUCHHOLTZ, K. P. and KUST, C. A. (1970). *Weed Sci.*, **18**, 101.

RIVERA, C. M. and PENNER, D. (1978). *Abstr. 1978 Mtg. Weed Sci. Soc. Amer.*, 70.

ROBINSON, D. E. and GREENE, D. W. (1976). *Weed Sci.*, **24**, 500.

ROETH, F. W. and LAVY, T. L. (1971). *Weed Sci.*, **19**, 98.

ROMANOVSKAYA, O. I. and STRAUME, O. P. (1975). *Agrokhimiya*, **12**, 117.

RUBIN, B. (1976). *Studies on Tolerance Mechanisms of Cotton (*Gossypium hirsutum *L.) to Herbicide Inhibition of Photosynthesis*, Ph.D. Thesis, Hebrew Univ. of Jerusalem, 122 pp.

RUBIN, B. and ESHEL, Y. (1978). *Weed Sci.*, **26**, 378.

SCHREIBER, J. D. (1972). *Diss. Abstr. Int. B.*, **32**, 4350.

SCHULTZ, D. P. and TWEEDY, B. G. (1972). *J. Agric. Fd. Chem.*, **20**, 10.

SHEETS, T. J. (1961). *Weeds*, **9**, 1.

SHIMABUKURO, R. H. and LINCK, A. J. (1967). *Weeds*, **15**, 175.

SHIMOTORI, H. and KUWATSUKA, S. (1978). *J. Pestic. Sci.*, **3**, 267.

SIEBER, K., SIEGEL, G. and GÖHLER, K. D. (1973). *Wissenschaftliche Beiträge der Martin-Luther - Universität Halle-Wittenberg*, **10**, 127.

SIKKA, H. C. and DAVIS, D. E. (1968). *Weed Sci.*, **16**, 474.

SIKKA, H. C., DAVIS, D. E. and FUNDERBURK, H. H. JR. (1964). *Proc. 17th S. Weed Conf.*, 340.

SIKKA, H. C., LYNCH, R. S. and LINDENBERGER, M. (1974). *J. Agric. Fd. Chem.*, **22**, 230.

SILVIA, E. L. and PUTNAM, A. R. (1973). *Abstr. Mtg. Weed Sci. Soc. Amer., Atlanta, Georgia*, 14.

SLATER, C. H., DAWSON, J. H., FURTICK, W. R. and APPLEBY, A. P. (1969). *Weed Sci.*, **17**, 238.

SMITH, D. and BUCHHOLTZ, K. P. (1962). *Science*, **136**, 263.

SMITH, D. and BUCHHOLTZ, K. P. (1964). *Pl. Physiol., Lancaster*, **39**, 572.

SMITH, J. W. and SHEETS, J. J. (1966). *WSSA Abstr.*, 39.

STILL, G. G. and MANSAGER, E. R. (1973a). *Abstr. 166th Nat. Mtg. Amer. Chem. Soc., Pestic.*, 8.

STILL, G. G. and MANSAGER, E. R. (1973b). *J. Agric. Fd. Chem.*, **21**, 787.

STRANG, R. H. and ROGERS, R. L. (1971a). *Weed Sci.*, **19**, 355.

STRANG, R. H. and ROGERS, R. L. (1971b). *Weed Sci.*, **19**, 363.

STRANG, R. H. and ROGERS, R. L. (1974). *J. Agric. Fd. Chem.*, **22**, 1119.

SÜSS, A., FUCHSBICHLER, G. and SIEGMUND, H. (1978). *Z. Pflanzenernährung Bodenkunde*, **141**, 275.

SWANN, C. W. and BEHRENS, R. (1969). *WSSA Abstr.*, 222.

TAKEMATSU, T. and YANAGISHIMA, S. (1963). *Riso*, **12**, 37.

TAYLOR, T. D. and WARREN, G. F. (1968). *WSSA Abstr.*, 11.

THIELE, G. H. and ZIMDAHL, R. L. (1976). *Weed Sci.*, **24**, 183.

THOMPSON, R. P. and SLIFE, F. W. (1973). *Abstr. 1973 Mtg. Weed Sci. Soc. Amer., Atlanta, Georgia*, 64.

UHLIG, S. K. (1968). *Arch. PflSchutz.*, **4**, 215.

VANDEVENTER, J. W., MEGGITT, W. F. and PENNER, D. (1979). *Abstr. Mtg. Weed Sci. Soc. Amer.*, 95.

VANSTONE, D. E. and STOBBE, E. H. (1978). *Weed. Sci.*, **26**, 389.

VERLOOP, A. and NIMMO, W. B. (1969). *Weed Res.*, **9**, 357.

VERLOOP, A. and NIMMO, W. B. (1970). *Weed Res.*, **10**, 65.

VOSS, G. and GEISSBÜHLER, H. (1966). *Proc. 8th Brit. Weed Control Conf.*, 266.

VOSTRAL, H. J., BUCHHOLTZ, K. P. and KUST, C. A. (1970). *Weed Sci.*, **18**, 115.

WALKER, A. (1972a). *Ann. Rept. Nat. Veg. Res. Sta., 1971, Warwick, U.K. Wellesbourne*, 80.

WALKER, A. (1972b). *Proc. 11th Brit. Weed Control Conf.*, 800.

WALKER, A. (1973a). *Weed Res.*, **13**, 407.

WALKER, A. (1973b). *Weed Res.*, **13**, 416.

WALKER, A. and FEATHERSTONE, R. E. (1972). *Ann. Rept. Nat. Veg. Res. Sta., 1971, Warwick, U.K. Wellesbourne*, 81.

WALKER, A. R. and FEATHERSTONE, R. M. (1973). *J. Exp. Bot.*, **24**, 450.

WAX, L. M. and BEHRENS, R. (1965). *Weeds*, **13**, 107.

WEIRICH, A. J., NELSON, Z. A. and APPLEBY, A. P. (1977). *Weed Sci.*, **25**, 27.

WERNER, G. M. and PUTNAM, A. R. (1980). *Weed Sci.*, **28**, 142.

WICKMAN, J. R. and BYRNES, W. R. (1975). *Weed Sci.*, **23**, 448.

WILKINSON, R. E. and KARUNEN, P. (1976). *Ann. Bot.*, **40**, 1043.

WILLS, G. D., DAVIS, D. E. and FUNDERBURK, H. H. (1963). *Weeds*, **11**, 253.

WU, C. H., BUEHRING, N. and SANTELMANN, P. W. (1974). *Proc. 27th Ann. Mtg. S. Weed Sci. Soc.*, 336.

YAKLICH, R. W., KARCZMARCZYK, S. J. and DEVLIN, R. M. (1974a). *Weed Res.*, **14**, 261.

YAKLICH, R. W., KARCZMARCZYK, S. J. and DEVLIN, R. M. (1974b). *Weed Sci.*, **22**, 595.

YAMAGUCHI, S. (1961). *Weeds*, **9**, 374.

YIH, R. Y., MCRAE, D. M. and WILSON, H. F. (1968). *Science*, **161**, 376.

ZURQUIYAH, A. A., JORDAN, L. S. and JOLLIFFE. V. A. (1976). *Pestic. Biochem. Physiol.*, **6**, 35.

CHAPTER FIVE

Biochemical Mechanisms of Action

In this chapter, the major mechanisms of action of the various groups of herbicides will be examined with particular reference to action on respiration and intermediary metabolism, nucleic acid and protein synthesis, lipid synthesis and photosynthesis. The primary site of action is often difficult to assess due to multi-action effects and to different responses to the range of herbicide concentration especially under *in vitro* conditions. The biochemical modes of action of pesticides including herbicides have been discussed in an excellent review by Corbett (1974).

Respiration

THE RESPIRATORY PROCESS

The mechanism of respiratory metabolism is well known (fig. 5.1). Sugars are broken down to the three-carbon pyruvic acid (glycolysis) which is subsequently degraded by a series of oxidative steps (Krebs cycle) with the release of CO_2 and electrons and H^+ ions which unite with oxygen to form water. The electrons are transferred along an electron transport system from compounds of low reduction potential to those of higher reduction potential, O_2 being the ultimate electron acceptor. ATP synthesis from $ADP+P_i$ is coupled to this electron transfer, the process being termed oxidative phosphorylation. Apart from the glycolytic steps, these reactions occur under aerobic conditions in the mitochondria and many studies of the action of herbicides on respiratory metabolism have been carried out using isolated mitochondria. It is important that such mitochondria exhibit 'tight-coupling' of oxidation and phosphorylation and should possess high RC (respiratory control) and P/O (phosphorylation/oxidation) ratios.

UNCOUPLERS AND INHIBITORS OF OXIDATIVE PHOSPHORYLATION

Chemicals which interfere with ATP synthesis can be classified as

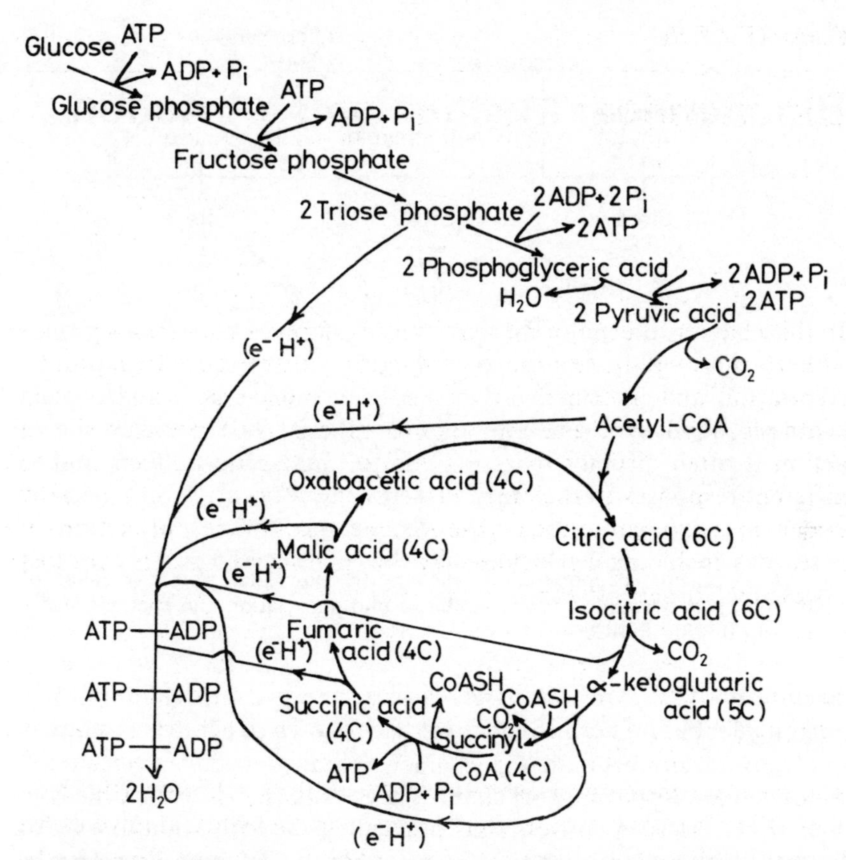

Fig. 5.1 A simplified view of the steps involved in glycolysis and the Krebs cycle leading to ATP synthesis.

uncoupling agents, inhibitors of energy transfer or electron transport, or inhibitory uncouplers (Moreland, 1980) (fig. 5.2). Characteristically uncouplers completely inhibit ATP synthesis without interfering with electron transport. They stimulate state 4 respiration, induce ATPase activity, circumvent oligomycin-imposed inhibition of state 3 respiration and lastly inhibit the various exchange reactions catalysed by mitochondria in the absence of substrates which involve ADP, P_i, ATP and water (Slater, 1963, 1970). At present the actual mechanism of uncoupling is uncertain though a number of hypotheses have been proposed (Moreland, 1980).

Energy-transfer inhibitors inhibit phosphorylating electron transport (state 3 respiration). When the energy-conserving functions of the mitochondria are intact the inhibition is circumvented by uncouplers. Such inhibitors are believed to combine with an intermediate in the energy-coupling chain and thus inhibit the phosphorylation process

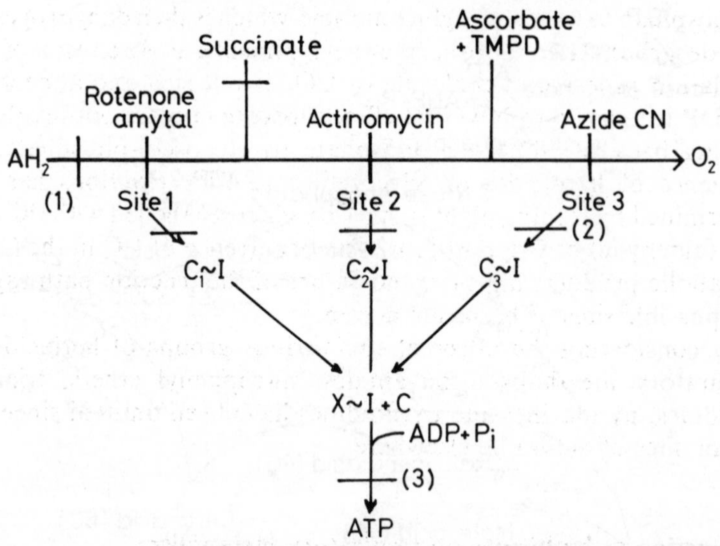

Fig. 5.2 A schematic representation showing the postulated points of action of inhibitors (1) or uncouplers (2) of oxidative phosphorylation and energy-transfer inhibitors (3) (after Matlib, 1970).

leading to ATP synthesis. The classical energy-transfer inhibitor is oligomycin but no herbicide has yet been shown to act in this manner.

Electron transport inhibitors characteristically interrupt electron flow along the electron transport chain. They are believed to combine with one of the electron carriers thus preventing the formation of a redox couple. Such action is manifest as an inhibition of state 3 respiration and coupled phosphorylation reactions. Herbicides which act as electron transport inhibitors include pyrichlor (2,3,5-trichloro-4-pyridinol) (Killion and Frans, 1969) and certain diphenylethers (Moreland *et al.*, 1970).

The term inhibitory uncouplers is applied to the many herbicides which produce a range of responses with oxidative phosphorylation. At low molar concentrations they may act as uncouplers while at higher molar concentrations they act in a manner typical of inhibitors of electron transport. The characteristics of such compounds has been described by Moreland (1980). Herbicides known to act as inhibitory uncouplers include acylanilides, benzimidazoles, dinitroanilines, dinitrophenols, halogenated benzonitriles, imidazoles, phenylcarbamates and thiadiazoles.

In addition to effects associated with electron transport and ATP synthesis, herbicide effects on glycolysis and the pentose phosphate pathway (PPP) have been reported. The PPP operates when O_2 is limiting. The initial step of the PPP involves the oxidation of glucose-

6-phosphate to 6-phosphogluconic acid which is then dehydrogenated and decarboxylated to form ribulose-5-phosphate. As a result of these oxidation processes a molecule of CO_2 is released and at each step NADP is reduced to $NADPH_2$. The subsequent reactions involve the cycling back of ribulose-5-phosphate to glucose-6-phosphate. The influence of herbicides on glycolytic and PPP reactions has been determined by feeding plant tissue with glucose labelled with ^{14}C in the C-1 (aldehyde) or C-6 positions. The occurrence of ^{14}C in the CO_2 or metabolic products indicates the nature of the reaction pathway and the possible sites of herbicide action.

In considering the effect of the various groups of herbicides on respiratory metabolism the amides, nitrophenyl ethers, triazines, pyridines, pyridazines and pyrimidines have been omitted since their major sites of action lie elsewhere.

The action of herbicides on respiratory metabolism

HALOALKANOIC ACIDS

These compounds appear to have relatively little effect on respiratory metabolism though conflicting views are evident from the literature. Foy and Penner (1965) found that of the chloro-aliphatic compounds which they investigated, only **TCA** had a noteworthy effect on succinate oxidation by cucumber mitochondria; **dalapon** failed to inhibit succinate oxidation even at concentrations of 10^{-2} and 10^{-3} M. Similar findings for dalapon have been reported by other workers (e.g. Ingle and Rodgers, 1961; Ross, 1966; Jain *et al.*, 1966; Lotlikar *et al.*, 1968).

The effect of dalapon (500–1500 mg/l *in vitro* and 16–46 kg/ha *in vivo*) on oxidative phosphorylation in mitochondria isolated from lupin, pea and cabbage has been investigated by Deeva *et al.* (1972). Greatest inhibitions of phosphorylation were noted in the more susceptible pea and lupin and less in the more resistant cabbage. Dalapon increased the ATPase activity of mitochondria in all species and the authors concluded that the activity of this herbicide could be explained by the uncoupling of oxidative phosphorylation combined with the inhibition of synthesis of pantothenic acid which participates in the synthesis of coenzyme A (CoA). This compound is an important intermediate in respiratory metabolism.

The effect of dalapon on glycolysis is somewhat controversial. Bourke *et al.* (1964) found that dalapon inhibited glycolysis in peas; Jain *et al.* (1966), however, found no such effect. The latter studied the effect of dalapon on the utilisation of 1-^{14}C-glucose, 6-^{14}C-glucose and

U-^{14}C-glucose by shoots and roots of one-week-old barley seedlings. No differences were found in the C-6/C-1 ratio or $^{14}CO_2$ release due to dalapon treatment; they concluded that the PPP was not favoured to the exclusion of the Krebs cycle. Jain *et al.* (1966) and Ross (1966) also concluded that dalapon did not inhibit activity; however these results are at variance with the findings of Oyolu and Huffaker (1964) (who found evidence to suggest that dalapon reduced the level of CoA in wheat) and also the results of Deeva *et al.* (1972) noted earlier.

PHENOXYALKANOIC ACIDS

There is considerable evidence that these compounds act as uncouplers and inhibitors of oxidative phosphorylation (Kirkwood, 1976) and Moreland (1980) has tentatively placed them in the group known as inhibitory uncouplers.

Uncoupling activity has been reported for **2,4-D** using mitochondria isolated from rat liver (Brody, 1952), lupin (Dow, 1952), etiolated soybeans (Switzer, 1957), cabbage (Lotlikar, 1960), cauliflower and red beet (Wedding and Black, 1962) and maize (Bottrill, 1965). Conversely, Baxter (1967) found no effect of 10^{-3} M 2,4-D on respiration or phosphorylation of mitochondria isolated from dark-grown soybean hypocotyls.

While most work has been carried out with 2,4-D the effect of certain other phenoxyalkanoic acid compounds on respiratory metabolism has also been reported. For example Stenlid and Saddik (1962) demonstrated uncoupling in cucumber (*Cucumis sativus*) mitochondria treated with **dichlorprop** and Smith and Shennan (1966) found that **MCPA** and **MCPB** both inhibit endogenous respiration of *Aspergillus niger*, MCPB being the more inhibitory. The effect of a range of phenoxyacetic and phenoxybutyric acid compounds on oxidative phosphorylation of mitochondria isolated from hypocotyls of broad bean (*Vicia faba*) has been studied by Matlib (1970). Uncoupling at relatively low concentrations was found in the case of **2,4-DB**, **2,4,5-T** and **4(2,4,5-TB)** and at high concentrations all the test herbicides inhibited respiration (Matlib *et al.*, 1972). *In vitro*, phenoxybutyric acid compounds were found to be more toxic as uncouplers than the corresponding acetic acid analogues and uncoupling activity was enhanced by chlorine substitution. Bovine serum albumin (BSA) was used to maintain mitochondrial integrity during these studies and there was evidence of a higher degree of binding of butyric than acetic compounds to BSA (Matlib *et al.*, 1971); the possible significance of this finding is discussed elsewhere.

The reaction of mitochondria and other organelles of the epidermal cells of *Rhoeo discolor* and *Eichhornia crassipes* to 2,4-D has been examined *in vivo* by George (1971). Extreme vesiculation and swelling was accompanied by aggregation. These changes are irreversible and are attributed to dispersal of lipid micelles from the mitochondrial membrane allowing increased intrusion of water. Following dispersal of the fatty materials from the membrane, a protective protein layer prevented further toxic activity.

Mitochondrial respiration is also susceptible to phenoxypropionic acid compounds. Olofinoba and Fawole (1976) found that **mecoprop** inhibited O_2 uptake and ATPase activity of mitochondria isolated from the cotyledon, hypocotyl or root of 17-day-old seedlings of *Theobroma cacao*.

The activity of 2,4-D on glycolysis and associated metabolic processes has been examined by a number of workers. For example Humphreys and Dugger (1957a) found that 2,4-D stimulated the respiration of etiolated pea (*Pisum sativum*) seedlings and enhanced the metabolism of 1-^{14}C-glucose via the PPP (Humphreys and Dugger, 1957b), possibly reflecting blockage of the synthetic metabolic pathways (Humphreys and Dugger, 1958, 1959). Similar results were found following treatment of 3-day-old etiolated corn (*Zea mays*) seedlings with 2,4-D (Black and Humphreys, 1962). Bourke *et al.* (1962), investigating the effect of a range of phenoxy compounds on absorption of ^{14}C-glucose by pea root tips, concluded that effects on oxidation were either due to a stimulation of glycolysis by low concentrations of 2,4-D, or to a reduction in glycolysis and PPP activity at higher concentrations. Kim and Bidwell (1967) found that 5×10^{-5} M 2,4-D had no effect on pyruvic acid decarboxylation but inhibited carboxylation of pyruvic acid and the entry into the Krebs cycle of acetate derived from pyruvate. They concluded that 2,4-D had an inhibitory effect on co-carboxylase or α-lipoic acid metabolism but not on CoA.

AROMATIC ACIDS

In general these compounds appear to act as inhibitors of oxidative phosphorylation. For example, when using succinate as substrate, **2,3,6-TBA** at a concentration range of 10^{-5}–10^{-2} M, inhibited O_2 uptake by mitochondria isolated from 5-day-old cucumber cotyledons by 30–100% respectively; greater inhibitions were obtained using α-keto-glutarate as substrate (Foy and Penner, 1965). Similarly, using citrate as substrate and cabbage mitochondria, Lotlikar *et al.* (1968) found that inhibition of oxygen uptake and phosphorylation increased to 82% and 100% at 1.2×10^{-3} M 2,3,6-TBA. The relevance *in vivo* of these findings achieved with relatively high concentrations applied *in vitro* must be a matter of conjecture.

NITRILES

The hydroxybenzonitrile group of herbicides which includes **ioxynil** and **bromoxynil** possess activity as uncouplers of oxidative phosphorylation; **dichlobenil**, the other major compound in the group, has little or no activity as an uncoupler though the 3-hydroxy and 4-hydroxy derivatives may be active.

While *p*-hydroxybenzonitrile *per se* is not an effective uncoupler, substitution of an iodine atom in the 3-position or dihalogenation in the 3,5-positions completely alters the situation. Kerr and Wain (1964) found that di-iodo, dibromo and dichloro analogues were all more active as uncouplers than were DNC (dinitrocresol) or DNP and Parker (1965) working with rat-liver mitochondria confirmed that the relative activity of these compounds was di-iodo>dibromo>dichloro, coinciding with their relative activity. Evidence of uncoupling activity of bromoxynil was also reported by Moreland and Blackmon (1970) though Foy and Penner (1965) found that while dichlobenil appeared to act by disrupting the phosphorylation mechanism of cucumber mitochondria, ioxynil inhibited TCA oxidases leading to a reduction rather than increase in O_2 uptake.

In general, however, the evidence is strongly in favour of uncoupling activity. The effect of ioxynil on the respiratory electron transport chain of mitochondria isolated from roots of *Vicia faba* was studied by Paton and Smith (1965, 1967). Oxygen uptake was stimulated at low concentrations (5×10^{-8}–5×10^{-7} M) and inhibited at high values (5×10^{-6}–5×10^{-3} M). They found evidence to suggest that ioxynil acted at a site between flavoproteins and cytochrome b, possibly at ubiquinone. Uncoupling activity was also reported by Ferrari and Moreland (1967, 1969) who investigated the effect of several hydroxybenzonitriles on the respiration of white potato tubers (*Solanum tuberosum*). While uncoupling activity was confirmed as I>Br>Cl only 10% separated ioxynil from its analogues indicating that halogen substitution may play only a minor role.

Dichlobenil is not generally regarded as an uncoupler *per se* (Wit and van Genderen, 1966), though the 3-hydroxy and 4-hydroxy derivatives may act as uncouplers in plants which can degrade dichlobenil into these metabolites (Verloop, 1972). Moreland *et al.* (1974) found that unlike dichlobenil the 3-hydroxy and 4-hydroxy products interfere strongly with electron transport and phosphorylation in both isolated spinach chloroplasts and mung bean mitochondria.

ANILIDES

The effect of **propanil** on respiration has been examined by several workers and while this herbicide appears to inhibit respiration, the results are somewhat variable. For example, Sierra and Vega (1967) reported that treatment of barnyard grass with 100 p.p.m. propanil increased respiration after 1 h, while 1000 p.p.m. propanil reduced respiration. Further evidence of inhibition of oxidative phosphorylation was obtained following propanil treatment of soybean mitochondria (Hofstra and Switzer, 1968), yeast (Inoue *et al.*, 1967), excised soybean hypocotyls (Gruenhagen and Moreland, 1971), rice and water foxtail (*Alopecurus geniculatus*) (Shirakawa and Ueki, 1974). Respiration appears to be affected by **alachlor** (Bardzik and Marsh, 1974) or **propachlor** (Duke *et al.*, 1975), but the effects do not appear to represent an important mechanism of action.

NITROPHENOLS

Most work on uncoupling of oxidative phosphorylation has been carried out using nitrophenols. Dinitrophenol (**DNP**) is generally regarded as a classical uncoupler though Moreland (1980) points out that nitrophenols should more properly be regarded as inhibitory uncouplers. At low molar concentrations they act as uncouplers while at higher molar concentrations they behave as inhibitors of electron transport.

The stimulatory effect of nitrophenols on respiration has been known since the mid-1930s (Ehrenfest and Ronzoni, 1933; Dodds and Greville, 1934) and the differential effect of low and high concentration was probably first reported by Bonner (1949) using *Avena* coleoptiles. Concentration-dependent responses on O_2 uptake by plant mitochondria have been reported by Foy and Penner (1965) and Bottrill and Hanson (1969). Similar responses have been found for other substituted phenols. For example Gaur and Beevers (1959) studied the effect of nitro-, chloro- and bromo-substituted phenols on the respiration of carrot tissue and found the response to be highly concentration dependent. Further, disubstitution was more effective than monosubstitution in inducing uncoupling, the order of effectiveness being di>*para*>*meta*>*ortho*. The uncoupling action of **dinoseb** on respiration of tomato leaf discs was later reported by Wojtaszek *et al.* (1966).

Knowledge of the site and mechanism of action of nitrophenols results largely from work with DNP. Most workers have favoured the view that DNP promotes the hydrolysis of an intermediate nonphosphorylated high-energy compound (Borst and Slater, 1961; Stoner *et al.*, 1964). Racker (1961) believes that DNP dissociates the process of phosphorylation from electron transport, a view supported by Pinchot (1963, 1967). They envisaged that DNP prevents the association of the uncoupling enzyme with coupling particles, thus preventing the formation of an enzyme–particle complex. The hypotheses currently proposed to explain uncoupling have been reviewed by Moreland (1980). They include the classical uncoupler-induced hydrolysis of a high-energy intermediate; conformational change and altered function of coupling proteins; energy-linked transport of the uncoupler anion into, and passive diffusion of the protonated species out of the mitochondria; acid- or base-catalysed hydrolysis by uncouplers of an intermediate of oxidative phosphorylation in a lipophilic region of the membrane and increase of membrane conductance and collapse of the transmembrane proton gradient by uncouplers.

The action of **PCP** (pentachlorophenol) on rat-liver mitochondria has been studied by Weinbach and Garbus (1964, 1965). They found that PCP bonds have intact mitochondria and the bonding is not readily broken except by treatment of the protein complex with 0.1 N NaOH. They found that BSA restored the morphological integrity of mitochondria swollen by treatment with PCP and restored the capacity for oxidative phosphorylation, possibly by tightly binding the uncoupling phenol and removing it from the mitochondria. They believe that mitochondrial proteins may possess a similar capacity to bind lipophilic reagents, the degree of binding being influenced by pH, concentration of phenols, mitochondria or mitochondrial protein. They concluded that the uncoupling effect of phenols may be due to such protein/phenol interaction.

The cause of selectivity of substituted phenols to tomato plants has been examined by Wojtaszek (1966) using twenty-four crop and weed species. Considerably greater levels of ^{32}P were accumulated in leaf discs isolated from plants which were resistant rather than susceptible to dinoseb. Wojtaszek suggested that susceptibility to dinoseb depends upon the level of ATP in tissues and this in turn, depends on its formation or storage in the plant.

NITROANILINES

Various authors appear to have different views on the importance of respiratory metabolism as a site of action of compounds of this group. In a number of cases the concentrations of **trifluralin** which inhibit growth fail to affect respiration (Talbert, 1967; Negi *et al.*, 1967;

Hendrix and Muench, 1969). However Negi *et al.* (1968) reported that 10^{-4} M trifluralin (and **nitralin**) inhibited O_2 uptake and P_i esterification by mitochondria isolated from several species. Gruenhagen and Moreland (1971) found however, that 0.2 mM trifluralin (*in vitro*) did not affect the ATP level of soybean hypocotyls. Subsequently Moreland, Farmer and Hussey (1972a) reported on the effects of **benfluralin**, **isopropalin**, nitralin, **oryzalin** and trifluralin on mitochondrial activities of mung bean (*Vigna radiata*). All of these compounds except nitralin markedly inhibited phosphorylating electron transport (state 3 respiration) with malate as substrate and most were strong inhibitors of NADH and succinate oxidation. Certain compounds may have interfered with the energy-generating pathway, but an accurate evaluation of their effects was not possible. They concluded that provided that these compounds partitioned into the organelles *in vivo*, ATP synthesis could be a major mechanism of action of the 2,6-dinitroanilines.

Wang, Grooms and Frans (1974) reported on the response of soybean mitochondria to the substituted dinitroanilines. They found that oryzalin and nitralin uncoupled succinate oxidation at the ADP-limited second state 4 (IV_2) while trifluralin inhibited the second state 3 (III_2) oxidation, suggesting that it acted as an energy-transfer inhibitor. They found that DNP overcame inhibition by trifluralin indicating that the locus of inhibition of trifluralin is at or near to, the site of action of DNP in the energy-transfer system. Wang *et al.* (1974) also concluded that oryzalin acted at a site on the energy-transfer system though closer to the formation of H_2O than did DNP or trifluralin; nitralin appeared to act earlier than trifluralin on the respiratory chain. **Profluralin** (1×10^{-5} M) also strongly inhibited state 3 respiration of isolated bean hypocotyl and rat-liver mitochondria, suggesting considerable herbicidal potential (Pillai and Davis, 1974).

CARBAMATES

It is unlikely that inhibition of oxidative phosphorylation is the primary site of action of these compounds, but rather that such activities may indirectly affect RNA and protein synthesis which is generally regarded as the primary mechanism. However, carbamates such as **propham** and **chlorpropham** can be highly effective inhibitors of ATP synthesis. Lotlikar *et al.* (1968) using mitochondria isolated from cabbage, found that propham (5×10^{-4}–5×10^{-3} M) inhibited O_2 uptake by 13–73% respectively and P_i esterification by 53–100% respectively; chlorpropham was even more effective. Gruenhagen and Moreland (1971) reported a reduction in the ATP synthesis following treatment of excised soybean hypocotyls with 0.6 mM chlorpropham.

The effect of chlorpropham and its metabolites on the respiration and phosphorylation activities of plant mitochondria isolated from maize, cucumber and soybean has been studied by Still, Rusness and Mansager (1974) and Rusness and Still (1974). They found that 0.1 mM chlorpropham and its 2-hydroxy metabolite inhibited NADH oxidation by 30% whereas only the latter inhibited NADH-linked ATP formation (85-100%); the 4-hydroxy chlorpropham had no effect on respiration, phosphorylation or ATPase activity. DNP and the 2-hydroxy metabolite had similar effects on respiration, phosphorylation and the ATPase activity.

In vitro and *in vivo* treatment with 25, 250 p.p.m. and 1000 p.p.m. **barban** inhibited O_2 uptake and esterification of P_i by mitochondria isolated from 2-day-old etiolated seedlings of wild oat; there was also evidence of uncoupling activity (Ladonin, 1966, 1967a; Ladonin and Svittser, 1967). However, Kobayashi and Ishizuka (1974) found no reduction in respiration 1 day after treatment of oats (susceptible) or wheat (resistant) and concluded that the selective action of barban was due to a marked inhibition of protein and RNA syntheses in the susceptible species.

A similar picture emerges from studies of the mode of action of **asulam** in bracken. RNA and protein synthesis are apparently primary target sites but there is evidence that effects on respiratory metabolism may also be important (Veerasekaran, Kirkwood and Fletcher, 1977). *In vitro* treatment of excised bud tissue with 10–1000 p.p.m. asulam inhibited respiration by up to 50% after 72 h while 5 and 10 p.p.m. asulam stimulated O_2 uptake. *In vivo* treatment with 2,4,6 and 8 kg/ha resulted in inhibition of up to 100% in the respiration of rhizome buds 6 and 12 months after treatment. Similar results were obtained by Yukinaga *et al.* (1973).

THIOCARBAMATES

These compounds may affect a number of physiological processes in plants including photosynthesis, nucleic acid and protein synthesis, and respiratory metabolism. The importance of effects on respiration appear to differ according to compounds and species and their relevance to the overall herbicidal action is difficult to assess.

Ashton (1963) reported that relatively low concentrations of **EPTC** stimulate respiration of excised embryos of maize and mung bean (expressed on a fresh weight basis), while relatively high concentrations severely inhibit O_2 and P_i uptake. Lotlikar *et al.* (1968) found that 1×10^{-3} M EPTC inhibited O_2 uptake (77%) and P_i esterification

(100%), though Gruenhagen and Moreland (1971) reported that EPTC (0.6 mM) and CDEC (sulfallate) (0.2 mM) failed to significantly reduce ATP levels in excised soybean hypocotyls pointing out that the herbicides may not have penetrated the tissues. They also suggest that lack of sensitivity of the techniques used may have failed to detect small but possibly important changes in ATP levels.

The mode of action of benzyl *N,N*-dialkyldithiocarbamates has been examined by Wakamori (1973) who found that they did not increase ATPase activity in cauliflower mitochondria but did increase the respiration of rice and barnyard grass (*Echinochloa crus-galli*); oxidative phosphorylation in cauliflower mitochondria was strongly inhibited. He concluded that the main mode of action may be to increase respiration via the uncoupling of oxidative phosphorylation. Uncoupling activity in mung bean mitochondria has also been attributed to the effect of **di-allate** but an additional primary site may lie at the malate–cytochrome c reductase step (Tetley, Ikuma and Hayashi, 1975).

UREAS

Many phenylurea herbicides are known to interfere with ATP synthesis by oxidative phosphorylation (Moreland, 1974) but the effects vary according to compound and concentration. For example Foy and Penner (1965) reported that **diuron** (1.46×10^{-6}–1.46×10^{-4} M) had little effect on O_2 uptake while Pillai and Davis (1974) found that this herbicide (5×10^{-7} M) increased respiration by nearly 50% after 80 min. **Monuron** appears to be more inhibitory, Lotlikar *et al.* (1968) reporting that at 5×10^{-4}–3×10^{-3} M, O_2 uptake and P_i esterification by cabbage mitochondria were inhibited by up to 25 and 63% respectively. Likewise incorporation of 10 p.p.m. **fluometuron** in the nutrient solution inhibited respiration of *Abutilon theophrasti* seedlings by 75% after 7 days (Potter and Wergin, 1974).

The action of substituted 1,2,3-thiadiazidylphenylureas on isolated mitochondria has revealed uncoupling activity at a concentration of 9 μM, while at higher concentrations they may act as energy-transfer inhibitors (Hauska *et al.*, 1975).

However, these effects might almost be regarded as incidental to the major mechanism of action of the phenylurea compounds, namely on the electron transport systems associated with photophosphorylation.

TRIAZINES

The effect of triazine herbicides on respiratory activity also appears to vary according to a number of factors including concentration, treatment duration, species, stage of growth and the specific compound under test. Increases in respiration have been reported following treatment of bean plants with **atrazine** (10 p.p.m.) (Ashton, 1960) and a variety of crop plants (1 kg/ha for 1–24 h) (Funderburk and Davis, 1963); in the latter case inhibition of respiration was noted in the long term (7–11 days). Similar inhibition has been noted for atrazine by Olech (1966).

Many authors have failed to find any effect of triazine compounds on respiration. For example Ashton and Uribe (1962) reported no effect on respiration of bean embryos by 10 p.p.m. atrazine for 48 h following an 8 h treatment of bean seeds. Treatment of 5–6-day-old seedlings of bean, sunflower, barley and *Agrostemma githago* in the dark with atrazine (0.5, 0.65 and 1.6×10^{-7} mol/dm^2 of the leaf surface or to the roots) failed to inhibit respiration. Similarly no effects on respiration were reported by Krishning (1965), Davis (1968), or Gruenhagen and Moreland (1971). Treatment of cotton seedlings with **prometryne** resulted initially in inhibition of respiration but recovery occurred later (Nasyrova *et al.*, 1968).

Variation is also evident in the response of isolated tissue or mitochondria to triazine treatment. Stimulation of respiration of excised barley roots by **simazine** was noted by Palmer and Allen (1962) though these authors later reported differential effects with various substrates (Allen and Palmer, 1963). Atrazine (4×10^{-4} M) had no effect on O_2 uptake or the P/O ratio of corn shoot mitochondria (Davis, 1968), though 1×10^{-5}–6×10^{-5} M simazine inhibited O_2 uptake and P_i esterification by cabbage mitochondria by up to 10% and 21% respectively (Lotlikar *et al.*, 1968). Results similar to the latter were obtained for atrazine (Voinilo *et al.*, 1967), the site of action occurring between cytochromes b and c (Voinilo *et al.*, 1968).

Further evidence of the nature of the inhibitory action of simazine comes from the work of Tluczkiewicz (1974) who found that *in vivo* treatment of barley (0.15% simazine) at the heading stage, decreased succinate oxidase activity and slightly increased catalytic ATPase activity in the mitochondria of germinating embryos (cv. Damazy). Metcalf and Collin (1978) treated celery cell suspension cultures with a range of concentrations of simazine (0.01–10 mg/l) and obtained results which suggested that it may be acting as a respiratory uncoupler.

Several workers have reported on the inhibitory effect of prometryne on mitochondrial respiration. Truelove and Davis (1969) found that state 3 respiration of maize shoot and rat-liver mitochondria was inhibited by prometryne, the effect being relieved by uncoupling agents. McDaniel and Frans (1969) reported similar effects of prometryne on soybean hypocotyl mitochondria, inhibition being overcome by the uncoupling action of DNP. Thompson *et al.* (1969) found that both state 3 and state 4 respiration were inhibited by prometryne, the former effect being partially relieved by uncoupling agents. They suggested that the active site occurred in the sequence of reactions leading to ATP synthesis, i.e. that prometryne acted as an energy-transfer inhibitor. **Ametryne** has also been reported to inhibit respiration of isolated mitochondria (Thompson *et al.*, 1970).

Application of prometryne and the insecticide phorate on bean (*Phaseolus vulgaris*) resulted in the inhibition of state 3 respiration of mitochondria isolated from 6-day-old bean hypocotyls, inhibition increasing with concentration (0.1–10 μM) (Parks, Truelove and Buchanan, 1971, 1972). However, simultaneous application of these compounds failed to produce interactions affecting state 3 respiration.

It has been suggested that inhibition of respiration by triazine compounds is an indirect effect of a reduction in the level of assimilates resulting from an inhibition of photosynthesis. This view has been expressed by Olech (1966) reporting on the inhibitory effect of atrazine. Similarly Merezhinskii and Lapina (1974), using a number of species, noted that 2,4-D (5 μM) inhibited ATP synthesis in pea mitochondria, the effect being reinforced by sublethal concentrations of simazine. They concluded that simazine reduced the accumulation of reduction products in the light, hindering electron transport, while 2,4-D intensified this effect and disrupted electron transport in mitochondria.

HETEROCYCLIC NITROGEN COMPOUNDS (UNCLASSIFIED)

Variable effects of **aminotriazole** on respiratory metabolism have been reported and these appear to depend on the concentration applied and the duration of treatment. For example Hall *et al.* (1954) found that *in vivo* treatment of cotton plants with aminotriazole (10^{-3}–10^{-1} M) resulted in increases of 20–60% in the level of O_2 uptake within 3 h of treatment; after longer periods respiration was inhibited by the two higher concentrations. Miller and Hall (1957) reported a similar pattern following spraying of cotton leaves with 4.8×10^{-2} M aminotriazole. Respiration was enhanced by application of aminotriazole to grass seedlings (Russell, 1957), wheat or bean plants (Wort and Shrimpton, 1958), maize (McWhorter, 1959), or maize (via roots only) (McWhorter and Porter, 1960). On the other hand, absence of effect (Lotlikar *et al.*, 1968) or inhibition of oxidation (Foy and Penner, 1965) or ATP synthesis (Gruenhagen and Moreland, 1971) have also been reported.

ORGANOARSENIC COMPOUNDS

There is some evidence that **MSMA** (monosodium methylarsonate) may act as an inhibitory uncoupler since a low concentration (10^{-7} M) increased respiration of rat-liver mitochondria (but not bean mitochondria) while a relatively high concentration (10^{-3} M) inhibited state 3 respiration of bean and rat-liver mitochondria (Pillai *et al.*, 1973).

Inhibition of photosynthetic systems

THE PHOTOSYNTHETIC PROCESS

The synthesis of carbohydrates from CO_2 and H_2O in the presence of light is unique to green plants. The process takes place in chloroplasts which in higher plants, are disc-shaped organelles (4–10 μm thick) found in the palisade and mesophyll cells of the leaf. The chlorophyll is located in structures within the chloroplasts called grana (0.4 μm in diameter) which are embedded in a matrix known as the stroma. Photosynthesis involves two types of reactions, the light reactions which take place in light and the dark reactions which can take place in the absence of light. The former takes place within the grana while the enzymes involved in CO_2 fixation (the dark reactions) are believed to be confined to the stroma. The products of the light reactions, $NADPH+H^+$ and ATP are required to reduce CO_2 to carbohydrate in the dark reactions. The processes are summarised in the following unbalanced equations, each of which involves a complex series of reactions:

$$H_2O+ADP+P_i+NADP^+ \xrightarrow[\text{energy}]{\text{light}} O_2+ATP+H^++NADPH \qquad (1)$$

$$CO_2+ATP+NADPH+H^+ \longrightarrow (CH_2O)+ADP+P_i+NADP^+ \qquad (2)$$

The process of phosphorylation associated with photo-induced electron transport is shown schematically in fig. 5.3. The products of these light reactions are O_2 produced by the oxidation of water and ATP and NADPH, which result from the transmission of electrons from the ultimate donor (H_2O) to the ultimate acceptor $NADP^+$ via a series of intermediate electron acceptors and involving Photosystems I and II (PS I and PS II respectively) (non-cyclic photophosphorylation). Phosphorylation associated with the cyclic flow of electrons involving PS I results in the production of ATP only and is termed cyclic photophosphorylation.

Herbicides may inhibit photosynthesis in several ways (Moreland and Hilton, 1976):

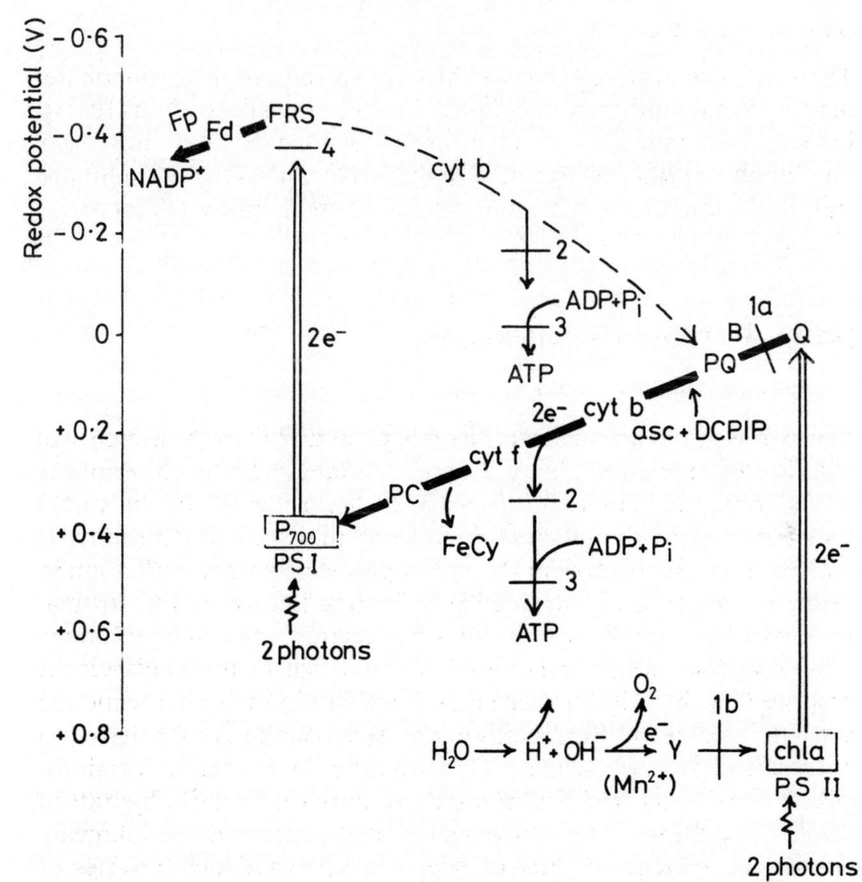

Fig. 5.3 A schematic representation of non-cyclic and cyclic photophosphorylation reactions which are believed to occur in chloroplast lamellae. Open arrows indicate light reactions, solid arrows indicate dark reactions and the narrow dashed line represents the cyclic pathway. The abbreviations used: Y (postulated electron donor), PS I (Photosystem I), PS II (Photosystem II), Q (unknown primary electron acceptor for PS II), B (secondary acceptor for PS II), chla (chlorophyll a), PQ (plastoquinones), cyt b (b-type cytochromes), cyt f (cytochrome f), PC (plastocyanin), P_{700} (reaction centre chlorophyll of PS I), FRS (ferredoxin-reducing substance), Fd (ferredoxin), Fp (ferredoxin–NADP oxido-reductase), FeCy (ferricyanide), asc (ascorbate) and DCPIP (2,6-dichlorophenolindophenol). The postulated sites of action by herbicides are indicated as 1a and b (electron transport inhibitors), 2 (uncouplers), 3 (energy-transfer inhibitors) and 4 (electron acceptors) (after Moreland and Hilton, 1976).

(1) by interference with the reproduction, development, structure and integrity of the chloroplasts
(2) by the light reactions involved in the conversion of light energy to chemical energy
(3) by the biosynthetic pathways concerned with the production of photosynthetic products.

UNCOUPLERS AND INHIBITORS OF PHOTOPHOSPHORYLATION

The action of herbicides as inhibitors of the light reactions has received considerable attention due largely to the successful development of a technique using isolated chloroplast systems. Various types of inhibitory action have been identified depending on their effects as follows: (a) electron transport inhibitors, (b) uncouplers, (c) energy transfer inhibitors, (d) inhibitory uncouplers (multiple forms of inhibitory action) and (e) electron acceptors (Moreland and Hilton, 1976). These authors have defined the terms which can be used to describe the photochemical reactions of isolated chloroplasts.

The term 'basal electron transport' describes the reductive reaction in which water serves as the electron donor and $NADP^+$ or ferricyanide acts as the electron acceptor; the reaction is carried out in the absence of phosphorylating agents (i.e. ADP, P_i and magnesium). The degree of reduction of the electron acceptor can be assessed using a spectrophotometer and effects on oxygen evolution determined polarographically. When these reactions are carried out in the presence of the phosphorylating agents, electron transport is coupled to phosphorylation and is termed non-cyclic photophosphorylation. Variations in the choice of electron acceptor or donor enable the effect of inhibitors on various mechanisms to be studied and inhibition sites to be by-passed. Effects on cyclic photophosphorylation are carried out using phenazine methosulphate (PMS) as the electron mediator under an argon gas phase. Under these conditions ATP only is produced due to electrons emitted from PS I being passed from the ferredoxin-reducing substance (FRS) to cytochrome b rather than to ferredoxin and $NADP^+$.

Electron transport inhibitors

Compounds of this type act by removal or inactivation of one or more of the intermediate electron transport carriers. Herbicides which act in this way include chlorinated phenylureas, bis-carbamates, chlorinated triazines and substituted uracils; these compounds appearing to act at a single site on the photochemical pathway (Moreland, 1967, 1969). Other inhibitors of electron transport include pyridazinones (Hilton *et al.*, 1969), diphenyl ethers (Moreland *et al.*, 1970), 1,2,4-triazinones (Draber *et al.*, 1969), azido-s-triazines (Gabbott, 1969), cyclopropane carboxamides (Hopkins *et al.*, 1967), *p*-alkylanilides (Bellotti *et al.*, 1968) and *p*-alkylthioanilides (Baruffini *et al.*, 1970). Their action includes inhibition of basal electron transport, methylamine-uncoupled electron transport and non-cyclic electron transport with water as electron donor and ferricyanide or $NADP^+$ as electron acceptor. Coupled photophosphorylation is inhibited by these herbicides to the same degree as the reductive reaction, but cyclic photophosphorylation is not inhibited to the same extent. Photoreduction of $NADP^+$ with electron acceptors which by-pass PS II is not affected. The site of action of such compounds (exemplified by diuron) may be at sites 1a and 1b.

Uncouplers

These are defined as compounds which dissociate electron transport from ATP formation, thus inhibiting phosphorylation but not electron transport. Characteristically uncouplers, in the absence of a component(s) required for photophosphorylation (ADP, P_i or magnesium), should stimulate the rate of electron flow to at least the rate obtained on addition of the missing component. They are believed to act by promoting the breakdown of high-energy intermediates and the postulated sites are shown in fig. 5.3.

Energy-transfer inhibitors

Energy-transfer inhibitors act on the phosphorylation process, by inhibiting both electron transport and ATP synthesis in coupled systems. Typically the inhibition of electron flow, though not ATP formation, is released by addition of an appropriate uncoupler. They are believed to act by preventing the formation of high-energy intermediates and thus inhibiting electron flow. The 1,2,3-thiadiazolylphenylureas alone have been found to act as energy-transfer inhibitors under certain conditions (Hauska *et al.*, 1975).

Inhibitory uncouplers

Such compounds have more than one site of action, since they may inhibit electron transport or act as uncouplers. Characteristic inhibitory uncoupler effects include (1) inhibition of basal, methylamine-uncoupled and coupled electron transport with ferricyanide acting as electron acceptor and water as electron donor, (2) inhibition of coupled non-cyclic and cyclic photophosphorylation occurs in a similar fashion to that associated with uncouplers.

Herbicides which act in this way include the *N*-phenylcarbamates, acylamides, 3,5-dihalogenated 4-hydroxybenzonitriles, substituted imidazoles and substituted benzimidazoles (Moreland, 1967, 1969), certain heterocyclic compounds (Büchel and Draber, 1969), substituted 2,6-dinitroanilines (Moreland *et al.*, 1972b), pyriclor (Meikle, 1970) and substituted 1,2,4-thiadiazoles (Bracha *et al.*, 1971). Moreland and Hilton (1976) suggest that dinitrophenols and alkylated dinitrophenols which are normally regarded as uncouplers may justifiably be considered as inhibitory uncouplers.

Electron acceptors

Such compounds are able to compete with some component of the electron transport pathway and subsequently undergo reduction. The bipyridylium herbicides, such as diquat and paraquat, are able to compete with FRS for electrons. In intercepting electrons from PS I they essentially shunt the electron transport chain. Phytotoxicity is evident in quaternary salts which possess nitrogen atoms in the 2,2′-, 2,4′- or 4,4′-positions; bipyridyls joined at the 2,3′- or 3,3′-position are not herbicidally active.

STRUCTURE–ACTIVITY RELATIONS

Structure–activity correlations of herbicides which inhibit photo-induced electrons have recently been reviewed by Moreland (1980). The value of such studies lies in elucidating information on (1) the substituents required for maximum herbicidal activity, (2) the relationship of chemical and physical properties of the herbicide with the inhibitory action, (3) the environment in which herbicides operate most efficiently and (4) the existence and nature of interactions between substituents of inhibitors and the presumed receptors in the chloroplasts.

In vivo studies using isolated chloroplasts have revealed not only a range of possible herbicide actions but also a variation in their efficiency as indicated by the wide range of toxicity (I_{50} values) found in the literature. However, these studies suffer from the shortcomings that they give no indication of herbicidal efficiency under *in vivo* conditions, when the herbicide may fail to accumulate at the specific site of action, and further that the herbicide may become associated with cell components which are not involved in the photochemical reactions and thus not be available. It should be borne in mind that the results of such studies do not provide an estimate of the herbicide concentration within the chloroplast if intact organelles are used or within the thylakoid if chloroplast fragments are used. Penetration to the active site will depend on a number of characteristics including partitioning properties (hydrophilic/lipophilic balance), molecular configuration (steric relations), resonance, keto–enol tautomerism, *cis* or *trans* relations of the amide hydrogen and carbonyl oxygen and the critical charge properties of particular substituent groups which participate in intermolecular interactions at active centres (Moreland, 1969).

Most herbicides can be readily eluted from chloroplasts, reversing their inhibitory action and indicating that only weak bonds are involved in the interaction between the compound and the reactive components of the electron transport system. It has been demonstrated that structural requirements may include the need for a free and sterically unhindered amide or imino hydrogen (e.g. phenylamide herbicides), suggesting the involvement of H bonds (Moreland, 1969) while the requirement of a N+H group attached to an electron-deficient sp^2 carbon atom has been noted by Hansch (1969) (fig. 5.4) for a number of herbicides (acylanilides, uracils, benzimidazoles,

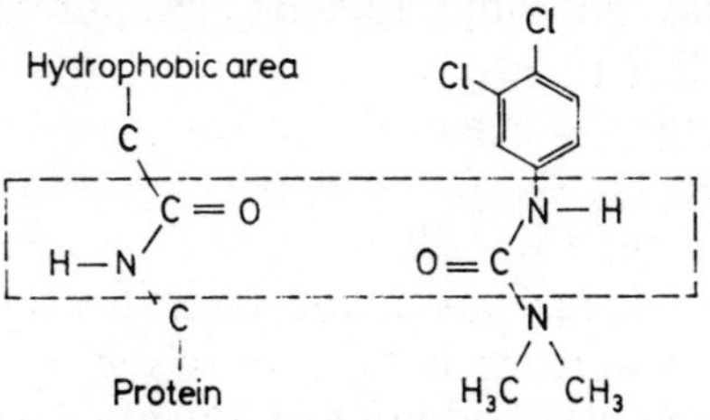

Fig. 5.4 The possible interaction between diuron and chloroplast membrane protein suggested by the scheme of Hansch (1969).

imidazoles and triazines). Binding is believed to occur between the lone-pair electron of the herbicide nitrogen and the electron-deficient carbonyl of the amide group of a strategically located chloroplast protein. This binding postulate was later modified to involve consummation by a hydrophobic interchange or a charge-transfer interaction between the free electron pair of the amide or heterocyclic nitrogen and the active centre (Draber *et al.*, 1974; Trebst and Harth, 1974). This alteration took into account herbicides which possessed heterocyclic rings, but lacked a free amide hydrogen (e.g. triazinones, pyrolidones, pyridazinones, pyrazolones, triazolones and oxadiazolinones).

The interpretation of results obtained in structure–activity studies has been facilitated by multiple regression analysis, now called quantitative structure–activity relationship analysis (QSAR). It has also been referred to as the 'Hansch approach' after Hansch who first used the analysis with phenylurea and acylanilide derivatives (Hansch and Deutsch, 1966). QSAR analysis has also been used to correlate inhibitory responses obtained with acylanilides (Hansch, 1969), azidotriazines (Gabbott, 1968, 1969) benzimidazoles, imidazoles

(Büchel and Draber, 1969), bis-carbamates (Schultz, 1969), nitrophenols (Trebst and Draber, 1979), phenylureas (Dicks, 1978), pyrimidinones (Gibbons *et al.*, 1976), s-triazines (Gabbott, 1968), triazinones (Draber *et al.*, 1969, 1974; Schmidt *et al.*, 1975) and thiadiazoles (Bracha *et al.*, 1971).

In most regression analysis with different classes of electron transport inhibitors, inhibition has been related to the lipophilicity of the ring substituents; usually electronic contributions are considered to be of minor importance. However, in the case of imidazoles and benzimidazoles, electronic effects were included in terms of p*K* (Büchel, 1972).

Regression analyses have demonstrated that inhibitory action against the Hill reaction can be correlated with physicochemical parameters within a particular group of herbicides; no mathematical correlation has been possible between different chemical families (Büchel, 1972). Failure to develop these correlations may be related to suggestions that while the herbicides have a common moiety that may bind them to the same or closely related site in the shielding protein of PS II, the various hydrophilic side chains, characteristic of the different herbicide families, may associate with different regions of the shielding protein (Moreland, 1980).

The basis of resistance to various herbicide groups has suggested that resistance is not related to differences in uptake, translocation or metabolism but rather to alterations associated with the chloroplasts. Results of comparative studies suggest that the herbicidal binding site is modified with subsequent alterations in the level of affinity for different classes of herbicide. Thus chloroplasts from triazine-resistant plants are relatively resistant to the s-triazines, strongly resistant to triazinones, partially resistant to pyridazinones and uracils, only slightly resistant to ureas and amides and even less to nitrophenols, phenols and benzothiadiazoles (Pfister *et al.*, 1979a). It has been postulated that resistance may be due to modifications or deletions of certain amino acids or segments of the PS II shielding protein which have been detected in at least two species (Arntzen *et al.*, 1979; Pfister and Arntzen, 1979; Pfister *et al.*, 1979b); such polypeptide modification may affect the redox poising of Q and B and the specific binding of herbicides (Moreland, 1980).

INHIBITION OF PHOTOSYNTHESIS – *IN VIVO* STUDIES

All herbicides which inhibit the Hill reaction of isolated chloroplasts also inhibit photosynthesis of intact plants. The technique which has been used in many studies is that developed by van Oorschot and

Belksma (1961). The level of CO_2 utilisation of plants maintained in an enclosed chamber is monitored by infrared gas analysis following application of the herbicide to the nutrient solution. The different types of inhibitory response obtained are shown in fig. 5.5 (Moreland

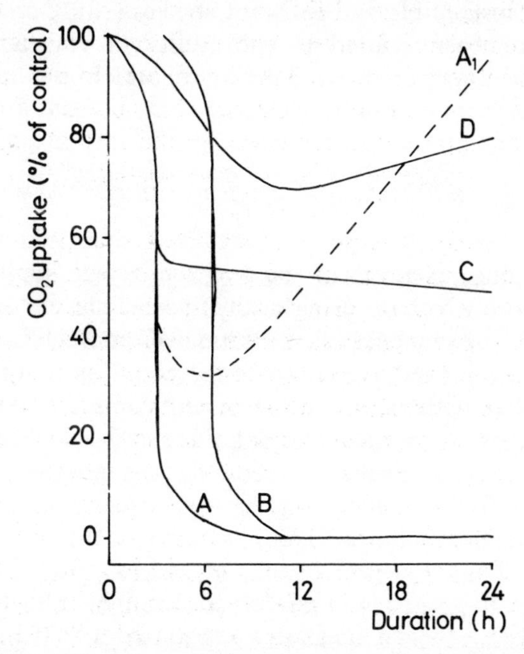

Fig. 5.5 Inhibition of photosynthesis, measured as CO_2 fixation, by inhibitors of chloroplast electron transport. Root exposure to herbicides was continuous (solid line) or temporary (dashed line) (after Moreland and Hilton, 1976). See text for details.

and Hilton, 1976). If the herbicide is readily root absorbed and xylem translocated, photosynthetic inhibition can be detected in 1–2 h, complete inhibition generally being obtained within 6–24 h. Rapid inhibition is typical of s-triazine, phenylurea and uracil herbicides (curve A). Delay in absorption may lead to delay in complete inhibition (e.g. pyridazinone) (curve B). Curves A and B are typical of herbicides classified as electron transport inhibitors which have a single site of action associated with PS II. Curve C is typical of that obtained with ioxynil and PCP when applied to the root. The initial inhibition of CO_2 uptake is not complete possibly due to arrested uptake of these herbicides which are also inhibitory uncouplers of ATP synthesis. If these herbicides are applied directly to the foliage, complete inhibition of CO_2 results. Curve D is representative of the response of resistant plants usually to low concentrations of the herbicide, increase in concentration leading to curve A response. However, in the case of certain herbicides such as propanil, application to root systems even in high concentration results in the D-type curve. It is known that propanil is rapidly hydrolysed in the roots of some species and it may be metabolised as rapidly as it is absorbed. The technique can also be used to examine the selective action of herbicides in susceptible and resistant species (van Oorschot, 1965). Selectivity has been related to the ability of resistant plants to metabolise the herbicides to derivatives which are non-toxic to photosynthesis.

EFFECTS ON THE STRUCTURE, COMPOSITION AND DEVELOPMENT OF CHLOROPLASTS

Herbicides induce changes in the leaf which are similar in many respects to those which occur naturally in senescence (Moreland and Hilton, 1976). The sequence of events has been characterised by Butler and Simon (1971). First a decrease in the population of ribosomes is evident together with the initiation of chloroplast breakdown. The chloroplast stroma disappears, thylakoids swell and disintegrate and there is generally a marked increase in the number and size of osmiophilic globules. Secondly, changes also occur in the structure of the mitochondria, endoplasmic reticulum and nuclear membrane.

The nature of alterations in the fine structure of chloroplasts due to herbicide action has been reviewed by Anderson and Thomson (1973). First the intergranal and then the granal thylakoids begin to swell until the whole lamellar system is disorganised. Subsequently the tonoplast and thylakoid membranes are ruptured with mixing of plastids and cytoplasmic contents being detected within 4 h in the atrazine-treated plants (Hill *et al.*, 1968). These effects may occur several days before external symptoms of injury are evident. They have been reported for a number of herbicides including bromacil, haloxydine (3,5-dichloro-2,6-difluoro-4-hydroxypyridine), pyrazon, pyriclor, 2,4-D, monuron, diuron, dichlormate (3,4-dichlorobenzyl methylcarbamate), metflurazone, paraquat and diquat (Anderson and Thomson, 1973).

Some herbicides interfere with normal chloroplast development, resulting in the formation of bleached or albino leaves after chemical treatment; such leaves do not become green even after the inhibitory levels of herbicide are dissipated. Compounds which produce this effect include aminotriazole (Miller and Hall, 1957), fluometuron (Bruinsma, 1965), dichlormate (Herrett and Berthold, 1965), haloxydine (Slater, 1968; Dodge and Lawes, 1972), pyriclor (Burns *et al.*, 1971), norflurazon, metflurazone and Sandoz 9774 (5-amino-4-chloro-2-[α,α,α-trifluoro-*m*-tolyl]-3(24)-pyridazinone) (Hilton *et al.*, 1969). No essential common structural feature is evident among these inhibitors (Moreland and Hilton, 1976).

The action of herbicides on photosynthesis

The haloalkanoic, phenoxyalkanoic and aromatic acids and amides, are generally regarded as ineffective on photosynthetic mechanisms except at high concentrations and their primary mechanism of action is regarded as lying elsewhere. However, the hydroxybenzonitriles inhibit the Hill reaction and this together with uncoupling of oxidative phosphorylation appears to be their primary mechanism of action.

NITRILES

The inhibitory effect of **ioxynil** on the Hill reaction was first reported by Wain (1964). Paton and Smith (1965) using chloroplasts, chloroplast fragments or grana isolated from the leaves of *Vicia faba* demonstrated that the presence of ioxynil in the culture medium resulted in inhibition of electron transport and non-cyclic phosphorylation (50% at 8×10^{-7} M). These workers also found that ioxynil was a more effective inhibitor of the Hill reaction than was 2,4,5-T, 2,4,5-TBA and CIPC (isopropyl *m*-chlorocarbanilate) and was comparable in activity to diuron (Smith, Paton and Robertson, 1966); they believe the site of action to be near plastoquinone. Similarly they found evidence that low concentrations of ioxynil (e.g. 10^{-8} M) uncoupled oxidative phosphorylation, the probable site of action being in the region of ubiquinone. The inhibition of photosynthetic electron transport by ioxynil was also reported by Friend and Olsson (1967) using chloroplasts isolated form broad bean (*V. faba*) and sugar-beet; they believe that ioxynil inhibits plastoquinone photoreduction at the same site as *o*-phenanthroline and diuron. Ioxynil inhibited the Hill reaction when 2,6-dichlorophenol-indophenol (2,6-DCIP) or NADP were used as electron acceptors, but had little effect on the photoreduction of NADP using ascorbate and indophenol as electron donors. They conclude that ioxynil inhibits the oxygen-evolving step in photosynthetic electron transport. Reimer, Link and Trebst (1979) reported that ioxynil had a relationship of inhibitory potency to chemical structure which differed from that of diuron and metribuzin but inhibited photosynthetic electron flow at the same functional site. They found that labelled metribuzin was displaced from the thylakoid membrane by ioxynil (and bromonitrothymol), indicating identical binding sites. In addition to effects on electron transport, there is evidence that ioxynil destroys chlorophyll (Davies *et al.*, 1968).

ANILIDES

Inhibition of the Hill reaction by **propanil** was first reported by Good (1961) and Moreland and Hill (1963); later Nishimura and Takamiya (1966) pinpointed its site of action as the reduction of cytochrome 553 by PS II. Nishimura *et al.* (1975) using spinach chloroplasts, found that propanil (and certain other Hill reaction inhibitors) uncoupled cyclic phosphorylation mediated by PMS (10^{-4} M) and accelerated PS I-dependent electron flow. The activity of these compounds was related to their partition coefficient in an octanol–water system, indicating the importance of efficient penetration of the chloroplast membrane. Earlier Monaco (1968) had shown that absorption and accumulation of propanil by isolated spinach chloroplasts was influenced by their lipid, but not protein, content. Inhibitory activity against the Hill reaction in spinach chloroplasts has also been found for the *cis* isomers of five 2,5-dimethyl-1-pyrrolidine carboxanilides (Holm and Stallard, 1974).

Pentanochlor (Solan) (3–6 kg/ha) applied post-emergence to carrots in the field, reduces the chlorophyll and carotenoid content of leaves and reduces the photochemical activity of chloroplasts (Sivtsev and Kuznetsova, 1973a, b).

NITROPHENOLS

In addition to affecting oxidative phosphorylation, the nitrophenols also inhibit photosynthesis, but only concentrations high enough to inhibit respiration also inhibit photosynthesis (Kandler, 1958). Wessels (1959) found that **DNP** was able to catalyse the generation of ATP by chloroplasts in light, the optimum concentration being 0.6 μM; a concentration of 3.0 μM inhibited ATP synthesis. Light reduced the action of DNP in bean leaves suggesting that the protective action may be due to a restoration of ATP by photophosphorylation (Ross and Salisbury, 1960). This view was substantiated by the findings of Colby *et al.* (1965) and Wojtaszek (1966).

Subsequently **DNOC** was shown to act as a potent inhibitor of the Hill reaction in isolated spinach chloroplasts (van Rensen *et al.*, 1977), acting at the same site of PS II as diuron. In addition, at higher concentrations, DNOC uncouples electron transport in the PS I-dependent Mehler reaction. Van Rensen, Wong and Govindjee (1978) later reported that both herbicides inhibit electron transport between the primary electron acceptor of PS II and the plastoquinone pool, causing a closing of the reaction centres of PS II. However, the interaction with the inhibited molecule differs for the two herbicides. Subsequently van Rensen (1979) concluded that DNOC acts as an inhibitory uncoupler. Structure–activity aspects associated with the inhibition of photosynthetic electron flow by alkyl-substituted nitrophenols and certain other herbicides have been discussed by Trebst and Draber (1978).

NITROPHENYL ETHERS

These compounds generally inhibit non-cyclic electron transport and uncouple photophosphorylation. This has been demonstrated for **nitrofen** in isolated spinach chloroplasts (Pereira, Splittstoesser and Hopen, 1971). Study of the selective action of **chlornitrofen** in rice and barnyard grass (*Echinochloa crus-galli* var. *oryzicola*) suggests that the susceptibility of the latter to chlornitrofen may be due to interference with cyclic phosphorylation, stimulating PS I and decreasing O_2 uptake on exposure to light (Hyakutake and Ishizuka, 1976). In the case of **oxyfluorfen** however, preliminary evidence suggests that photosynthesis is affected only after membrane integrity is disrupted (Vanstone and Stobbe, 1979). The effect of a number of diphenyl ether herbicides on photosynthetic electron transport has been reported by Bugg *et al.* (1979). Using isolated pea seedling and spinach chloroplasts, they found that all of the diphenyl ethers tested appear to block linear electron flow in the plastoquinone–cytochrome f region of the electron transport chain; cyclic electron flow mediated either by diaminodurene or pyocyanine was partially inhibited. They propose a site of inhibition for these compounds either at the dibromothymoquinone inhibition site or between this site and cytochrome f.

DINITROANILINES

Photosynthesis does not appear to be a major site of action of nitroaniline compounds. Moreland, Farmer and Hussey (1972a) examined the effect of twelve dinitroaniline herbicides on photosynthesis of spinach leaf disc tissue. While both *in vivo* and *in vitro* treatments interfered with photosynthesis (and respiration), effects on growth and metabolism were only partially explained by interference with oxidation and photoproduction of ATP. It seems that the importance of interference with ATP production to the phytotoxic action of these herbicides remains to be established.

CARBAMATES

Carbamate herbicides normally inhibit the Hill reaction of photosynthesis, but it is unlikely that in practice, this mechanism is a major factor in their toxicity since they are mainly pre-emergence herbicides; those which are foliage-applied appear to inhibit meristematic activity.

Using the Hill reaction, Moreland and Hill (1959) screened a series of alkyl esters of *N*-(3-chlorophenyl)carbamic acid and found the s-butyl to be most active followed by the n-butyl>n-propyl>isopropyl >amyl esters. They propose that molecules of ethyl *N*-phenylcarbamates may form association polymers by bonding between imino hydrogens and carbonyl oxygens. In inhibiting the Hill reaction, the imino hydrogen may take part in hydrogen bond formation with some electronegative constituent located at or near the reactive centre of the chloroplast.

The effects of a range of herbicides including chlorpropham on light-activated, magnesium-dependent ATPase of isolated spinach chloroplasts have been studied by Alsop and Moreland (1975). They found that ATPase activity was stimulated by low and inhibited by high molar concentrations. The light-induced synthesis of ATP mediated by dithiothreitol (DTT) and PMS was inhibited strongly by **chlorpropham** and a number of the other herbicides. Alsop and Moreland concluded that chlorpropham (and most of the other compounds tested) had a site of action on the ATP generating pathway that is separate from the site involved in the inhibition of the Hill reaction.

THIOCARBAMATES

There is little evidence that photosynthesis is a major mechanism of action of thiocarbamate herbicides, though some reports of inhibitory effects do exist. For example, **EPTC** inhibits photosynthesis and thus sugar accumulation in the roots of sugar-beet (Slovtsov, Arabi and Dorozhkina, 1976), while in combination with certain other herbicides this compound inhibits photosynthesis in sugar-beet grown in the greenhouse but not in the field (Merezhinskii, Mitrofanov and Ivanischev, 1975). Again, examination of quinone accumulation in EPTC-treated wheat suggests that transmembrane electron transport systems are inhibited, though growth in size is reduced more than is synthesis of the various pigments and quinones (Wilkinson, 1978) suggesting that other mechanisms of inhibition may be involved. Previously Weinbach and Castelfranco (1975) found that EPTC increased chlorophyll accumulation in germinating cucumber cotyledons.

UREAS

Urea herbicides are normally root absorbed, translocated in the xylem and exert their toxic action on the leaf tissues. Early studies showed that **monuron** inhibits the Hill reaction (Cooke, 1956; Wessels and van der Veen, 1956; Spikes, 1956), almost completely blocks the pathway of photosynthetic CO_2 fixation but does not affect the dark CO_2 fixation pathway (Ashton *et al.*, 1961). Subsequently the effect of monuron and other urea compounds on photosynthesis was reported by a number of workers and many of these findings have been included in reviews by van Overbeek (1962), Black and Meyers (1966), Moreland (1967), Zweig (1969), Caseley (1970) and Hoffmann (1971). The effect of urea herbicides on the light reactions of photosynthesis have received particular attention from many investigators including Jagendorf and Avron (1959), Ashahi and Jagendorf (1963), Bamberger *et al.* (1963), Izawa and Good (1965), Izawa (1968), Moreland and Blackmon (1968) Nishida (1968), Zweig *et al.* (1968), Petkova (1972) and Floyd (1972). The primary site of action is apparently located in PS II in or near the oxygen-evolving step; inhibition of PS I appears not to be a significant factor.

The site of inhibition of PS II by **diuron** has been investigated using thylakoids of the cyanobacterium *Anabaena cylindrica* (Codd and Cossar, 1978). Diphenylcarbazide-supported PS II activity is completely insensitive to diuron, indicating that the site of action lies on the donor side of PS II, before the site of electron donation by diphenylcarbazide. Further information on the specific site of action of PS II inhibitors has resulted from studies by Radosevich, Steinback and Arntzen (1979). They examined the effect of diuron (and bromacil and several s-triazines) on the thylakoid membranes of two common groundsel (*Senecio vulgaris*) biotypes. Each herbicide inhibited photosynthetic electron transport on the reducing side of PS II in thylakoid membranes from the susceptible biotype. Thylakoids isolated from the resistant biotype were 60–3200 times more tolerant to the inhibitors than those from the susceptible biotype. Specific structural or configurational changes associated with the reaction site may account for the differences in Hill reaction inhibition between biotypes (Radosevich *et al.*, 1979). Further evidence of a common molecular target for the PS II inhibitors including *N*-phenylureas (also *N*-phenylcarbamates, cyclic ureas, *N*-acylanilides, 1,3,5-triazines, triazinones, uracils, pyridazines and substituted thiadiazoles) has been presented by van Assche (1979). These diuron-type inhibitors are believed to induce a negative potential shift of B which is a secondary electron carrier on the acceptor side of PS II; the existence and role of an allosteric protein remains to be clarified. Diphenyl ethers, dinitroanilines and halogenated benzonitriles do not appear to affect photoreactions at the same site. The long-term treatment of the photosynthetic membrane system of the unicellular alga *Bumilleriopsis filiformis* with sublethal doses of diuron (or triazines or pyridazinones) suggests that the electron transport system is not impaired though pyridazines damage the photosynthetic redox system (Boger and Schlue, 1976).

The rapid degradation of carotenoid pigments may have a role to play in the action of certain urea compounds. For example, Stranger (1972) showed that *in vitro* treatment of spinach chloroplasts with diuron results in the degradation of carotenoid pigments more rapidly than chlorophyll. Specific carotenoids normally function through oxidation–reduction reactions to protect chlorophyll from photo-oxidation in the presence of reduced NADP. Pallett and Dodge (1976) also examined the cause and implications of rapid breakdown of carotenoid pigments following treatment of flax cotyledons with monuron. They believe that it is due to an overloading of a protective system whereby when electron flow is inhibited, excitation energy is initially channelled from PS II to PS I and to the carotenoids. The subsequent degradation of chlorophylls may be due to generation of excited singlet oxygen and the induction of lipid peroxidation.

Degradation of chlorophyll due to bleaching and lipid peroxidation results from treatment of isolated intact chloroplasts with certain substituted ureas (and uracils, s-triazines, benzonitriles and bipyridyls) (Giannopolitis and Ayers, 1978). Photo-oxidation is prevented by addition of NADH or NADPH, supporting the view that depletion of NADPH is responsible for chloroplast photo-oxidation and death. On the other hand, Watanabi and Kiuchi (1975) previously reported that monuron significantly retards chlorophyll degradation in light. They believe that O_2 evolved in photosynthetic reactions cause light-induced degradation of chlorophyll and monuron suppresses such degradation by inhibiting O_2 evolution in the light reactions of photosynthesis.

The effect of light on the phytotoxicity of diuron has been examined by Ladonin (1972) who found that growth is more inhibited in light than in shade. He concluded that shading helps the plant either to inactivate the herbicide or in some way to become more resistant to it; shade also reduces transpiration and herbicide translocation.

There is also evidence that phenylurea herbicides tend to reduce transpiration (Lay and Ilnicki, 1971). When applied to excised shoots via the nutrient solution, compounds with methyl groups are more inhibitory than those with a methoxy group in the 1-*N* position; when applied to the foliage the position is reversed. Transpiration is also reduced when the herbicide is applied to the roots, the rate of H_2O loss decreasing with reduction in diuron concentration (8×10^{-6}–2×10^{-6} M). Lay and Ilnicki (1971, 1972a, b) suggest that reduced transpiration may reflect stomatal closure due to photosynthetic inhibition.

Studies of the influence of monuron on PS II and CO_2 fixation in isolated cells of C_3 and C_4 plants suggest that selectivity of monuron may involve differing responses of cells from different species together with the differing photosynthetic pathways in C_3 and C_4 plants (Malakandaiah and Fang, 1978). Again, examination of a range of phenylurea herbicides on a lettuce chloroplast test system reveals variation in the level of activity (I_{50} values), the compounds falling into two classes on the basis of the relationship between pI_{50} and water solubility (Dicks, 1978).

The need for care in the interpretation of the results of *in vitro* studies is indicated by the findings of Santakumari and Das (1978). They examined the mode of action of diuron (and atrazine) on the photochemical activities of chloroplasts (*in vivo* and *in vitro*) isolated from the leaves of certain crop and weed species. Hill reaction activity, as indicated by DCPIP photoreduction of chloroplasts isolated from herbicide-sprayed plants, was unaffected by sublethal or intermediate levels of diuron (or atrazine), while chloroplasts treated *in vitro* were severely inhibited. Ferricyanide-catalysed non-cyclic photophosphorylation was markedly reduced in both *in vitro* and *in vivo* systems while PMS-mediated cyclic photophosphorylation was inhibited *in vivo* but enhanced *in vitro*; $NADP^+$ photoreduction was severely inhibited *in vitro* but unaffected *in vivo*. Further, it should be borne in mind that inhibition of Hill reaction activities by technical diuron may be due partly to the inhibitory effect of an impurity (1,3-bis-[3,4-dichlorophenyl]urea) which is also an active uncoupler of oxidative phosphorylation and modifies the permeability of liposome membranes.

Compared with diuron and monuron, other members of the phenylurea group have received less attention. The mechanism of action of **fluometuron** has been examined by Wergin and Potter (1975) and Richardson and Frans (1975). The former germinated seedlings of velvet leaf (*Abutilon theophrasti*) in the presence of 0.5, 10 or 80 μg/ml fluometuron. Increasing concentration of this herbicide reduced the number and length of stroma lamellae and promoted more elongate grana with fewer compartments. Chlorophyll content was increasingly reduced with increased concentration, albinistic seedlings being produced at 80 μg/ml; net photosynthesis was inhibited by all fluometuron treatments. Richardson and Frans (1975) found that 0–40 p.p.m. fluometuron reduced cell number and chlorophyll content, I_{50} values being 22 and 5 p.p.m. respectively.

Methabenzthiazuron also inhibits photosynthetic activity particularly under high light conditions (Fedtke, 1973, 1974). Application of this herbicide to wheat seeds prior to germination or to 13-day-old wheat plants results in a reduction in the concentration of reducing sugars, enlarged chloroplasts and a decrease in the ratio of chlorophyll a:b. However increases are found in the total amount of chlorophyll (due to delayed breakdown), ATP concentration and nitrogen metabolism (Fedtke, 1974). It was later reported that fixation of $^{14}CO_2$ by wheat plants is reduced by methabenzthiazuron and while the proportion of ^{14}C found in amino acid and insoluble fractions increased, less was found in the neutral sugar fraction (Fedtke, 1975). These results are also discussed in a later report (Fedtke, Deischgräber and Schnepf, 1977). Similar effects on the levels of chlorophyll and other pigments were found in barley (*Hordeum vulgare*) seedlings (Kleudgen, 1978). He suggests that inhibition of photosynthetic electron transport, like light deficiency in shade leaves, may be compensated by metabolic reactions leading to greater production of PS II particles with an enhanced accumulation of the chlorophylls and carotenoids.

Little difference is evident in the action of **siduron** on PS II in isolated chloroplasts of spinach, blue grass (*Poa pratensis*) and Bermuda grass (*Cynodon dactylon*); suggesting that the basis of selectivity does not depend upon differential inhibition of the Hill reaction (Mur, Swader and Youngner, 1972). The implication is that selectivity depends on differences in the amount reaching the sites of action. This appears to be the case for the differential action of **linuron** (and prometryne) on the pigment system of potato and wild radish (Timofeev, 1978). These herbicides accumulated differently in resistant and susceptible plants and differently according to fertiliser level. Linuron treatment of maize reveals that peroxidative breakdown of the polyunsaturated membrane lipids leads to destruction of the pigments, changes in energy transfer, disorganisation of the chloroplast membranes and accumulation of toxic organic peroxides (Muschinek *et al.*, 1979). Earlier Ladonin, Khachatryan and Gal'per-Blichenko (1973) had reported the effects of linuron (and monuron) on certain crop and weed species and conclude that inhibition of photosynthesis could not be the sole mechanism of action of these compounds; they also found effects on the protein:RNA ratio and amino acid synthesis.

Methurin has produced results which are at variance with the classical diuron or monuron pattern. It has been found to interrupt electron transfer from water to NADP in the transition from PS II to PS I in potato chloroplasts (Luzhnova, Shekhtman and Sharkov, 1972). Later work with isolated spinach and pea chloroplasts showed that the compound had no effect on PS I and PS II; the Hill reaction was inhibited only when plants were directly exposed to the herbicide (Stonov *et al.*, 1974). However, Luzhnova and Shekhtman (1975), using isolated pea and spinach chloroplasts, found that methurin had no effect on PS I, but unlike monuron and diuron had only a very weak inhibitory effect on PS II. This probably results from transformation of methurin into such Hill reaction inhibitors as *N*-methyl-*N*-phenylurea.

Certain other phenylurea herbicides are effective as inhibitors of the Hill reaction including compounds of the *N*-aryl-carbamoyloxyphenylureas group (Arndt, Boroschewski and Kotter, 1971), *N*-methyl-*N*-propargyl-*N'*-(3,4-dichlorophenyl)urea (Pyne, Szabo and Holm, 1979) and *N*-methoxy-*N*-methyl-*N'*-(3-t-butylacetamidophenyl)urea (Swader and Howe, 1978, 1979). The site of action of the last compound is believed to be located in PS II between water splitting and PS I.

The mode of action of **buthidazole**, one of the most recent of this group of compounds has been examined by two groups of workers. York, Arntzen and and Slife (1978) examined the effects of buthidazole on the photochemical reactions of isolated pea chloroplast thylakoids. Inhibition of PS I-mediated electron transport was insignificant, with either whole chain electron transport or PS II-mediated DCPIP reduction, however, the effects of buthidazole on electron transport were time dependent on I_{50} of 4×10^{-7} M being obtained after 3 h dark incubation. Further study indicated that the major site of inhibition of electron transport occurs on the reducing side of PS II, at or near the site of action of diuron, a secondary site of inhibition appearing to occur on the oxidising side of PS II (York and Arntzen, 1979).

Inhibition of photosynthesis by buthidazole has also been demonstrated by Hatzios and Penner (1978). Pre-emergence application of 2.24 kg/ha inhibits maize photosynthesis and electron microscopic examination of plant tissues revealed reduced amounts of starch in the bundle sheath chloroplasts and some disruption of mesophyll chloroplasts at lower concentrations. Further evidence of the action of buthidazole (10^{-6}–10^{-4}M) as a photosynthetic inhibitor was presented by Hatzios and Penner (1979) who found that sensitive species (*Amaranthus retroflexus* and *Agropyron repens*) were affected to a greater degree than were tolerant species (maize and lucerne).

Using isolated spinach chloroplasts they too demonstrated that the site of inhibition of the electron transport system was primarily at the reducing side of PS II though a secondary site may also occur on the oxidising side of PS II (Hatzios, Penner and Bell, 1979).

TRIAZINES

The inhibition of photosynthesis by triazine herbicides is well documented, having been demonstrated by many workers using a variety of techniques on a variety of test systems including isolated chloroplasts, unicellular algae and intact higher plants. In two of the earlier studies by Exer (1958) and Moreland *et al.* (1958, 1959) it was shown that **simazine** inhibits the Hill reaction in isolated chloroplasts. Subsequently the effect of other triazines on the Hill reaction was reported by Gysin and Knüsli (1960), Exer (1961) and Moreland and Hill (1962).

Moreland and Hill (1962) demonstrated that the inhibitory action of simazine on the Hill reaction was reversible, and that the degree of inhibition by chloroplasts treated with simazine from various species was similar although their *in vivo* susceptibility was quite different. They concluded that the selectivity of simazine in these species could not be due to differences in the chloroplasts. Bishop (1962) presented evidence to suggest that triazine herbicides inhibit the oxygen-evolving system of photosynthesis; photophosphorylation is also inhibited when flavin mononucleotide (FMN) is used as electron acceptor (Good, 1961; Shimabukuro and Swanson, 1968).

Several investigations have been carried out to pinpoint the specific site of action of these compounds. For example, Merbach and Schilling (1977) found evidence to suggest that the major site of action of the s-triazines (and several other herbicide groups) is the acceptor site of PS II. Goss *et al.* (1978) using pea chloroplasts found that both **atrazine** and diuron have a single and specific site of electron transport inhibition which is located on the reducing side of PS II in the Q–PQ complex; both herbicides inhibit non-cyclic but not cyclic photophosphorylation. Further studies by the Illinois group reveal that the primary site of inhibition for atrazine and **cyanazine** is on the reducing side of PS II, at the electron transfer step between the primary electron acceptor and the plastoquinone pool of the electron transport chain; I_{50} values of approximately 2×10^{-7} M were obtained for each herbicide (Brewer, Arntzen and Slife, 1979). Previously it had been found that cyanazine reduces cyclic photophosphorylation of isolated pea chloroplasts (Pillai and Davis, 1975).

The photochemical properties of isolated chloroplasts and thylakoids from atrazine-susceptible and resistant strains of pigweed (*Amaranthus retroflexus*) were characterised by Arntzen (1978). Alterations in both PS II electron transport processes were found to be correlated to differences in a protein of the PS II reaction centre complex. The mechanism of resistance to atrazine of biotypes of *Chenopodium album*, however, was found to exist at the chloroplast level (Souza Machado *et al.*, 1978a, b). Diuron blocked the photochemical activity of isolated chloroplasts of the resistant biotype while atrazine did not; there was no difference in chloroplast membrane permeability to atrazine in the two biotypes. Other studies have involved the evaluation of binding of radiolabelled diuron and atrazine to stroma-free chloroplast thylakoids isolated from triazine-susceptible and resistant biotypes of *Senecio vulgaris* (Pfister, Radosevich and Arntzen, 1979b). They found evidence of a specific chloroplast membrane constituent which binds both atrazine and diuron, these compounds interacting with different segments of the constituent. They suggest that resistance to triazine herbicides results from a selective alteration in the segments such that there is a selective loss in binding of triazine. Arntzen, Ditto and Brewer (1979) carried out electrophoretic analysis of chloroplast membrane proteins of redroot pigweed (*Amaranthus retroflexus*) biotypes exhibiting resistance and susceptibility to triazines and showed small differences in two polypeptide species. They conclude that differences in resistance are due to genetically controlled modification of the herbicide target site.

Inhibition of the light reactions to the triazine compounds must indirectly affect the efficiency of CO_2 fixation and this has indeed been reported. For example CO_2 fixation in beans is blocked by root-applied simazine (Ashton and Zweig, 1958; Ashton *et al.*, 1960) and **trietazine** (Ashton *et al.*, 1960), inhibition of CO_2 fixation by all compounds being almost complete 48 h after treatment (1 p.p.m.). Similar effects on CO_2 fixation have been obtained using **simetone** (2,4-bis[ethylamino]-6-methoxy-5-triazine) in beans (van Oorschot, 1964), atrazine in beans ((Funderburk and Carter, 1965), atrazine in maize, cotton and soybean (Sikka and Davis, 1969), atrazine separately or with 2,4-D in beans (Dorozhkina *et al.*, 1973) and with simazine, atrazine (and 2,4-D) in *Pinus nigra* var. *calabrica* (Shaukat, 1975; Shaukat, Moore and Lovell, 1975).

In addition to affecting the light and dark reactions of photosynthesis, the triazine herbicides are also known to affect the synthesis of chlorophyll and other pigments. For example Litvinov (1972) found that atrazine (4 and 10 kg/ha) selectively inhibited photosynthesis and chlorophyll synthesis in *Cirsium arvense* and *Sonchus arvensis*, but not in maize. Copping and Davis (1972) studying the effects of atrazine on chlorophyll a retention in maize leaf discs, observed a decrease in retention in the dark but not in light. The combination of small amounts of kinetin with atrazine accentuated the deleterious effect of high concentrations of atrazine on chlorophyll a retention, but the highest concentrations (150 and 1500 μg/ml) masked the effects of atrazine. Application of atrazine (0.1, 0.5 or 1%) to seedlings of wheat (cv. Giza 155) also accelerated the breakdown of chlorophyll a and b in shoots and chlorophyll/protein/phospholipid complexes in leaf blades; carotenoids, however, are more resistant to atrazine than is chlorophyll (Sharaky and El-Gandour, 1975). Degradative effects of atrazine on chlorophyll content have also been reported in *Phaseolus vulgaris* (Bolha'r-Nordenkampf, 1975) and maize (Muzychenko and Borodycheva, 1976), while synergistic effects of atrazine and alachlor in reducing chlorophyll protein synthesis have been noted for *Echinochloa crus-galli* var. *frumentacea* (Akobundu *et al.*, 1975).

The effect of other triazine compounds on chlorophyll content has also been examined. For example, **prometryne** (1.5–3.0 kg/ha) applied to onion (*Allium cepa*) reduces chlorophyll and carotenoid content (Sivtsev and Kuznetsova, 1973b). Treatment of *Echinochloa crus-galli* seedlings with prometryne (1–10 p.p.m.) (or **simetryn** 5–50 p.p.m.) results in abnormal chloroplast formation and other effects which included inhibition of transpiration (Nakayama, 1974). Application of simazine (1.5–3.0 kg/ha) to seedlings of maize (Susheela Deevi and Perur, 1978) or barley (Kleudgen, 1979) results in a reduction in the chlorophyll content. A detailed study of the effect of simazine (10 and 100 μM) applied via the nutrient solution, on the formation of plastid prenyllipids in barley seedlings, showed that metabolic changes in chloroplasts could be related to a common site of regulation perhaps involving the endogenous cytokinins (Kleudgen, 1979). Changes include accumulation of chlorophylls and the dependence of carotenoids on age and the concentration of herbicide applied. The ratio of chlorophyll a/b decreased in the older plants (10 and 13 days) while the ratio of xanthophylls/β-carotene and chlorophyll a/prenylquinones increased. The manner in which these prenyllipid ratios were changed in these older plants is characteristic of a shade-type adaptation which is typical of other PS II-inhibiting herbicides.

A number of studies indicate that triazine herbicides also affect the transpiration rate. For example **desmetryne** (as 0.2% 'Semeron') inhibits the transpiration of cabbage, pigweed (*Amaranthus* sp.) and fat hen (*Chenopodium album*), complete recovery within 8 days occurring in cabbage but not in the other species (Pochinok, 1971).

Transpiration is also reduced in fall panicum (*Panicum dichotomiflorum*) treated at the 3- to 4-leaf stage in solution culture with 5 to 20 μM simazine, atrazine or mixtures of both (Smith and Ilnicki, 1972). Das and Santakumari (1975) found a correlation between stomatal response to atrazine (and diuron, alachlor, butachlor, EPTC and molinate) and transpiration indicating that the stomatal mechanism is a determining factor in the selectivity of these herbicides to a range of weed species. However, West, Muzik and Witters (1976) found evidence to suggest that reduction of photosynthesis in susceptible (S) and resistant (R) biotypes of pigweed (*Amaranthus retroflexus*) was due to a greater effect of atrazine on the mesophyll tissue rather than on the guard cells; photosynthesis and transpiration were both reduced to a greater extent in the S biotype. Radosevich (1976) and Radosevich and Devilliers (1976) also found that photochemical activity of 'S' chloroplasts (from a susceptible biotype of *Senecio vulgaris*) was severely inhibited, 'R' chloroplasts being unaffected.

A study of the effect of atrazine on Douglas fir (*Pseudotsuga menziesii*) seedlings showed that low levels of herbicide stimulated H_2O uptake (and transpiration) apparently acting as a cytokinin-like growth-promoting compound (Peterson, 1977).

The unsymmetrical triazines also affect photosynthesis. **Metribuzin** inhibits photosynthesis and respiration (Fedtke, 1972) and all Hill reactions when water is the electron donor (Trebst and Wietoska, 1975). The latter presented evidence to suggest that metribuzin inhibits photosynthesis electron flow between the primary and secondary electron acceptor of PS II (Q and plastoquinone). **Metamitron** also inhibits photosynthesis but recovery is rapid and complete in certain species due to inactivation of the herbicide (van Oorschot and van Leeuwen, 1979).

PYRIDINES

The photosynthetic mechanism is an important site of action of the bipyridylium herbicides. This was suggested in early studies by Mees (1960) who found that the action of **diquat** occurred more rapidly in the light than in the dark and that the presence of oxygen is necessary for development of the phytotoxic symptoms. Mees (1960) also showed that pretreatment of leaves with monuron delayed the appearance of phytotoxic symptoms of diquat suggesting that the site of action of the latter occurred later in the photosynthetic system than the Hill reaction which monuron inhibits. While CO_2 fixation is also inhibited (e.g. Couch and Davis, 1966; van Oorschot, 1964; Zweig, 1969), this is regarded as a secondary effect resulting from inhibition of the light reactions with consequent reduced availability of NADPH and ATP. It was later demonstrated that diquat could act as an electron carrier (Davenport, 1963; Zweig and Avron, 1965) and could competitively inhibit NADP reduction by isolated chloroplasts (Davenport, 1963). Diquat also competitively inhibits photophosphorylation catalysed by PMS suggesting that it acts at the same site as PMS in the electron transfer pathway (Zweig *et al.*, 1965). Diquat is also known to shunt electrons from ferredoxin to form the diquat-free radical thus preventing the normal reduction of NADP to NADPH (Ashton and Crafts, 1973).

The re-oxidation of the free radical of the bipyridylium compounds has also been demonstrated (Kok *et al.*, 1965; Zweig *et al.*, 1965; Black, 1965; Black and Meyers, 1966). In light, rapid reduction of the bipyridylium ion to the free radical occurs while in darkness re-oxidation to the ion takes place. In the absence of oxygen, reduction occurs rapidly in light but is relatively slight in the dark, demonstrating the catalytic nature of these compounds. Consequently the bipyridylium herbicides remove electrons from the electron transport system associated with PS I, thus inhibiting reduction of NADP to NADPH. However, the relatively slow death resulting from this action is not consistent with the rapid development of phytotoxic symptoms.

It has been proposed that hydrogen peroxide liberated during the oxidation of the bipyridylium free radical in the toxic agents causes this rapid injury (Mees, 1960; Calderbank and Crowdy, 1962; Black and Meyers, 1966; Calderbank, 1968) though reactive hydroxyl radicals may also be involved (Calderbank, 1968). Since catalases and peroxidases are widely distributed in higher plants, it might reasonably be assumed that the bipyridylium herbicides produce enough hydrogen peroxide to exceed the capacity of these enzymes to destroy it.

The effect of **paraquat** on the fine structure of flax cotyledon leaves has been examined by Harris and Dodge (1972a, b). Six hours after treatment with paraquat under light conditions, tonoplast deterioration was followed by disruption of the mitochondria and chloroplast thylakoid membrane structure; after 30 h in darkness, however, only limited breakdown of the chloroplast membrane was observed (Harris and Dodge, 1972a).

These ultrastructural changes are correlated with the inhibition of photosynthetic CO_2 uptake and electron flow in isolated chloroplasts, increased membrane permeability and changes in the level of the major lipid components and malondialdehyde (Harris and Dodge, 1972b).

In contrast, Davis and Shimabukuro (1973) using isolated

groundnut cells, found no evidence that diquat, paraquat or perfluidone (1,1,1-trifluoro-4-[phenylsulphonyl]methane-sulphono-*o*-toluidide) altered the sub-cellular components during a 2 h incubation period. However, diquat and paraquat (10^{-5} M) prevented net photosynthetic oxygen evolution after 23 min, perfluidone having no effect; $^{14}CO_2$ fixation was also inhibited by diquat (10^{-5} M and 5×10^{-7} M). It would appear that the photosynthetic inhibition caused by these bipyridylium herbicides was not caused by gross structural alterations of the cellular membranes. On the other hand, van Rensen (1975) found that addition of diquat to *Scenedesmus obtusiusculus* caused lipid peroxidation and a decrease in chlorophyll a though chlorophyll b was unaltered.

Farrington (1976) calculated that superoxide ions (O_2^-) should be formed inside photosynthesising chloroplasts in paraquat-treated plants and these should be sufficiently stable to diffuse to the chloroplast envelope which is the first site to reveal damage following paraquat treatment. Superoxide ions should initiate peroxidation of unsaturated lipids in the chloroplast membrane and cause their partial oxidation to malondialdehyde, which has been noted in treated plants. Farrington believes that this compound is the most likely cause of phototoxicity attributable to paraquat treatment.

The effects of paraquat on carbon assimilation and transport by sugar-cane leaves in white light and darkness have been studied by Alexander and Biddulph (1976). Assimilation is almost completely blocked when paraquat is applied prior to $^{14}CO_2$ and this is attributed to its known effect on the light reactions.

The phytotoxic action of the bipyridyls on certain microalgae has also been investigated. For example the action of paraquat on the photosynthetic apparatus of *Bumilleriopsis filiformis* appears to centre primarily on the photosynthetic membrane system and redox chain (Boger and Kunert, 1978). Severe chlorophyll bleaching accompanied by formation of malondialdehyde and 50% inactivation of partial redox reactions was observed after 20–160 h in the region of PS II and PS I, except for the terminal portion of PS I (pigment 700→ $NADP^+$). Sub-toxic concentrations of diquat enhance cupric ion inhibition of *Chlorella* growth and photosynthesis and electron transport in isolated chloroplasts (Swader, 1978a, b). Cupric ion movement through the outer chloroplast envelope is stimulated by the presence of diquat, suggesting that the latter influences membrane permeability.

Fluridone also is an effective photosynthetic inhibitor. For example, Anderson (1978) exposed winter buds of American pondweed (*Potamogeton nodosus*) and Sago pondweed (*P. pectinatus*) to a range of concentrations of fluridone. Light was necessary for herbicide activity which appeared to involve chlorophyll synthesis. Similarly, fluridone (e.g. 10 μM w/v) caused a marked reduction in the chlorophyll content of wheat (cv. Mericopa) and maize (cv. Merit) seedlings grown under high but not low, light intensity (Devlin *et al.*, 1978) suggesting that inhibition of carotenoid production represents the main mode of action. Devlin, Kisiel and Kostusiak (1979) suggest that a second, related site of action may involve GA sensitivity, while Rafii, Ashton and Glenn (1979) found that in addition to inhibiting photosynthesis (the major site), fluridone blocked RNA and protein synthesis in *Phaseolus vulgaris* (susceptible) and cotton (cv. Acala, SJ-1) (resistant); differences in response were believed to contribute to herbicide selectivity.

PYRIDAZINES

Several sites of action have been proposed for the substituted pyriazinones including inhibition of the Hill reaction and CO_2 fixation, inhibition of carotenoid biosynthesis accompanied by photodestruction of chlorophyll and interference with the formation of chloroplast membrane polar lipids. The primary action which appears to be on carotenoid biosynthesis was probably first reported by Bartels and McCullough (1972) who found that **metflurazone** inhibited the accumulation of carotene and xanthophyll chloroplast pigments in wheat seedlings grown in light and dark conditions. They indicated that the site of action on carotenoid biosynthesis occurred at a point prior to the intermediate phytofluene. Porter and Bartels (1977) subsequently reported that metflurazone inhibited RNA and lipid synthesis in single leaf cells of cotton and soybean but suggested that these effects were an indirect result of inhibited photosynthesis.

The action of several chlorosis-inducing herbicides including metflurazone on the biogenesis of chloroplasts and leaf microbodies in wheat and rye has been examined by Feierabend and Schubert (1978). In light, all herbicides prevented the accumulation of carotenoids and of chloroplastic rRNA, the concentrations used causing complete chlorosis. Specific effects on peroxisomal enzymes could not be explained by a hypothesis involving herbicide-induced photodestruction in carotenoid-deficient plastids.

Norflurazon is also inhibitory to chlorophyll production. For example Devlin and Karczmarczyk (1975a) found that low concentrations of this compound reduced the chlorophyll content of *Potamogeton pectinatus* under high light conditions (1000 ft-candles), the effect being greatly reduced at a light intensity of 10 ft-candles. They suggest that norflurazon inhibits carotenoid synthesis thus exposing chlorophyll to photo-oxidation. Devlin and Karczmarczyk (1975b) also found that under high light intensity (1000 ft-candles), 1 p.p.m. norflurazon almost completely bleached maize and wheat seedlings, while lucerne was even more sensitive, 0.1 p.p.m. causing almost complete chlorosis. At low light intensity (1 ft-candle) the effect of norflurazon on chlorophyll production was greatly reduced and they explained the results on the basis of herbicide-induced photoreduction in carotene-deficient plastids. Norflurazon also caused progressive deterioration of the chloroplast lamellar system (Devlin, Kisiel and Karczmarczyk, 1976).

Carotenoid synthesis was also blocked by norflurazon (20–1000 μg/ml) in growing and resting cells of *Euglena gracilis*, inhibition occurring prior to the synthesis of the intermediate phytoene (Vaisberg and Schiff, 1976). In addition, chlorophyll synthesis and membrane formation were also blocked but these effects appeared to be secondary to the inhibition of carotenoid synthesis which normally protects chloroplasts from photodestruction. St. John (1976) reported effects by various pyridazones, including norflurazon, on fatty acid composition of the major lipids of wheat chloroplast membranes. The structure of the parent compound pyrazone accounted for inhibition of the Hill reaction and CO_2 fixation, but it lacked activity at the membrane polar lipid site (St. John and Hilton, 1976). Both trifluoromethyl substitution of the phenyl ring and monomethyl substitution of the amine were required for maximum inhibition of carotenoid biosynthesis, while dimethyl substitution of the amine was related to certain changes in the fatty acid of the membrane lipids. Bartels and Watson (1978) also found that disruption of chloroplasts and loss of chlorophyll from wheat seedlings were related to the absence of carotene and accumulation of the carotenoid precursors phytoene and phytofluene. They suggested that dehydrogenation reactions following phytoene formation were inhibited.

The effect of norflurazon applied to *Sinapis alba* on carotenoid biosynthesis has been examined in infrared, red and white light (Frosch *et al.*, 1979). The self-photo-oxidation and photosensitising action of chlorophyll appear to occur after carotenoid protection of chlorophyll against such photo-oxidation is lost. In addition to blocking carotenoid and chlorophyll synthesis, Kunert and Böger (1979) found substantial reduction of α- and β-carotenes under aerobic conditions.

The enhancement of gibberellic acid (GA) sensitivity by certain carotenoid-inhibiting enzymes including norflurazon has been reported by Devlin, Kisiel and Kostusiak (1979). Increase in shoot length following application of GA to imbibing maize seeds was greatly enhanced by simultaneous application of norflurazon (and fluridone), this activity paralleling their ability to block carotenoid biosynthesis.

PYRIMIDINES

The pyrimidines are effective inhibitors of the Hill reaction and their activity as herbicides appears primarily to be due to this action; combination of these compounds with urea herbicides results in additive effects (Hoffman, 1971).

HETEROCYCLIC NITROGEN COMPOUNDS (UNCLASSIFIED)

This group includes **aminotriazole** and **difenzoquat**, which like metribuzin, metamitron and buthidazole, affect photosynthesis either directly by influencing the efficiency of photochemical reactions or indirectly by impairing chlorophyll development.

The most striking symptom of aminotriazole is leaf chlorosis leading to the production of albino leaves, apparently due to chlorophyll destruction and/or inhibition of chlorophyll synthesis (Hall *et al.*, 1954); at sublethal doses, chlorosis may be temporary with subsequent recovery to the normal green colour. The action of aminotriazole on plastid structure has been reviewed by Ashton and Crafts (1973). They point out that mature leaves treated with moderate rates of aminotriazole normally do not become albinised while young expanding leaves do, indicating that pigment synthesis rather than pigment destruction is involved. Cytological examination of the plastids from treated plants has revealed gross alteration of their structure in maize (Jacobson and Rogers, 1961), barley and potato (Pyfrom *et al.*, 1957) or the presence of proteinaceous structures, possibly deformed plastids, in pine (Shive and Hansen, 1958) and wheat (Weier and Imam, 1965). In tomato sublethal concentrations of aminotriazole prevent proplastid maturation (Putala, 1967). Vesicles which would normally differentiate into granal discs are retained in the proplastids, while existing chloroplasts form disorganised granal discs along with scattered vesicles.

Aminotriazole (10^{-4} M) blocks light-induced plastid development in wheat leaf tissue, while in dark-grown plants, membrane formation in proplastids is unaffected (Bartels, 1965). Subsequent work showed that plastids from light-grown wheat treated with aminotriazole lack normal 70S chloroplast ribosomes and 18S fraction I protein (Bartels *et al.*, 1967). These findings were confirmed by Bartels and Weier (1969) who showed that light-treated plants lack grana-fret membrane systems and chloroplast ribosomes and suggested that aminotriazole may interfere with chloroplast nucleic acid metabolism.

Aminotriazole is also known to influence aspects of leaf metabolism associated with leaf chlorosis. In oats, chlorosis, induced by aminotriazole (10^{-3} M), is associated with increased concentrations of amino acids and free ammonia (Baumann and Günther, 1976a), inhibition of formation of sucrose from glycolate (Baumann and Günther, 1976b) and an increased rate of chloroplast rRNA degradation (Baumann, Baumann and Günther, 1977).

Aminotriazole is also known to induce rapid degradation of chlorophyll a and b in leaves (Misra, Kapoor and Choudhri, 1974) and to inhibit $^{14}CO_2$ fixation (Kulandaivelu and Gnanam, 1975). The latter believe that aminotriazole has a generalised non-specific inhibitory action on all the partial reactions of photosynthesis, contrasting with growth regulator compounds such as IAA, GA and 2,4-D which appear to act on the electron transport chain connecting PS II and PS I.

Bentazone also arrests net photosynthesis, causing depletion of carbohydrate reserves and loss of chloroplast membrane integrity (Wergin and Potter, 1974). Treatment of cocklebur (*Xanthium pennsylvanicum*) with bentazone (0.25 kg/ha) resulted in leaf necrosis within 36 h of treatment. Potter and Wergin (1975) treated cocklebur with 0.01–1 kg/ha bentazone and exposed the treated plants to a range of light (21–86 klx). Their results show that light is required for development of necrosis, the higher the illumination, the more rapidly necrosis develops. They believe that changes in shape and distribution of the chloroplasts are not directly due to bentazone but propose that photo-induced toxic by-products result from the inhibition of photosynthesis.

The effect of bentazone on photosynthesis of rice and certain weed species was examined by Mine and Matsunaka (1975). Generally, the herbicidal effect developed slowly, after translocation of the herbicide had occurred, though high concentrations caused more rapid effects. They suggest that the slow effect may be due to inhibition of photosynthesis and cite in evidence the fact that (a) bentazone inhibits the Hill reaction in isolated chloroplasts, (b) that it rapidly inhibits photosynthetic CO_2 fixation in *Cyperus serotinus* (susceptible) and other species and (c) that bentazone injury is prevented by endogenously or exogenously applied carbohydrate.

Retzlaff and Hamm (1976) reported that the reduction in the rate of CO_2 fixation by wheat plants after treatment with bentazone was correlated with the uptake of the active ingredient by the leaf tissue. A subsequent increase in assimilation was associated with a decline in herbicide level within the leaf resulting from metabolism to hydroxy bentazones. The reversibility of action of bentazone on the alga *Bumilleriopsis* was found to depend on the duration of exposure

(Retzlaff and Hamm, 1976). Chloroplast material, isolated from cells, grown for several days in the presence of the herbicide showed inhibition of PS II, though PS I was not affected. Recovery from these long-term treatments was slow and incomplete compared with that of short-term treatments when recovery was rapid and complete. They believe that irreversibility of long-term inhibition could result from a metabolic process which brings about the binding of bentazone (or derivative) to the thylakoids following *in vivo* but not *in vitro* treatment with this herbicide. Potter (1977) monitoring the movement of biologically active levels of bentazone following application to roots, stems, petioles and leaves of cocklebur (*X. pennsylvanicum*) found that photosynthesis was inhibited by application of the herbicide to any part of the plant.

Another herbicide of this group, **oxadiazon**, also affects chloroplast development. This effect on the ultrastructure of barnyard grass chloroplasts has been examined by Kawamura, Miyake and Maeda (1976). Herbicidal action occurred only in light, greening and elongation of leaves of etiolated seedlings being markedly inhibited while leaf tissues were distorted and vascular bundles disintegrated. The affected chloroplasts were characterised by prolamellar bodies and a circular lamella structure though rupture of the chloroplast membrane and change in chloroplast shape did not occur. Oxadiazon treatment also affected the nucleus and cytoplasm.

HETEROCYCLIC COMPOUNDS

Ethofumesate also inhibits photosynthesis, though recovery can occur in more resistant species. For example, this herbicide inhibits photosynthesis (and respiration) in redroot pigweed (*Amaranthus retroflexus*), wild mustard (*Sinapis arvensis*) and sugar-beet at the 2–4 leaf stage within 4 h of treatment. However after 96 h, wild mustard (R) and sugar-beet (R) recover, but pigweed (S) only partially, suggesting that metabolism is involved in selectivity (Duncan, Meggitt and Penner, 1978).

ORGANOARSENIC COMPOUNDS

The organoarsenic herbicides such as **DSMA** and **MSMA** affect the functioning of many organelles including chloroplasts (Ashton and Crafts, 1973). Application of MSMA to isolated chloroplasts of *Sorghum halepense* causes inhibition of the Hill reaction (I_{50}= 1.8×10^{-3} M) and root exposure to this compound (10^{-2} and 10^{-3} M) causes inhibition of photosynthesis (Spilsbury, 1972).

ORGANOPHOSPHORUS COMPOUNDS

Glyphosate can inhibit photosynthetic reactions and affect chloroplast ultrastructure. Incubation of cells of the green alga *Scenedesmus* sp. with glyphosate (7×10^{-4} M) results in a 50% inhibition of oxygen evolution after 1 h in light, the degree of inhibition increasing with higher light intensity and temperature (van Rensen, 1974). Glyphosate (<10 mM) has no immediate effect on electron transport in isolated spinach chloroplasts, but after 15 min pretreatment with the herbicide, inhibition is obtained at levels below 10 mM. Van Rensen concludes that the herbicide inhibits electron transport at or near PS II.

Effects on chloroplast ultrastructure resulting from glyphosate treatment have been reported by a number of workers. Campbell, Evans and Reed (1973) found that application of glyphosate (0.56–4.49 kg/ha) to *Agropyron repens* damages the chloroplast organelles. They subsequently reported partial to complete disruption of the chloroplast envelope and swelling of the rough endoplasmic reticulum with later formation of the vesicles; eventually the chloroplast becomes completely disrupted (Campbell, Evans and Reed, 1976). Examination of mesophyll chloroplasts from leaves of purple nutsedge (*Cyperus rotundus*) which had been immersed for 1 h in glyphosate (2.8×10^{-2} M) reveals some fretwork disarray of internal membranes, more diffuse stroma and progressive swelling (Hoagland and Paul, 1978b); after 36 and 48 h increasing numbers of spherical chloroplasts and rupture of some chloroplast envelopes are evident (Hoagland and Paul, 1978c).

Glyphosate causes significant reductions in the chlorophyll and carotenoid contents of leaves of *C. rotundus* (Abu-Irmaileh and Jordan, 1978), and in maize seedlings treated with glyphosate (1.12 kg/ha) (or aminotriazole) (Ali and Fletcher, 1978). Application of glyphosate to bean (*Phaseolus vulgaris*) decreases leaf conductance and net photosynthesis simultaneously (Shaner and Lyon, 1979a, b). In a constant environment (300 μE m^{-2} s^{-1} light, r.h. 50% and 21 or 27°C) spontaneous stomatal cycling occurs in response to glyphosate treatment while at higher temperature (37° C) or r.h. (80%), no cycling was recorded.

Nucleic acid and protein metabolism

The actions of herbicides on nucleic acid and protein metabolism have recently been reviewed by Ashton and Crafts (1973), Cherry (1976) and Moreland (1980).

Cell division, nucleic acid metabolism and protein synthesis are essential processes in plant growth and development; all are biosynthetic reactions powered by ATP. Herbicides may act directly or indirectly on these reactions by inhibiting ATP synthesis and thus energy-dependent reactions such as nucleic acid or protein synthesis.

Cell division is known to be affected by a number of herbicides. They may act by interfering with the synthesis or active transport into the nucleus of precursors required for DNA synthesis during interphase, by modifying the chemical and physical properties of DNA, by interfering with spindle formation or function or by inhibitory cell wall formation (Moreland, 1980). Compounds active in this way include *N*-phenylcarbamate, dinitroanilines, phenoxyalkane carboxylic acids, pronamide (*N*-[1,1-dimethylpropynyl]-3,5-dichlorobenzamide), di-allate and CDAA (*N,N*-diallyl-2-chloroacetamide) (Linck, 1976).

The effect of herbicides on DNA, RNA and protein synthesis by excised plant tissue or isolated mesophyll cells has been studied by assessing their effect on ^{14}C-precursor incorporation. The most inhibitory herbicides on non-photosynthetic tissues include ioxynil, dinoseb, pyriclor and chlorpropham, while in photosynthetic tissue, additional inhibitory herbicides include s-triazines, phenylureas and dinitroanilines (Moreland, 1980). In addition many of the strong inhibitors of these reactions also inhibit oxidative photophosphorylation and correlations have been established between inhibition of RNA and protein synthesis and ATP levels (Gruenhagen and Moreland, 1971).

The action of herbicides on nucleic acid and protein synthesis

HALOALKANOIC ACIDS

The haloalkanoic acids interfere with the meristematic activity of root tips and apical meristems, modify protein structure and thus enzyme and membrane permeability. Secondary effects include alteration of carbohydrate, lipid and nitrogen metabolism (Ashton and Crafts, 1973). Amino acid contents may be reduced as reported for **dalapon** (5–20 kg/ha) applied to *Cynodon dactylon* (Srinivasan and Sakharam Rao, 1973). Foliage treatments cause decreases in alanine, proline and tryptophan while aspartic and glutamic acids increase; DNA and RNA also increase in foliage but decrease in stolons. Nasyrova, Aripova and Babaev (1974) also report that dalapon treatment of cotton seeds (cv. Tashkent-1) reduces total nucleic acid and protein nitrogen contents.

PHENOXYALKANOIC ACIDS

The characteristic symptoms of phenoxy acid activity include the proliferation of tissue resulting in epinasty and swelling of leaves and stems, generally accompanied by changes in the nitrogen levels of the treated plants. Further, nitrogen, amino acid and protein contents of the stems increase (Sell *et al.*, 1949), possibly due to loss from leaves and roots (Weller *et al.*, 1950). However, there is evidence that increased protein levels in stems are paralleled by the increase in the level of proteolytic enzymes; in leaves the concentrations of protein and digestive enzyme decrease in proportion.

Many researchers have reported increases in RNA following **2,4-D** treatment (West, Hanson and Key, 1960; Key and Hanson, 1961; Basler and Nakazawa, 1961), though high concentrations may inhibit RNA formation (Shannon, 1963; Basler and Hanson, 1964). Key and Shannon (1964) found that the 2,4-D-induced increase in RNA results primarily in the ribosomal fraction with an apparent transfer of RNA from the nuclear to ribosomal fraction; messenger RNA also increases. It has been suggested that such excessive nucleic acid and protein synthesis would preclude normal cell functioning and may well be the primary biochemical mechanism of action (Key *et al.*, 1966). They suggest that 2,4-D action results in renewal of RNA and protein synthesis leading to massive tissue proliferation and disorganised growth. Where proliferation of this nature occurs in the phloem it may lead to distortion and disruption, resulting in starvation of the roots and subsequent death of the plants. Key *et al.* (1966) believe that uncoupling of phosphorylation cannot be the major mechanism of action since it is incompatible with the massive new synthesis of macromolecules and cell divisions which occur in response to lethal doses of 2,4-D.

Increased RNAase activity and microsomal RNA formation in pea seedlings treated with herbicides of **MCPA** and **mecoprop** (and 2,4-D) have also been reported (Okubo and Switzer, 1965). Herbicidal levels of picloram have similar activity to that of 2,4-D, DNA and RNA synthesis being enhanced in susceptible species (soybean and cucumber) but not in resistant species (barley and wheat) (Malhotra and Hanson, 1966). They found that the activity of DNAase and RNAase bound to membranes is inversely related to herbicide sensitivity.

It has been postulated that auxin compounds such as IAA and 2,4-D act as allosteric effectors (Monod, Changeux and Jacob, 1963), perhaps acting by derepressing RNA synthesis, thus triggering polypeptide chain synthesis. Work with maize mitochondria suggests that IAA acts as an allosteric effector, possibly acting on the CoA–acetyl-CoA complex (Sarkissian and McDaniel, 1966). Armstrong (1966) proposed that a specific sRNA fraction must be charged with auxin to derepress RNA synthesis, triggering polypeptide chain synthesis. Continued synthesis of RNA would take place where the auxin-specific sRNA is saturated and inhibition of growth at high auxin concentrations would be due to a failure in the specificity of the aminoacyl-RNA synthesis leading to premature chain termination. Such miscoding would prevent meaningful protein synthesis.

In their review of biochemical mechanisms of the phenoxy acids, Robertson and Kirkwood (1970) conclude that the main mechanism of action of 2,4-D and probably other phenoxy acids involves a complex series of reactions initiated by the derepression of the gene-regulating synthesis of the enzyme RNAase. The resultant synthesis of RNA and protein is accompanied by massive cell proliferation, provided a proper auxin–cytokinin balance exists. This balance presumably depends upon the dose of herbicide applied and effective movement to the sites of action. Since the newly synthesised RNA and protein apparently migrate to the stem tissue, proliferation leads to disruption of the transpiration and translocation systems resulting in assimilate accumulation in the shoots and starvation of the roots. Other mechanisms of action including inhibition of the Hill reaction and oxidative phosphorylation appear to be of secondary importance. However, their significance should not be underestimated since any inhibition of these processes must influence the efficiency of energy-dependent functions including mineral ion uptake, photosynthate translocation and indeed the absorption and translocation of the herbicide itself.

Since 1970, many papers have been published on the effect of 2,4-D and other phenoxyalkanoic herbicides on aspects of protein synthesis. Ezerzha *et al.* (1971) examined the localisation of ^{14}C-ring- or side-chain-labelled 2,4-D and its metabolites in the cell organelles of wheat (cv. Kizil Shark) and upland cotton (cv. 108-F), the rate of application being 2 kg/ha or 100 g/ha respectively. They analysed the radioactivity of leaf homogenates of cytoplasm, proteins, chloroplasts, mitochondria, nuclear and ribosome fractions and reported that differences in sensitivity could not be explained on the basis of change in location of ^{14}C-2,4-D which did not differ between species. They tentatively concluded that sensitivity of cotton was attributable to the longer exposure to herbicide of the nucleoproteins of the nuclei and ribosomes of cotton and suggest that the resistance of wheat is due to its containing more protein capable of inactivating the herbicide. It was later confirmed that the acid proteins of wheat may play a role in inactivating 2,4-D (Ezerzha and Kulikov, 1973). Ezerzha and his co-workers subsequently examined the interaction of 2,4-D with nucleotides from twelve crop species and postulated that an interaction may occur between 2,4-D and the DNA spiral with the formation of supermolecular complexes between 2,4-D and adenine (Ezerzha, Kulikov and Ibragimova, 1975).

The effect of 2,4-D on nucleic acid levels has been examined by several workers. For example Chen, Switzer and Fletcher (1972) obtained results which suggest that differential effects on protein synthesis may be the basis for selectivity of 2,4-D, **2,4,5-T**, dicamba and picloram in cucumber (susceptible) and wheat (tolerant). While DNA and protein levels were increased by all of the test herbicides in both species, RNA levels decreased in wheat but increased in cucumber. Nucleic acid synthesis was also affected by treatment of etiolated soybean seedlings with 2,4-D (5×10^{-4} M) (Chkanikov, Mikityuk and Makeev, 1973). The synthesis of nucleic acids was activated in differentiating cells of the basal portion of the hypocotyl but decreased in the apical part, the latter being attributed to the inhibition of oxidative phosphorylation by 2,4-D. Éidel'nant *et al.* (1974) found evidence to suggest that the accelerated turnover of RNA after treatment of upland cotton (cv. 108-F) with 2,4-D (30 mg/l) was mediated by stimulation of ethylene biosynthesis.

Effects on RNAase activity have also been reported. For example Nevzorova and Voevodin (1977) examined the effect of 2,4-D sodium on RNAase activity of resistant spring wheat (cv. Lyutestsens. 758) (0.8 kg/ha) and sensitive barley (cv. Viner) (0.036–1.2 kg/ha). In the resistant plants, 2,4-D (0.8–1.2 kg/ha) promoted RNAase activity, possibly due to *de novo* synthesis of RNAase in cells, or to a higher rate of circulation of RNAase. In the sensitive species stimulatory rates of 2,4-D (0.036 g/ha) reduced, while herbicidal rates (0.8–11.2 kg/ha) increased RNAase activity. These workers subsequently examined the effect of several years treatment with 2,4-D (0.8 kg/ha) on the content of nucleic acid and other phosphorus compounds in the endosperms of grains of spring wheat (cv. Minskaya and Zarya). There was no direct correlation between herbicide resistance and RNA/DNA contents (Voevodin, Nevzorova and Kazarina, 1977).

Treatments with 2,4-D have also caused changes in protein content and activity of certain enzymes. Treatment of barley grains (cv. Damazy) with 2,4-D (or simazine or linuron) results in inhibition of ribonucleases and α-amylases, deterioration of certain protein fractions in embryos and endosperms and a decline in heterogeneity of simple proteins (Grzesiuk, Login and Nowak, 1973). Treatment of spruce (*Picea abies*) seedlings with 2,4,5-T (100–300 mg/l) causes a small decrease in soluble protein level and marked disturbances in the activity of proteases and acid phosphatases (Tonecki, 1975).

Studies of nucleic acid and histone-precursor incorporation into nuclei show that 2,4-D may have a retarding effect on the process of transcription in differentiating nuclei and an accelerating effect on histone synthesis (Sen, 1975). Effects on transcription occur at the level of initiation of RNA polymerase activity and once transcription is initiated the herbicide is unable to exert its retarding effect (Sen, 1978).

Effects on protein and amino acid metabolism have been reported by a number of workers. For example, Krishchenko, Kovach and Gruzdev (1973) found that plants of spring wheat (cv. Krasnozernaya) treated with 2,4-D amine (1 mg/plant) show an initial increase in the rate of protein synthesis which, however, gradually declines as the growing season progresses. The free amino acid content of wheat foliage and roots was increased (histidine, arginine, proline and α-aminobutyric acid). Treated plants synthesise proteins with relatively high proportions of proline, alanine and aspartic acid and low proportions of lysine, leucine and phenylalanine. Subsequent work with this cultivar showed that after 24 h, 2,4-D (1 mg/plant) or chlormequat (2 mg/plant) inhibit protein synthesis and enhance the accumulation of free amino acids. While these affects disappear by the boot-to-heading stage in the case of 2,4-D, they persist in the case of chlormequat (Gruzdev, 1979). However, treatment of barley with 2,4-D, dicamba and picloram showed that herbicidal effects on amino acids and proteins are most pronounced at the tillering-to-boot stages and decline at the initiation of grain formation; dicamba and picloram cause greater changes in amino acids than does 2,4-D (Gruzdev, Poshitnaya and Gruzdev, 1979).

Foliage applications of MCPA (0.83 kg/ha) to *Raphanes raphanistrum* increases the protein and RNA contents of leaves and roots, though no significant changes were observed in the DNA content (Rensburg and de Villiers, 1979). There is no evidence that 2,4-D induces mitotic irregularities in suspension cultures of *Vicia hajastane* and *Haplopappus gracilis* treated with 2,4-D (0.1–10 and 0.1–5 μg/ml respectively) (Singh and Harvey, 1975). Indeed the higher the concentration of 2,4-D in the culture medium, the lower the frequency of mitotic irregularities. However, treatment of soybean hypocotyls with 2,4-D (5×10^{-4} M) (or IAA) inhibits apical meristem activity, while cells in the sub-apical zone undergo isodiametric elongation and dedifferentiation. RNA polymerase and ethylene synthesis are stimulated in the sub-apical zone (Mikityuk, Petelina and Chkanikov, 1977). They conclude that the cessation of cell division in the apical meristem of soybean caused by 2,4-D is not mediated by ethylene and is probably not directed towards RNA polymerase.

AROMATIC ACIDS

There is evidence that **2,3,6-TBA**, chlorthal-dimethyl and **dicamba** affect nitrogen metabolism in treated plants. Zhirmunskaya (1966) reported a two- to three-fold increase in total nitrogen of Canada thistle (*Cirsium arvense*) and perennial sowthistle (*Sonchus arvensis*), 3 to 24 months following treatment with 2,3,6-TBA. Intensive synthesis of protein and accumulation of non-protein forms of nitrogen occurred at the expense of amides and this increase was not due to translocation from the shoots.

Dicamba was reported to influence the DNA-precipitating properties of histone and it was suggested that it may thus interfere with the normal functioning of the genetic mechanism (Quimby, 1967). It inhibits gibberellic acid induced α-amylase biosynthesis in barley grains but has little effect on RNA or protein synthesis (Moreland *et al.*, 1969). On the other hand, Arnold and Nalewaja (1971) found that dicamba treatment increases the RNA and protein contents in wild buckwheat at the 5–8 cm growth stage and at flowering; similar increases were noted in wheat at the boot stage. It was suggested that a DNA–histone–dicamba complex may be formed.

Contrary to the findings of Moreland *et al.* (1969) it seems that dicamba may not affect enzyme synthesis to a sufficient degree to fully explain its phytotoxic effects. For example, Ashton *et al.* (1968) found that dicamba (10^{-3} M) had little effect on the development of proteolytic activity in germinating squash seeds, though it had a marked effect on seedling growth. Again, Penner (1970) reported that dicamba (10^{-4} M) had no significant effect on phytase activity in germinating barley grains.

More recently, Khristov (1974) reported that dicamba–dimethylamine (as Banvel-D) (1×10^{-5}–1×10^{-2} M) applied to germinating grains of maize, increased the proportion of aberrant cells due to its action as an inhibiting mutagen. Ladonin *et al.* (1974) found that dicamba (as 'Banvel-D' 0.02%) reduced the protein nitrogen content of seedlings and disrupted the metabolism of free amino acid in leaves and stems, causing an imbalance in their production and ratio and adversely affecting protein synthesis. However, there was little effect on the amino acid composition of the hydrolysates of total protein.

Chloramben does not appear to inhibit protein synthesis (Mann *et al.*, 1965a; Moreland *et al.*, 1969) nor RNA synthesis or gibberellic acid induced α-amylase synthesis in de-embryonated barley grains

(Moreland *et al.*, 1969). This is contrary to the results of Penner (1968) who had previously reported that chloramben inhibits RNA synthesis and α-amylase synthesis in intact grains of barley. He also found that chloramben increases the phytase level in the embryonic axis of 2-day-old squash seedlings (Penner, 1970).

Chlorthal-dimethyl (2–16 p.p.m. w.) applied to the soil, reduces the root and shoot growth of oats and green foxtail (*Setaria viridis*), causes hypertrophy of the meristems and irregular differentiation; it also reduces the levels of starch, protein and nucleic acid in seedling tissues (Shaybany and Anderson, 1972). Treatment of foxtail millet (*Setaria italica*) seedlings with chlorthal-dimethyl (100 mg/l) decreases levels of DNA and ribosomal RNA and increases soluble RNA (Chang and Smith, 1972a). Electron microscopy of meristematic cells from plants treated with 2 and 20 mg/l chlorthal-dimethyl reveals an interruption of cell division with the presence of giant nuclei and nucleoli (Chang and Smith, 1972b).

AMIDES

There is evidence to suggest that compounds of this group may interfere with RNA or protein synthetic mechanisms.

For example, **CDAA** (*N,N*-diallyl-2-chloroacetamide) (now superseded) was reported to inhibit ^{14}C-leucine incorporation into tissue segments of barley and hemp sesbania (*Sesbania exaltata*) (Mann *et al.*, 1965b), though protein synthesis in soybean (resistant) was not affected (Moreland *et al.*, 1969). In addition inhibition of certain hydrolytic enzymes by CDAA has been found to occur. For instance inhibition of proteolytic activity in squash cotyledons has been reported by Ashton *et al.* (1968) and of gibberellic acid induced α-amylase synthesis in barley seedlings by Moreland *et al.* (1969). It is uncertain whether such inhibitory effects result from interference with protein synthesis *per se* or with hormonal control mechanisms, or indirectly result from effects on ATP synthesis (Ashton and Crafts, 1973).

Diphenamid appears to be less effective on such mechanisms, failing to inhibit protein synthesis in oat roots (Briquet and Wiaux, 1967), only slightly affecting ^{14}C-leucine incorporation into protein in barley coleoptiles and hemp sesbania hypocotyls (Mann *et al.*, 1965b) and scarcely affecting α-amylase synthesis in barley (Penner, 1968) or proteolytic activity in squash cotyledons (Ashton *et al.*, 1968)). In contrast to those findings, Briquet and Wiaux (1967) reported that diphenamid (10 p.p.m.) completely inhibits RNA synthesis in barley roots, possibly reflecting the greater sensitivity to diphenamid of RNA synthesis in roots, or indicating a greater concentration of the herbicide at this site of action (Ashton and Crafts, 1973).

Propyzamide has been shown to markedly affect DNA, RNA and protein levels in rhizomes of couch grass (*Agropyron repens*) (Smith, Peterson and Horton, 1971), and inhibit mitosis in oat root tips (Carlson, Lignowski and Hopen, 1975). The latter found evidence of arrested metaphase, multi-nucleate cells, polyploidy and distorted nuclei; incorporation of ^{14}C-leucine into protein was increased by 20% following a 4 h incubation.

MDMP (2-[4-methyl-2,6-dinitroanilino]-*N*-methylpropionamide), was found to be a potent inhibitor of mRNA-dependent amino acid incorporation into wheat embryos (*in vitro*) (Weeks and Baxter, 1972). It appeared to prevent peptide chain inhibition by specifically blocking the interaction of 60S ribosomal sub-units with the 40S ribosomal sub-unit–mRNA–Met–tRNA complex. Exposure of wheat root tips to MDMP increased monomers and dimers at the expense of the high-ribosome-number polyribosomes (Baxter *et al.*, 1973).

ANILIDES

Moreland *et al.* (1969) reported that **propanil** strongly inhibited RNA and protein synthesis and repressed α-amylase synthesis in half-seeds of barley; Mann *et al.* (1965a) however, had previously found that propanil had no effect on protein synthesis. Effects on oxidative phosphorylation and membrane permeability (Hofstra and Switzer, 1968) may indirectly influence protein synthesis.

Propachlor has been found to inhibit protein synthesis (Duke *et al.*, 1967), GA-induced α-amylase synthesis in barley half-seeds (Smith and Jaworski, cit. Ashton and Crafts, 1973) and phytase development in squash (Penner, 1970). Dhillon (1971) examined the cytological effects and differential activity of propachlor in several species including squash (*Curcubita maxima*). He postulates that inhibition of amino acid activation is the primary mechanism of action thus preventing the formation of proteinaceous compounds (including enzyme complexes) which break down storage material and alter cell walls to enable cell elongation. Subsequently Dhillon and Anderson (1972) reported that propachlor (4 or 16 p.p.m.) in the growth medium inhibits root and shoot growth of oat or squash, inhibits cell division in onion tips and auxin-induced cell elongation of oat coleoptiles; such treatment also prevents the normal senescence of squash coleoptiles by inhibiting the autolysis of protein and lipid reserves.

A number of workers have examined the mode of action of **alachlor**, comparisons being made in certain instances with other acetanilide compounds. Chandler, Croy and Santelmann (1972) studied the influence of alachlor (0.005 and 0.05 mM) on nitrate reductase, nitrate, amino acid and water-soluble protein levels in wheat seedlings treated with alachlor for varying periods before or after addition of nitrate to the nutrient solution. The amino acid and protein levels were depressed by high but not by low concentration of alachlor. No significant effect was obtained on leaf nitrate reductase activity following treatment of soybean with alachlor, metolachlor or oryzalin (Lolas, 1980).

The effects of alachlor on protein synthesis appear to differ however, according to susceptibility. Rao (1974) found that in barley (susceptible) alachlor, propachlor and pyrnachlor inhibited hydrolytic enzyme production, reducing the levels of ribosomes and polysomes and peptide synthesis; in maize (resistant) protein synthesis was unaffected. Rao and Duke (1974) concluded that these acetanilides affect polyribosome and protein synthesis in susceptible but not resistant species and that the major effect occurs on nascent protein formation. Rao and Duke (1976) also found that these herbicides inhibit GA-induced production of protease and α-amylase in de-embryonated barley (cv. Schuyler) grain, production of the former being the more sensitive. They believe, however, that the effect of these compounds on seed germination and seedling development may not be restricted to these two effects.

The effect of alachlor (2.5×10^{-4} M) on growth of *Avena* seedlings and its interaction with gibberellic acid (GA) and indoleacetic acid (IAA) has been reported by Chang, Marsh and Jennings (1975). While IAA does not prevent inhibition by alachlor, treatment with GA (10^{-3} M), prior to the herbicide treatment, overcomes the inhibitory effect of alachlor. Narsaiah and Harvey (1977) also found evidence of an interaction between alachlor and GA. Treatment of maize epicotyl segments with GA in combination with alachlor results in growth comparable with that of untreated tissues; the magnitude of response depends on the concentrations of GA and herbicide and on the duration of treatment. Further evidence of the complex nature of the mode of action of such compounds is evident from the results of Marsh, Bates and Downs (1976). They found that rapid inhibition of oat seedling growth by alachlor was paralleled by reductions in root elongation, cortical cell enlargement and mitotic activity (10^{-6}–10^{-4} M alachlor). Uptake of a number of compounds including leucine by excised root tips was inhibited as was auxin transport in coleoptile sections. They attribute the growth-inhibitory effects of alachlor to general inhibition of the transport system.

Further evidence of the effect of 250–2000 μg/ml solutions of alachlor (as 'Lasso', alachlor 50% e.c.) and propanil (as 'Stam F-34', propanil 35% e.c.) in causing cytogenetic aberrations has been recorded by Das *et al.* (1978). These include bridges, fragments and laggards at meiotic anaphase and non-orientation, precocious separation and univalents at meiotic metaphase I and lagging chromosomes, bridges and fragments at anaphase I. The evidence of meiotic irregularities was relatively lower in sprayed plants than in those derived from the herbicide-treated seed.

Additional evidence of effects of alachlor and **metolachlor** on root growth, cell division and enlargement in peas (cv. Alaska) and oats (cv. Victory) was obtained by Deal and Hess (1980). Significant reductions occur in the mitotic indices of pea roots treated for 48 h with alachlor (5×10^{-6} M) or metolachlor (1×10^{-5} M) and of oat roots treated with 1×10^{-7} M metolachlor or 1×10^{-6} M alachlor. Significant reductions in the elongation of etiolated oat coleoptiles and pea hypocotyls were also reported. They conclude that growth inhibition of plants treated with these herbicides results from both an inhibition of cell division and cell enlargement.

The mode of action and selectivity of metolachlor on growth and germination of several crops and weeds has also been examined by Davis, Pillai and Truelove (1979). The ability of cucumber root tips to absorb ^{14}C-leucine was inhibited by metolachlor (10^{-5} M) but the percentage of ^{14}C-leucine absorbed which was actually incorporated into protein was unaffected. An earlier report had suggested that membrane damage may play a major role in the phytoxic action of metolachlor since 10^{-4} and 10^{-5} M concentrations caused leakage of previously absorbed ^{32}P from the roots of onion (S) and cucumber (MS) but not maize (R).

There is evidence that **benzoylprop-ethyl** may inhibit DNA synthesis and to a lesser extent RNA and protein synthesis (Radtsev *et al.*, 1975). This was confirmed by Bergmannova and Taimr (1980) who studied the effect of this herbicide on the uptake, transport and metabolism of ^{32}P in *Avena fatua* and *Triticum aestivum*; ^{32}P incorporation into all fractions (organic, lipidic and nucleic acid) was hampered.

NITROPHENOLS

Inhibition of RNA and protein synthesis by soybean hypocotyl tissue has been reported for **DNBP** (2-s-butyl-4,6-dinitrophenol) (Shokraii and Moreland, 1966), **dinoseb** and certain other herbicides (Gruenhagen and Moreland, 1971). However, these nitrophenol compounds strongly inhibit oxidative phosphorylation and the effects on protein synthesis may partly reflect the limited energy available for these biosynthetic reactions. Van Hoogstraten (1973), using cotyledons of 4-day-old pea seedlings found that dinoseb, ioxynil and PCB (1 μM) were the most active of the herbicides tested in inhibiting the incorporation of phenylalanine into protein, the esterification of tRNA with phenylalanine and the incorporation of phenylalanine from Phe–tRNA. The most inhibitory herbicide, **PCP**, had almost no effect on exogenous ATP *in vitro* indicating that this and possibly other herbicides at relatively high concentrations, are capable of interfering with protein synthesis *per se*.

NITROANILINES

The dinitroanilines are known to affect several biochemical mechanisms in plants causing changes in carbohydrate, lipid and nitrogen content, nucleic acid levels and oxidative phosphorylation; they also interfere with cell division.

Trifluralin, for example, causes swelling and distortion of root tips in *Zea mays*, *Triticum aestivum* (Lignowski and Scott, 1971), rice seedlings (Mercado and Talatala, 1973), *Vicia faba* (Vaverkova and Vancova, 1973), soybean (Hunyadi and Bogdan, 1976) and sorghum (Lin, 1978). It also inhibits mitosis in root tips of onion and wheat (Lignowski and Scott, 1972), cotton (Rizk, 1973) and *Allium sativum* (Delcourt and Deysson, 1976). There is evidence to suggest that inhibition is due to arrested metaphase reflecting disruption of the spindle apparatus (Lignowski and Scott, 1972; Delcourt and Deysson, 1976). Loss of cortical and spindle microtubules of wheat and maize roots treated with trifluralin (or **oryzalin** or pronamide) has been reported by Bartels and Hilton (1973), possibly reflecting interference with the synthesis of microtubular protein.

Cellular abnormalities in the swollen stem of treated bean (*Phaseolus vulgaris*) (cv. Tenderette) or soybean (cv. Corsoy) plants have also been observed in the form of thin-walled cells, elongated xylem elements, hypertrophy–hypoplasia of cells and anomalous rings of vascular tissue in the cortex (Struckmeyer, Binning and Harvey, 1976).

Upadhyaya and Nooden (1976) believe that colchicine and the dinitroanilines may have similar actions at the molecular level, producing similar patterns of inhibition of elongation, swelling in the elongation zone, depolarised cell enlargement, disruption of cell differentiation and polyploidisation in maize seedling roots. Swelling commenced in the case of oryzalin at about 10^{-8} M, trifluralin at 10^{-7} M and colchicine at 5×10^{-5} M. Further study revealed that oryzalin induced swelling and inhibited elongation of apical root segments in maize (susceptible) and peas (resistant); the latter occurred at a higher concentration level. Considerable binding occurred in the insoluble fraction, presumably to membranes (Upadhyaya and Nooden, 1978).

The effect of trifluralin on amino acid and amide content and amylase and peroxidase activity has been examined by Mercado and his co-workers. Trifluralin (50 p.p.m.), applied to rice seedlings for 6–48 h, reduces leucine, alanine, glycine, methionine, phenylalanine, valine, serine and histidine and increases arginine, asparagine and glutamic acid; aspartate or glutamate are reduced or absent (Lapade and Mercado, 1973). They postulate that trifluralin causes a shift in the metabolism of glutamate and aspartate towards the formation of arginine, glutamine and asparagine. Treatment of mung bean (*Phaseolus aureus*) seedlings with trifluralin (50 p.p.m.) increases the level of glutamic acid and asparagine and reduces that of arginine (Mercado and Baltazar, 1973). Trifluralin (5 p.p.m.) also affects amylase activity of rice seedlings (Robles and Mercado, 1973) and peroxidase activity of maize and horse-radish seedlings; peroxidase from maize roots is the more responsive (Begonia, Mercado and Robles, 1973). Effects of trifluralin on the nitrogen metabolism of tomato seedlings have been reported by Bandy and Krieg (1974). They found that the free amine concentrations of the leaves decreases and that in the stem increases as the level of nitrate in the solution increases; substitution of ammonium for nitrate greatly reduces the effect of trifluralin on seedling growth and morphology. When nitrate and ammonium are supplied together a reduction in herbicide effects occurs.

CARBAMATES

Carbamate herbicides have been shown to inhibit RNA synthesis and protein synthesis as well as oxidative phosphorylation and the Hill reaction. For example Mann *et al.* (1965a) found that propham, chlorpropham and barban reduced radioactive methionine incorporation into a mainly protein polymer fraction of etiolated seedling sections. Several plant species were used and the degree of inhibition by **chlorpropham** was related to the susceptibility of the test species. Mann *et al.* (1965b) also reported that chlorpropham strongly inhibits the incorporation of ^{14}C-leucine into protein using barley coleoptiles and hemp sesbania (*Sesbania exaltata*) hypocotyls. It was later shown that chlorpropham and barban blocked the gibberellin-induced synthesis of amylase in barley half-seeds and the decapitation-induced rise in the rate of protein synthesis in sesbania hemp hypocotyls (Mann *et al.*, 1967). While Moreland *et al.* (1969) confirmed that chlorpropham inhibits ^{14}C-leucine incorporation into protein, they did not find significant inhibition of gibberellin-induced amylase synthesis. They showed that chlorpropham inhibited RNA synthesis and suggest that this herbicide inhibits RNA and protein synthesis by interfering with ATP production and thus the energy required to drive these biosynthetic reactions.

The effect of **propham** on cell division of the green alga *Oedogonium cardiacum* was examined by Coss and Pickett-Heaps (1974). The results suggest that propham (5.5×10^{-4} M) prevents assembly and orientation of microtubules and increases the number of functional polar bodies. Propham has also been observed to inhibit cell division in pea roots, distorting nuclear reformation and inhibiting DNA, RNA and protein synthesis (Rost and Bayer, 1974). However, further investigation by these workers revealed that the effect was concentration dependent. Low concentrations of propham (10^{-5} M) applied to pea root tip meristems induce abnormal mitotic figures but do not inhibit DNA synthesis or reduce the number of dividing cells. Higher concentrations of propham (10^{-3} M) inhibit DNA, RNA, protein synthesis and mitosis. They conclude that the most likely mode of action of propham is the inhibition of certain mitotic specific proteins.

Chlorpropham also inhibits cell division and protein synthesis though there is evidence to suggest that in potato tissue, the latter does not occur by blocking mRNA transcription (van Winkle and Ewing, 1974). Inhibition of mitosis has also been reported to occur in cell suspensions of soybean and carrot following treatment with chlorpropham (0.1, 5 and 10 μM) (Davis *et al.*, 1977). Mitotic index studies of intact 3-day-old soybean roots showed that the 2-hydroxy derivative inhibited to a greater extent than did chlorpropham, whereas the 4-hydroxy derivative produced only a slight reduction. Chlorpropham is also known to inhibit elongation and increase radical expansion of root cells (Bystrova, 1974).

Barban also affects the metabolism of RNA, protein and nucleoproteins, either directly or indirectly due to action on oxidative phosphorylation. For example, barban treatment of wild oats increases the nucleotide content due to disruption of RNA and protein synthesis brought about by an ATP deficiency (Ladonin, 1967a); similar effects have been found in treated wheat seedlings (Ladonin, 1967b). Again, increase in DNA, protein and nucleotide contents of wild oats treated with barban has been attributed to disruption of RNA synthesis causing abnormal cell division (Ladonin and Svittser, 1967). Later, Ladonin and Beketova (1972) reported that *in vivo* treatment of wild oat with barban (0.5 and 1 kg a.i./ha) results in increased DNA and RNA contents; *in vitro* treatment results in an increased RNAase activity. An investigation of the effect of barban (0.0001% a.i.) (and tri-allate 0.0003%) on mitosis of barley (cv. Viner) root caps indicates that neither herbicide is a serious mutagen in barley crops.

Asulam also affects mitosis, there being good evidence from work with onion root tips (Sterrett and Fretz, 1974, 1975) and bracken rhizome tips (Veerasekaran, Kirkwood and Fletcher, 1977) that the effects included arrested metaphase and anaphase and chromosome bridges. The inhibitory effect of asulam on the incorporation of ^{14}C precursors into RNA and protein in bracken rhizome buds has also been reported by Veerasekaran *et al.* (1977); respiratory metabolism is also affected.

THIOCARBAMATES

These soil-applied herbicides appear to be readily absorbed by the shoot of emerging seedlings causing abnormal growth of the leaves emerging from the coleoptiles of grasses. Indeed at high rates the leaves may not emerge from the coleoptiles. It seems that they all affect mitosis in the young shoot with relatively little influence on root tissues (Ashton and Crafts, 1973).

Both **EPTC** and **sulfallate** are known to inhibit protein synthesis (Mann *et al.*, 1965b). Moreland *et al.* (1969) found that sulfallate (2×10^{-4} M) and EPTC (6×10^{-4} M) inhibited GA-induced *de novo* α-amylase synthesis in barley half-seeds and 8-^{14}C-ATP or 6-^{14}C-orotic acid incorporation into RNA by soybean hypocotyls or maize mesocotyl sections respectively. These herbicides had little effect on ATP levels (Gruenhagen and Moreland, 1971) suggesting either that they did not affect energy metabolism or they failed to penetrate to the active sites.

EPTC has been found to inhibit RNA synthesis in soybean and maize tissue, this effect being antagonised by addition of 2,4-D (Beste, 1971; Beste and Schreiber, 1972a). EPTC inhibits rRNA, mRNA and tightly bound RNA (TB-RNA) synthesis; analysis of rRNA revealed that EPTC preferentially inhibits the synthesis of 18S rRNA to a greater degree than 25S rRNA. 2,4-D antagonises the action of EPTC apparently due to enhanced synthesis of D-RNA and TB-RNA by 2,4-D in the presence of EPTC (Beste and Schreiber, 1972a). 2,4-D enhancement of growth was not prevented by incubation of soybean tissue in EPTC for 4 h prior to the addition of 2,4-D; however, EPTC inhibited growth when added after 4 h of 2,4-D-enhanced growth

(Beste and Schreiber, 1972b); there was no evidence that EPTC inhibited the synthesis of pyrimidine precursors. A further example of antidote action is found in the work of Harvey, Chang and Fletcher (1975) who showed that the EPTC-induced inhibition of growth and stimulation of peroxidase activity in maize seedlings is annulled by the herbicide antidote *N,N*-diallyl-2,2-dichloroacetamide; IAA or GA had no effect.

Morimoto, Shimizu and Ueki (1972) reported that EPTC inhibits the germination of barley and wheat but not barnyard grass (*Echinochloa crus-galli*) and reduces *de novo* synthesis of α-amylase in barley and wheat but not in *E. crus-galli*. Differing susceptibilities of these species appears to be due to the fact that *de novo* synthesis of α-amylase during germination occurs 1 day later in *E. crus-galli* than in the other species. The herbicidal action of molinate on *E. crus-galli* has also been studied by Chen, Seaman and Ashton (1968). While molinate has little effect on the total RNA content of the treated coleoptiles, the sRNA content is reduced by 70% compared with the controls.

In addition to affecting α-amylase synthesis and activity, thiocarbamate herbicides may affect certain other enzyme reactions. For example, the IAA oxidase activity of pea root (*in vitro*) was inhibited by tri-allate (10^{-6}–10^{-2} M) (also diuron, DNOC, MCPA and diquat) (Schneider and Günther, 1974a, b) and sodium diethyldithiocarbamate reduced aldolase activity in germinating seeds of *Phaseolus radiatus* due to specific inhibition of protein synthesis (Balasimha, Tewari and Ram, 1977).

UREAS

Herbicides of this group have been shown to interfere with a number of enzyme reactions. Application of **linuron** (as 'Afalon', 600 l/ha) to spring barley reduces α-amylase and proteinase activity, though the basic protein content of the grain harvested from the treated plants is increased (Rejowski, Tluczkiewicz and Login, 1973). Spraying of barley plants at the onset of ear formation with linuron ('Afalon', 1.5 kg/ha) inhibits growth, yield, starch content and α-amylase activity of the ripening grain (Grzesiuk, Januszewicz and Sójka, 1973). These workers found that repeated application of linuron reduces the germination ability of barley grains, promoting protease activity and reducing protein contents (Grzesiuk, Paliwoda and Sójka, 1973).

Diuron (and simazine and atrazine) increase the *in vitro* extractable nitrate reductase activity and nitrogen content of detached barley leaves and reduce the *in vivo* nitrate reductase activity, thus increasing the internal level of nitrates (Aslam and Huffaker, 1973). Geike (1974) reported that phenylurea (and phenylcarbamate compounds) inhibited phosphatase and β-amylase activity, while Klepper (1975) demonstrated inhibition of *in vivo* nitrate reductase by certain photosynthetic inhibitors. Substituted ureas (**fenuron**, **monuron**, diuron, linuron, **siduron** and **neburon**) act as urease inhibitors, the enzyme molecule possibly reacting with the oxygen atom of the carboxyl group in the substituted urea (Cervelli *et al.*, 1975).

The effect of pre-emergence application of diuron (1.5–2 kg/ha), monuron (2 kg/ha) and **fluometuron** (3–4 kg/ha) on the nucleic acid content of cotton at the 2-leaf stage was examined by Ali-Zade and Ismailov (1976). All herbicides increased the nucleic acid content though fluometuron and monuron reduced the DNA content.

Mutagenic effects have been reported to result from treatment of germinating barley by substituted ureas (Pusztai and Vegh, 1978). The lethal dose values were found to occur between 300 and 1200 p.p.m. and the LD_{50} values were estimated to be between 250 and 1000 p.p.m. Root tip metaphase analysis suggests that the mutagenic effects were different for each chemical, frequencies of aberrant cells ranging from 2.5 to 11%.

Differing susceptibility of strains of the green alga (*Chlamydomonas reinhardii*) to diuron (1×10^{-6}, 1×10^{-5} and 7×10^{-5} M) has been observed by Khakimov and Kvitko (1978). Forty strains of the 137c line were tested and the majority were found to be resistant. Two types of sensitivity were noted, however, one involving a change of colony colour and the other growth suppression. In a preliminary study by Kahlon and Banerjee (1979), monuron (as 'Telvar') was included in an evaluation of three pesticides for possible mutagenic properties in barley shoot tips. At the higher concentrations tested (500 and 1000 p.p.m.), monuron significantly increased chromosome breaks as evidenced by the frequency of dicentric bridges, though it was only moderately toxic to the barley seedlings.

TRIAZINES

Many studies have demonstrated that the triazine compounds increase the level of several nitrogen fractions. Some studies show that the amount of nitrogen on a per plant basis is actually increased while other studies have failed to demonstrate this effect. In some instances the increase in the level of particular nitrogen fractions has been due to a decrease in plant dry weight rather than to a change in nitrogen level *per se*. Apparently it is difficult to demonstrate unequivocally that sub-toxic levels of these herbicides increase the actual amount of nitrogenous compounds present on a per plant basis. Changes in nitrogen metabolism induced by triazine herbicides are determined by a number of factors including the amount of herbicide applied, the amount and form of nitrogen supplied to the plant, the growth conditions, plant species and the method of presentation of the results, e.g. whether the data has been calculated as a percentage by weight or on a per plant basis (Ashton and Crafts, 1973).

Several investigators have reported increases in various nitrogen components. For example, **atrazine** and **simazine** have been reported to increase the protein content of maize (Gast and Grob, 1960) and simazine to increase the nitrogen uptake by maize (De Vries, 1963; Ries and Gast, 1965), the protein content of apple leaf tissue (Solecka *et al.*, 1969) and the total nitrogen, protein nitrogen and total soluble nitrogen levels of apple seedlings (Lin and Anderson, 1967); in this last case the increase was attributable to a decrease in other constituents rather than to an increase in nitrogen absorption. However, other studies have supported the view that sub-toxic levels of the triazine herbicides do increase growth and nitrogen content of certain species (Ries *et al.*, 1963; Tweedy and Ries, 1967; Schweizer and Ries, 1969). This apparently results from an increase in nitrate reductase activity which occurs in plants grown on a nitrate rather than on an ammonia nitrogen source, the degree of response declining as the nitrate concentration approaches optimum levels.

Several groups of workers have studied the action of triazine herbicides on various aspects of plant physiology including nitrogen metabolism. For example, in Poland, Ploszynski and his co-workers have examined the phytotoxic action of triazine compounds on certain crop species. Pre-emergence treatment of flax, beet and buckwheat results in inhibition of plant growth, dry matter (d.m.) and fresh weight yields, transpiration, sucrose and hexose levels; free amino acid contents and amides were increased. The accumulation of asparagine, glutamine and basic amino acids was attributed to the inhibition of photosynthesis with the subsequent decline in sugars leading to the formation of excess NH_2 groups which combine with glutamic and aspartic acids to form the corresponding amides (Ploszyński and Zurawski, 1971). Pre-emergence application of simazine (2–8 kg/ha) to pot-grown peas resulted in growth inhibition or death. At high herbicide concentrations, total nitrogen and total free amino acid levels of the leaves increased, while chlorophyll, sugar, certain organic acids and protein levels declined. At low rates (0.2–0.5 kg/ha), however, yields increased slightly and the content of chlorophyll, sugar, certain organic acids, total nitrogen and protein nitrogen increased (Ploszyński, 1972). Pre-emergence treatment of spring rye with simazine and atrazine (0.5 kg/ha) reduces saccharide and chlorophyll contents, and increases total nitrogen, nitrate nitrogen, amino acids, phosphorus and potassium contents and slightly increases the percentage content of protein nitrogen (Ploszyński and Rola, 1974). Also in Poland, Lehmann, Turowski and Frieske (1971) examined changes in the nitrogenous and mineral contents of winter rye, spring wheat and yellow lupin (*Lupinus luteus*) treated with simazine (as 'Gesatop 50', 0.5–2 kg/ha). They found that simazine increases the total and protein nitrogen contents but also causes an increase in potassium and nitrate levels. Similar results have been reported for forage maize (Frieske, 1971).

At Utah State University, Singh, Salunkhe, Wu and their co-workers have studied various aspects of growth and chemical responses to s-triazine herbicides. Foliage treatment of bush beans (*Phaseolus vulgaris* cv. Tendergreen) with atrazine, simazine, **terbutryne** and **secbumeton** (0.1 and 0.5 p.p.m.) increases the protein and amino acid constituents of the leaves; these effects appear to be attributable to a reduction in the content of starch and total and reducing sugars (Singh, Salunkhe and Lipton, 1972b). These treatments also increase the number of protein-containing vesicles, the number of cytoplasmic ribosomes in the cotyledons (Singh, Campbell and Salunkhe, 1972a) and the iron content of the leaves (Singh, Vadhawa and Salunkhe, 1972c).

Foliage treatment of pea and sweet corn with simazine, terbutryne (or **prometon**) (2 or 5 mg/l) significantly increases the total nitrogen and soluble protein content of the seeds and reduces the starch and total sugar contents (Singh *et al.*. 1972b). Such treatments (2 mg/l) increase the activity of starch phosphorylase, pyruvate kinase, cytochrome oxidase and glutamate dehydrogenase in these species and may induce the greater utilisation of carbohydrates for nitrate reduction and synthesis of amino acids and proteins (Wu, Singh and Salunkhe, 1972a). The activities of nitrate reductase, glutamic-pyruvic transaminase but not ribonuclease are also significantly enhanced (Wu, Singh and Salunkhe, 1972b). Triazine-induced increases in protein synthesis and enzyme activity in pea and sweet corn are accompanied by changes in the ultrastructure of the parenchyma cells of the developing cotyledons (Wu, Singh and Salunkhe, 1972c).

Treatment of wheat and mung beans with secbumeton (and two insecticides, menazon and disulfoton) inhibits germination and seedling growth as a result of impaired respiration and starch and protein degradation; amylase and ATPase activities are impaired to a greater degree in wheat than in mung bean (Dalvi, Singh and Salunkhe, 1972).

At Michigan State University, Pulver and his co-workers have examined the mode of action of sub-toxic levels of simazine in increasing plant protein content. Sub-toxic concentrations of simazine increase the protein content of barley plants (cv. Coho) with concomitant secondary increases in nitrate assimilation and soluble carbohydrate degradation (Pulver, 1972). However, simazine does not alter the protein content under conditions unfavourable for the accumulation of soluble carbohydrates (high temperatures and high nitrogen levels), or in situations where ammonia is the source of nitrogen. In a time course study, Pulver and Ries (1973) found that simazine (10^{-8} M) increased ^{14}C-leucine incorporation into protein after 24 h but did not increase the protein content or nitrogen uptake until 48 h. At a concentration of 10^{-5} M, simazine increased ^{14}C-leucine incorporation by 30% whereas hydroxysimazine (10^{-5} M) had no effect on protein synthesis. Ries, Pulver and Bush (1974) reported that simazine increased ^{14}C-leucine incorporation into protein prior to increasing nitrate uptake or total protein, probably indicating that the herbicide has a direct effect on protein synthesis; atrazine (10^{-5} M) also increased protein synthesis in red kidney beans (*Phaseolus vulgaris*). Bush and Ries (1974) also reported that at a concentration of 10^{-8} M atrazine stimulates the elongation of the embryonic axis of red kidney beans accompanied by increased light-dependent protein synthesis.

Gräser and co-workers examined the effect of simazine on the growth and nitrogen content of maize or *Sinapis alba*. Gzik and Gräser (1972) reported that simazine-induced growth stimulation of etiolated seedlings of *Zea mays* was accompanied by an initial increase in the size of the amino acid pool followed by an increase in the protein content of treated seedlings. Later Paul *et al.* (1973) reported that effects of triazine treatments of etiolated *S. alba* and maize seedlings on nucleic acids were affected by the stage of growth at the time of treatment, site of application, herbicide concentration and environmental conditions. Iliev and Gräser (1975) later found that in sensitive *S. alba* plants, an initial stimulation of rRNA and TB-RNA synthesis by atrazine (5×10^{-5}–5×10^{-3} M) was followed by progressive inhibition; the latter did not occur in resistant maize.

While lethal concentrations of triazine herbicides appear to inhibit nitrogen uptake and protein synthesis, at least in susceptible species (Chesalin and Timofeeva, 1971, 1973), there is evidence that sublethal concentrations may increase the utilisation of soil inorganic nitrogen and nitrogen metabolism. For example, sublethal concentrations of simazine increased the nitrogen content and nitrate reductase activity in *Triticum durum* seedlings (Brunetti, Picciurro and Ferrandi, 1972) and crude protein levels of alfalfa (*Medicago sativa*) and cocksfoot

(*Dactylis glomerata*) (Currey, 1972); a similar response was obtained for low concentrations of atrazine (1 p.p.m.) in maize seedlings (Lay and Ilnicki, 1972a).

Increased total nitrogen and amino acid levels have also been reported following treatment of maize with atrazine (<3 p.p.m.) (Hiranpradit, Foy and Shear, 1972a), maize and soybean with atrazine (10^{-6}–10^{-4} M) (Penner and Early, 1972) and wheat with 4×10^{-7} M atrazine (Terrillon and Paynot, 1973). Stimulatory effects on growth, nitrate absorption and nitrate reductase activity have also been reported for sublethal concentrations (0.02 p.p.m.) of other triazine herbicides (and ureas) (Decleire, De Cat and Bastin, 1974). Nucleic acid biosynthesis is not stimulated however, by 0.02 p.p.m. atrazine or **prometryne** (Decleire, De Cat and Bastin, 1976). Reduction of absorbed nitrate to nitrite reductase is considered to be the rate-limiting step in nitrogen nutrition of most plants (Streeter, 1972) and it may be significant that nitrate reduction is inhibited by triazine herbicides in susceptible but not in resistant plants (Finke, Warner, and Muzik, 1977).

Increased nitrate and d.m. levels following simazine treatment of wheat (0.1–0.5 p.p.m.) and sorghum (0.5–1 p.p.m.) have been reported by Rottman, Tweedy and Kapusta (1974). Application of simazine, prometryne plus simazine, or lenacil to lupin (*Lupinus* sp.) increases the crude protein by >10% (Elek *et al.*, 1974), while simazine or terbutryne (12.5–50 kg/ha) increases protein production in the kernels of cereals and d.m. of grasses (Aamisepp, 1976). On the other hand, no difference in protein synthesis was reported following treatment with simazine of radish (Gnanarethinam, 1976) or southern peas (*Vigna unguiculata*) (Meredith and Langdale, 1978). Treatment of peas with 0.8 kg/ha simazine causes disruption of nucleic acid synthesis reducing DNA and RNA levels (Voevodin, Nevzorova and Kazarina, 1977). Initially when only small amounts of simazine have entered the plants, the *de novo* synthesis of nucleic acids is stimulated, but as the herbicide accumulates, so the process is inhibited.

Similarly, low rates of prometryne (1–3 kg/ha) (and monolinuron) stimulate protein synthesis in soybean (Lalova, 1973); application of higher rates (3 kg/ha) of prometryne to peas, however, increases the total free amino acid content in the foliage but reduces it in the roots (Vasyuta *et al.*, 1974). Prometryne (and trifluralin) at the normal rates used in agricultural practice caused chromosome aberrations and disrupted the mitotic process.

The effect of triazine herbicides on certain enzymes associated with protein synthesis and intermediary metabolism has been studied by Decleire and his co-workers. In wheat, 1 p.p.m. atrazine, prometryne,

terbumeton and certain other herbicides inhibits the activity of glutamate–pyruvate transaminase and glutamate–oxaloacetate transaminase systems, both of which are concerned with protein synthesis (Decleire, van Roey and Bastin, 1974). Again, while application of 1 p.p.m. atrazine, chlortoluron and other herbicides to cucumber seedlings has a variable effect on alkaline pyrophosphatase (Decleire and De Cat, 1978), the same treatments progressively inhibit the activity of glucose-6-phosphate dehydrogenase and 6-phosphogluconate dehydrogenase (Decleire and van Roey, 1979). Triazine-induced inhibitions of saccharase, dihydrogenase, urease, protease and phosphatase activity of soil (Stirban and Soreanu, 1975) and nitrogenase activity of *Anabaena cylindrica* have also been reported (Rohwer and Flückiger, 1979).

PYRIDINES

Nucleic acid and protein synthesis seem to be important sites of action of **picloram** and differences in the susceptibility of these reactions may be an important basis of selectivity of this herbicide. For example, Malhotra and Hanson (1970) found that picloram enhances DNA and RNA synthesis in soybean and cucumber (susceptible), but not in barley and wheat (resistant). Similar differences were reported by Chen *et al.* (1972) who found that picloram (and 2,4-D, 2,4,5-T and dicamba), increase DNA and RNA contents of wheat (resistant) and cucumber (susceptible), particularly the latter. They suggest that differential effects on RNA and protein synthesis may act as a basis for the selective activity of picloram in wheat and cucumber. Sharma and Vanden Born (1970) also conclude that picloram has a role in altering both RNA and protein synthesis and some processes in the degradative metabolism of these constituents. Baur and Bowman (1972) obtained results which indicated that picloram-enhanced protein synthesis reflects the enhanced activity of cytoplasmic ribsomes rather than the chloroplast or mitochondrial ribosomes. This view is substantiated by their findings that the concentrations of picloram recovered from the chloroplasts and mitochondria are consistently low in comparison with the high recovery from the cytoplasmic contents (Baur and Bowman, 1973).

Kalinin, Musiyaka and co-workers have examined the effect of picloram on the synthesis of ribonucleic acids in peas. They sprayed germinating peas with picloram (2.3×10^{-8} M and 3.9×10^{-7} M) and examined ^{14}C-uracil incorporation into total RNA, 6, 12 and 24 h after treatment. Significant incorporation of ^{14}C occurred in the different forms of RNA, reflecting serious disturbances in the processes of RNA synthesis at cellular and whole seedling levels (Kalinin *et al.*, 1975). Treatment of tobacco tissue culture with these concentrations of picloram (substituting for 2,4-D) promoted cell division and extension and nucleic acid and protein contents (Musiyaka and Kalinin, 1975). They conclude that picloram is suitable as an auxin source, the optimal concentration being 0.5 mg/l (3.9×10^{-7} M). These workers also examined the effect of picloram (0.6 kg/ha) on the nucleic acid and protein contents of rhizome buds of Russian knapweed (*Centaurea repens*) (Musiyaka and Kalinin, 1979). Four to five hours after penetrating aerial and sub-surface meristems, picloram stimulated cell division and broke bud dormancy, even at significant depths in the soil; there was a close correlation between the nucleic acid and protein contents of the buds. Subsequent to the period of picloram-induced activation, the buds died.

Several workers have examined the ultrastructural effects of picloram with particular reference to meristems. For example Mikhno, Musiyaka and Kalinin (1972) reported that picloram (2.5×10^{-7}–4.1×10^{-6} M) decreases cell division in the meristem of pea seedlings, with predominant division occurring in radial and tangential planes causing thickening and stunting of the root. They also found that treatment of pea seedlings with picloram (as 'Tordon 22K') (0.0001%) causes intensified starch formation in meristematic cells, deformation of mitochondria, dispersion of chromatin in the nucleus, thickening of the cell membrane and alteration of the configuration of the nucleus (Musiyaka *et al.*, 1972). Musiyaka and Kalinin (1972) also recorded disruption of protoplasm movement and cell structure in hair cells of anther filaments of *Tradescantia paludosa* and *Centaurea repens* treated with picloram (0.001%). These authors subsequently reported the effects of foliage-applied picloram on meristems of the latter species, intensified cell division and extension in an isodiametric dimension resulting in death of the roots and thus the whole plant (Mikhno, Kalinin and Lesnevich, 1975). In the long term (5–10 days in shoots, 3–12 months in roots), picloram retarded cell division in the procambium and cambial zones, causing cell elongation in some tissues and destruction of secondary cell walls leading to death of the plants (Mikhno, Kalinin and Lesnevich, 1978).

The effect of picloram (1–40 p.p.m. w) on organogenesis in detached leaves of *Echeveria elegans* has been studied by Raju and Grover (1976). Low concentrations (1–15 p.p.m. w) did not affect shoot organogenesis but stimulated rhizogenesis, both, however, were inhibited at concentrations of 20 p.p.m. w and above; while root number increased at concentrations of less than 40 p.p.m. w, root elongation was markedly suppressed. However, they concluded that the apical meristems were not severely affected by these picloram treatments.

Treatment of the leaves of *Pinus radiata* with 'Tordon 50D' (picloram+2,4-D) results in ultrastructural changes, including effects on the nucleus and cell membranes and swelling of the internal chloroplast; eventually the membranes disintegrated (Ayling, 1976).

The mechanism of action of **diquat** and **paraquat** involves the formation of the free radical by reduction of the ion and subsequent autoxidation to yield the original ion. The OH^- radical or H_2O_2 which is formed during this autoxidation process appears to be the primary toxic agent. Indeed Hassan and Fridovich (1978) have described paraquat toxicity as largely 'a superoxide toxicity' which is diminished by superoxide dismutase. Harper and Harvey (1978) working with paraquat-sensitive and tolerant strains of ryegrass (*Lolium perenne*), have proposed that tolerance to paraquat is due to destruction of O_2^- by elevated concentrations of superoxide dismutase and detoxification of the resulting hydrogen peroxide by increased levels of catalase and peroxidase. Stimulatory effects on catalase and peroxidase (also amylase and diastase) in nutgrass (*Cyperus rotundus*) tubers treated with 'Gramoxone' (paraquat 20%, 2.5 l/ha) had previously been reported by Thangarat and Sakharam Rao (1973).

Effects on growth and nucleic acid content have also been reported following treatment with the bipyridyliums. Schneider and Günther (1974a, b) found that root-applied diquat (10^{-4}–10^{-2} M) inhibits the growth of etiolated pea seedlings and severely reduces the level of rRNA, probably reflecting damage to the ribosomes; the DNA–RNA fraction was little changed. Paraquat (10^{-6}–10^{-3} M) treatment of *Vicia faba* root tips, results in inhibition of DNA synthesis and cell division (Bell, Schwarz and Hughes, 1976). Goeltenboth (1978) found that treatment of *Eichhornia crassipes* with diquat or paraquat (1.5–6 p.p.m.) inhibits the onset of mitosis at interphase and spindle formation at late prophase.

Effects of paraquat on the integrity of the plasma membrane have also been demonstrated. For example Boulware and Camper (1972) exposed isolated protoplasts of immature tomato fruits to paraquat (and other herbicides) and found that an initial segregation of cytoplasm into isolated areas on the inner membrane surface was followed by rupture of the plasma membrane and collapse of the cell.

3,6-Dichloropicolinic acid may also have a marked effect on growth pattern. Treatment of soybean hypocotyls with 'Dowco 290' inhibits cell division in the meristem and cell elongation while radial enlargement of cells occurs even in mature zones due to increases in the size of nucleolae. Treated tissues of soybean hypocotyls contain about twice as much RNA and protein as untreated controls.

PYRIDAZINES

Certain of these compounds exhibit auxin-like activity. For example Carter and Camper (1973), using the *Avena* coleoptile and cucumber radicle elongation tests, compared the responses obtained from **norflurazon** and **metflurazone** with those obtained with 2,4-D and IAA. Both herbicides inhibit cucumber radicle elongation, while norflurazon stimulates oat coleoptile elongation; combination of norflurazon with 2,4-D results in an additive effect. They conclude that norflurazon, but not metflurazone, exhibits auxin-like activity, this being located at the nucleic acid and protein levels. However, Takeuchi, Konnai and Takematsu (1972), examining the inhibition of auxin activity by certain 3-phenoxypyridazine derivatives, find that they reverse the inhibitory effects of excess auxin on radish. Growth inhibition resulting from pyridazine action could be overcome by addition of low concentrations of auxin. Since the pyridazines removed apical dominance, Takeuchi *et al.*, concluded that their mode of action involved interference with auxin.

PYRIMIDINES (URACILS)

Uracils affect the meristematic region of root tips, disrupting cell wall formation, resulting in the occurrence of multiple nucleii, thickening of cell walls and other effects (Ashton *et al.*, 1969). Nitrogen metabolism is also affected. For example, treatment of soybeans with **bromacil** (0–0.045 p.p.m.) increases the levels of nitrogen and phosphorus in the shoots (Hiranpradit, Foy and Shear, 1972b). In treated shoots, accumulation of nitrate nitrogen, protein nitrogen, nucleic acid nitrogen and amino acids occurred. Similarly Lalova (1973) reports that treatment of fodder beet with 'Venzar' (**lenacil** 80%) (and benzthiazuron) influences the synthesis of free amino acids and increases protein synthesis.

ORGANOPHOSPHORUS COMPOUNDS

In recent years a considerable literature has developed concerning the effect of **glyphosate** on nitrogen metabolism. Roisch and Lingens (1974) presented evidence which showed that inhibition of growth of *E. coli* B. by glyphosate (2×10^{-3} M) is partially alleviated by addition of phenylalanine or tyrosine and to a greater extent by a combination of both. No effect was observed by addition of histidine, valine or leucine. Roisch and Lingens also found that glyphosate inhibited certain synthetase enzymes, these effects being countered by the addition of CO_2.

Reversibility of glyphosate inhibition was also demonstrated by Haderlie, Widholm and Slife (1974) who found that additions of penylalanine±tyrosine+tryptophan or casein hydrolysate result in a significant reversal of the inhibitory effect of the herbicide on growth of carrot and tobacco cells. Glyphosate increases the level of free amino nitrogen and the induction of nitrate reductase within 6 h of treatment (Haderlie, 1976). However, glyphosate does not appear to alter phenylalanine, leucine or protein synthesis. Reversal of glyphosate-induced inhibition was reported by Baur (1979) who found that addition of cyclohexamide or L-phenylalanine significantly reduces the incidence of basal stem swelling of sorghum (cv. Tophand) seedlings.

Precursor studies indicate that protein synthesis is a major site of action. For example, Tymonko (1979) examined the influence of glyphosate on the metabolism of separated soybean leaf cells. The incorporation of ^{14}C-leucine into acid-insoluble products was more sensitive to glyphosate than were other metabolic pathways examined (respiration, photosynthesis, RNA and protein synthesis) (Tymonko and Foy, 1978a). They conclude that protein synthesis is the primary site of action of glyphosate. Tymonko and Foy (1978b) also reported that 1 and 5 mM concentrations of glyphosate inhibited ^{14}C-leucine incorporation into protein by 36 and 57% respectively; additions of phenylalanine or phenylalanine combined with tyrosine or tryptophan reduced the inhibition by glyphosate to 21%. The shikimic acid pathway appears to be little involved. Similarly, glyphosate (10^{-4} M), applied to the cut ends of single-node fragments of *Agropyron repens* rhizome, inhibited the incorporation of ^{14}C-leucine and to a lesser degree ^{14}C-phenylalanine (Cole, Dodge and Caseley, 1979). They suggest that glyphosate may inhibit the rate of protein synthesis by reducing the level of free phenylalanine.

Nucleic acid synthesis is also affected by glyphosate. For example, Pillai, Davis and Truelove (1978) using soybean roots, found that 10^{-4} M glyphosate (and MSMA, profluralin and metribuzin) inhibits the uptake and incorporation of ^{14}C-thymidine into DNA ^{14}C-uridine into RNA and ^{14}C-leucine into protein; DNA synthesis was the process most affected. Again, treatment of bean leaf discs with glyphosate inhibits the incorporation of ^{14}C-uracil into RNA but not ^{14}C-leucine into protein, nor $^{14}CO_2$ assimilation into photosynthetic processes or $^{14}CO_2$ expiration from ^{14}C-glucose (Brecke and Duke, 1977). They conclude that glyphosate directly inhibits ion transport.

In a series of papers, Hoagland, Duke and co-workers examined the hypothesis that glyphosate exerts its effect in dark-grown seedlings of maize and soybean through induction of phenylalanine ammonia-lyase (PAL) which inhibits growth by altering the metabolism of phenolic compounds (Duke and Hoagland, 1978; Hoagland and Duke, 1978). Their results suggest that glyphosate may exert only some of its effects through reduction of phenylalanine levels (Duke and Hoagland, 1979a) and that complex changes in phenolic metabolism may vary with species (Hoagland and Duke, 1979). Protein synthesis was not affected by concentrations of glyphosate which inhibit growth (Duke and Hoagland, 1979b). Subsequently evidence was presented which indicated that glyphosate could exert its effect through induction of PAL activity and/or inhibition of aromatic amino acid synthesis (Hoagland, Duke and Elmore, 1978; Hoagland, Duke and Elmore, 1979). Increases in the levels of glutamine and other amino acids are believed to result from amination using the excess ammonia generated by the enhanced PAL activity (Duke, Hoagland and Elmore, 1979).

Ultrastructural effects of glyphosate (10^{-4} M) on *Lemna gibba* include progressive damage of chloroplasts, mitochondria and cell walls with increasing exposure for 12–24 h. Stunting of roots and chlorosis of frond tissue was evident after 48 h (Hoagland and Paul, 1978a, b, c). Exposure of excised leaf tips of *Cyperus esculentus* to glyphosate (2×10^{-2} M) results in reduced starch levels in the chloroplasts and disruption of membranes (Hoagland and Paul, 1979).

Davis and Harvey (1979) postulated that glyphosate may act by chelating the zinc co-factor of chlorismate mutase, an enzyme involved in the biosynthesis of aromatic amino acids. If this is indeed the case, then addition of the amino acids at the end of the biosynthetic sequence containing chlorismate mutase should reverse glyphosate inhibition. In the event there was only some reversal of glyphosate (1 mM) toxicity by the amino acids and this incompleteness may be due to the presence of three isoenzymes of chlorismate mutase with different activation energies, two of which are affected by phenylalanine and tyrosine.

It has been suggested that the inhibitory action of glyphosate on transpiration of excised bean (*Phaseolus vulgaris*) shoots may be due to some form of interaction with the aromatic amino acid levels in the plant (Shaner and Lyon, 1979a, b). Partial protection of *P. vulgaris* shoots from glyphosate (6×10^{-5} M) results from continuous feeding with tyrosine (10^{-3} M), while a combination of phenylalanine and tyrosine (5×10^{-4} M) completely protects the plant from inhibition of transpiration caused by glyphosate. Increased cellulase activity in this species may be attributable to glyphosate-enhanced ethylene production (Abu-Irmaileh, Jordan and Kumamoto, 1979).

Lipid metabolism

Lipids occur in plants in the form of true fats, phospholipids and waxes. Fats and phospholipids are synthesised by esterification of glycerol with fatty acids while waxes are esters of fatty acids and long-chain monohydroxy-alcohols or dihydroxy alcohols. Glycerol is synthesised from dihydroxyacetone phosphate while fatty acids are formed from acetyl-CoA which arises from glycolytic reactions or is derived from pantothenic acid. Herbicidal effects on lipid biosynthesis may affect the physicochemical properties of the epicuticular or cuticular waxes and the permeability of cell or organelle membranes; they may also influence the storage of fat materials.

The action of herbicides on lipid synthesis

HALOALKANOIC ACIDS

TCA and **dalapon** are both known to affect lipid metabolism and cuticle-wax deposition. For example, Kolattukudy (1968) found that these herbicides alter the cuticular wax of peas and maize rendering the leaves more wettable to subsequent sprays. Mashtakov *et al.* (1967) reported that TCA reduces the thickness of the cuticle and lamina of leaves of sensitive varieties of *Lupinus luteus*; these workers also found that TCA reduces the level of pantothenic acid (Mashtakov *et al.*, 1967). However, TCA appears to have relatively little effect on wax deposition in peas (*Pisum sativum*) compared to certain carbamates (Still *et al.*, 1970).

While these effects may be due to direct herbicidal action on lipid biosynthetic mechanisms, they may also reflect indirect effects on ATP metabolism. Ross (1966) found that dalapon applied to bean leaf discs, causes an increase in ATP and a reduction in an unidentified sugar phosphate. He concluded that if dalapon did have a specific site of action it was probably phosphorus metabolism and the increase in ATP probably occurred because it was not utilised for phosphorylation of some compound required for synthetic processes.

Mann and Pu (1968) reported that **2,4-D** and **2-4,5-T** stimulated lipogenesis, though the increase declined over the concentration range tested (1–20 p.p.m.); similar effects have been found for 2,4-D in

Chlorella (Sumida and Ueda, 1973). These effects may be related to reports that these herbicides may cause disintegration of the cell and organelle membranes. For example, White and Hemphill (1972) found that treatment of tobacco (cv. Samsun N.N.) leaves with 2,4-D (500 mg/l) caused rupturing and disintegration of the tonoplast, plasmalemma and membranes of chloroplasts and mitochondria. Again, loss of chlorophyll from *Chlorella pyrenoidosa* incubated with 2,4,5-T (10^{-3} M) was believed to be caused by defects in the membrane system of the chloroplasts or cell (Geike, 1972). Artamonov (1975) reported that high concentrations of 2,4-D (250 mg/l) caused irreversible disruption of the cell membrane; a lower concentration (25 mg/l) caused changes which were not lethal. Damage to cell membranes by 2,4-D was also reported by Haunold and Zsoldos (1976) and Zsoldos *et al.* (1978) who examined 2,4-D-induced changes in the uptake of K^+ by wheat roots. They suggest that 2,4-D exerts a non-specific effect on the lipid–protein interactions, giving rise to a general alteration of the transport barrier properties of the plasmalemma even at low concentration (0.01 mM) (Zsoldos *et al.*, 1978).

NITRILES

There is evidence that **ioxynil** and **dichlobenil** influence lipogenesis, though the nature of the effect may be species or concentration dependent. For example, Mann and Pu (1968) reported that 10 and 20 p.p.m. ioxynil and **dichlobenil** inhibited the incorporation of ^{14}C from 2-^{14}C-malonate into lipid by 60–65% and 30–40% respectively. On the other hand, Sumida and Ueda (1973) found that dichlobenil (5 mg/l) stimulated lipid biosynthesis in *Chlorella*.

ANILIDES

Increases in the percentage of unsaturated fatty acids in the plasmalemma and mitochondria were recorded for soybean plants treated with **alachlor** (2.24 kg/ha) and other herbicides (Penner and Rivera, 1977). In contrast, exposure of cotton root tips to **metolachlor** (10^{-4} M) reduced total lipid synthesis by about 25%; synthesis of phosphatidyl choline was almost completely inhibited while synthesis of other phospholipids was unaffected (Diner, Truelove and Davis, 1978).

NITROPHENOLS

Dinoseb (and perfluidone) have been found to inhibit glyceride synthesis in wheat (cv. Mediterranean) seedlings resulting in the accumulation of free fatty acids and a decrease in neutral and polar lipids (St. John and Hilton, 1973). They suggest that both compounds alter membrane structure, functioning through an inhibition of membrane lipid synthesis. Increased cell membrane permeability of *Lemma minor* has been reported for dinoseb and several other herbicides (O'Brien and Prendeville, 1979).

THIOCARBAMATES

A considerable literature is associated with the action of certain of these compounds, particularly **EPTC**, as inhibitors of lipid biogenesis. Still *et al.* (1970) found that **di-allate**, EPTC and **sulfallate** all have a significant effect on the biosynthesis of epicuticular lipids in leaves of pea (*Pisum sativum*). Introduction of di-allate via the roots reduced epicuticular lipids by 50%, while exposure of the leaves to vapour resulted in an 80% reduction. There was no change in the ratio of wax lipid components in di-allate-treated plants, with the exception of the primary alcohols which were reduced. EPTC altered wax structure but not composition. Still *et al.* suggested that the carbonyl group may be the functional part of the di-allate and EPTC molecules.

The inhibition of fatty acid synthesis by these two herbicides has been studied by Wilkinson and his co-workers. Treatment of isolated spinach (*Spinacia oleracea*) chloroplasts with di-allate and EPTC resulted in increased incorporation of 1-^{14}C-acetate into fatty acids at low concentrations of di-allate (0.24 and 2.4 p.p.m. w) and inhibition at high concentration (24 p.p.m. w) (Wilkinson and Smith, 1973). They also reported that EPTC (33 μM) and di-allate (90 μM) inhibited the incorporation of 1-^{14}C-acetate and 1-^{14}C-palmitic acid into fatty acyl moieties associated with the acyl carrier protein (ACP) (Wilkinson and Smith, 1974). Similar inhibition of 1-^{14}C-palmitate and 1-^{14}C-oleate incorporation into chloroplast lipids and 2-^{14}C-malonate into dienoic fatty acids has also been observed (Wilkinson and Smith, 1975a). The EPTC-induced (10^{-5} M) inhibition of acetate incorporation into fatty acids was counteracted by 1,8-naphthalic anhydride and *N,N*-diallyl-2,2-dichloroacetamide (R 25 788) (10^{-7} M) both of which are important as antidotes for EPTC injury in maize crops (Wilkinson and Smith, 1975b). A similar reversal of EPTC inhibition of acetate incorporation has been reported for 1,8-naphthalic anhydride in red beet tissue (Wilkinson and Smith, 1976a). **Butylate**, **pebulate** and **vernolate** also inhibit the incorporation of 2-^{14}C-acetate into fatty acids of isolated spinach chloroplasts (Wilkinson and Smith, 1976b).

Treatment of sicklepod (*Cassia obtusifolia*) with di-allate (0–1.12 kg/ha) showed that fatty acid content declined with increasing concentrations interacting with photoperiod; each lipid class responded differently according to the combination of treatments (Wilkinson, 1974). Treatment of navy bean (*Phaseolus vulgaris*) with EPTC (10^{-6} M) greatly reduced wax deposition of the hypocotyl which may partly explain the greater permeability (Wyse, Meggitt and Penner, 1974a) and enhanced disease injury (Wyse *et al.*, 1974b).

Inhibition of fatty acid synthesis by EPTC, di-allate and **tri-allate** has also been reported. For example, EPTC (0.25–1 mg/kg air dry sand) applied to winter wheat (*Triticum sativum* cv. Nisu) caused a marked inhibition of total phospholipid production and a slight alteration in the general quality of fatty acids in the phospholipid fraction (Karunen and Wilkinson, 1975). Again, EPTC, di-allate and tri-allate (10^{-4} M) inhibits the formation of very long chain fatty acids by aged potato discs (Bolton and Harwood, 1976). Incorporation of ^{14}C-acetate into total fatty acids was inhibited by 24% by EPTC, 50% by tri-allate and 55% by di-allate. Bolton and Harwood conclude that these herbicides reduce cuticular wax by inhibiting fatty acid elongation.

EPTC applied as a soil drench (2.24 kg/ha) in developing leaves of cabbage (*Brassica oleracea*) resulted in increased cuticle permeability as demonstrated by greater uptake of ^{14}C-NAA and increased cuticular transpiration (Flore and Bukovac, 1974). Wettability and retention of water applied to the leaf surface were greater following EPTC application, this being associated more with a change in epicuticular wax topography than the quantity or chemistry of the wax (Flore and Bukovac, 1975). In the epicuticular wax fraction esters increased while ketones and alkanes decreased. Increased polarity of the cuticular wax was indicated by an increase in acid and alcohol constituents; no qualitative differences in cutin acids were evident (Flore and Bukovac, 1974).

Similarly, Leavitt and Penner (1978) found that EPTC significantly decreased the amount of epicuticular wax on the surface of cabbage leaves and increased cuticular transpiration. However, EPTC did not affect wax deposition on maize leaves, though scanning electron microscopy showed that the epicuticular wax was arranged in large aggregations on EPTC-treated leaves. This aggregation was prevented by combining R 25788 (*N,N*-diallyl-2,2-dichloroacetamide) with EPTC, protecting maize against the deleterious effects of post-emergence applications of paraquat (Leavitt and Penner, 1979).

HETEROCYCLIC COMPOUNDS

Inhibition of epicuticular wax deposition on leaves of cabbage (cv.

Market Prize) has been reported following soil treatments with **ethofumesate** (and EPTC) (Leavitt *et al.*, 1978). Separation of the epicuticular wax into its major components by g.l.c. showed that ethofumesate reduced the deposition of n-nonocosane and n-nonocosan-15-one but increased the deposition of long-chain waxy esters, a relatively minor component. In comparison EPTC was less inhibitory to n-nonocosan-15-one deposition than was ethofumesate, and it did not increase long-chain waxy ester deposition. Scanning electron microscopy revealed that ethofumesate almost completely eliminated cuticle wax on cabbage leaves, while EPTC only diminished it, suggesting that the former was the more potent inhibitor of epicuticular wax deposition.

References

AAMISEPP, A. (1976). *Växtodling*, **30**, 41 pp.

ABU-IRMAILEH, B. E. and JORDAN, L. S. (1978). *Weed Sci.*, **26**, 700.

ABU-IRMAILEH, B. E., JORDAN, L. S. and KUMAMOTO, J. (1979). *Weed Sci.*, **27**, 103.

AKOBUNDU, I. O., DUKE, W. B., SWEET, R. D. and MINOTTI, P. L. (1975). *Weed Sci.*, **23**, 43.

ALEXANDER, A. G. and BIDDULPH, O. (1976). *Pl. Cell Physiol.*, **17**, 601.

ALI, A. and FLETCHER, R. A. (1978). *Can. J. Bot.*, **56**, 2196.

ALI-ZADE, M. A. and ISMAILOV, A. A. (1976). *Dokl. Akad. Nauk Azer. SSR.*, **32**, 75.

ALLEN, W. S. and PALMER, R. D. (1963). *Weeds*, **11**, 27.

ALSOP, W. R. and MORELAND, D. E. (1975). *Pestic. Biochem. Physiol.*, 5.

ANDERSON, J. L. and THOMSON, W. W. (1973). *Residue Rev.*, **47**, 167.

ANDERSON, L. W. J. (1978). *Abstr. 1978 Mtg. Weed Sci. Soc. Amer.*, 50.

ARMSTRONG, D. J. (1966). *Proc. Natl. Acad. Sci. U.S.A.*, **56**, 64.

ARNDT, F., BOROSCHEWSKI, G. and KOTTER, C. (1971). *Mitt. Biol. Bund. Land. Fort. Bert-Dahlem.*, No. 146, 171.

ARNOLD, W. E. and NALEWAJA, J. D. (1971). *Weed Sci.*, **19**, 301.

ARNTZEN, C. J. (1978). *Abstr. 1978 Mtg. Weed Sci. Soc. Amer.*, 75.

ARNTZEN, C. J., DITTO, C. L. and BREWER, P. E. (1979). *Proc. Natl. Acad. Sci.*, **76**, 278.

ARTAMONOV, V. I. (1975). *Dokl. Akad. Nauk SSR.*, **220**, 1230.

ASHAHI, T. and JAGENDORF, A. T. (1963). *Arch. Biochem. Biophys.*, **100**, 531.

ASHTON, F. M. (1960). *Pl. Physiol., Lancaster*, **35**, (Suppl.) XXVIII.

ASHTON, F. M. (1963). *Weeds*, **11**, 295.

ASHTON, F. M. and CRAFTS, A. S. (1973). *Modes of Action of Herbicides*, Wiley–Interscience, 504 pp.

ASHTON, F. M., CUTTER, E. G. and HUFFSTUTTER, D. (1969). *Weed Res.*, **9**, 198.

ASHTON, F. M., PENNER, D. and HOFFMAN, S. (1968). *Weed Sci.*, **16**, 169.

ASHTON, F. M. and URIBE, E. G. (1962). *Weeds*, **10**, 295.

ASHTON, F. M., URIBE, E. G. and ZWEIG, G. (1961). *Weeds*, **9**, 575.

ASHTON, F. M. and ZWEIG, G. (1958). *Pl. Physiol., Lancaster, Abstr.*, **33**, XXVI–XXVII.

ASHTON, F. M., ZWEIG, G. and MASON, G. W. (1960). *Weeds*, **8**, 448.

ASLAM, M. and HUFFAKER, R. C. (1973). *Physiol. Pl.*, **28**, 400.

ASSCHE, C. J. VAN (1979). *4th IUPAC Symp., Zurich, Switzerland*, 494.

AYLING, R. D. (1976). *Weed Res.*, **16**, 301.

BALASIMHA, D., TEWARI, M. N. and RAM, C. (1977). *Biochem. Physiol., Pflanzen*, **171**, 43.

BAMBERGER, E. S., BLACK, C. C., FEWSON, C. A. and GIBBS, M. (1963). *Pl. Physiol., Lancaster*, **38**, 483.

BANDY, J. T. and KRIEG, D. R. (1974). *Pl. Physiol., Lancaster*, **53**, (Suppl.) 57.

BARDZIK, J. M. and MARSH, H. V. (1974). *Pl. Physiol., Lancaster*, **53**, (Suppl.) 57.

BARTELS, P. G. (1965). *Plant Cell Physiol.*, **6**, 227.

BARTELS, P. G. and HILTON, J. L. (1973). *Pestic. Biochem. Physiol.*, **3**, 462.

BARTELS, P. G. and MCCULLOUGH, C. (1972). *Abstr. Weed Sci. Amer.*, 93.

BARTELS, P. G. K., MATSUDA, K., SIEGEL, A. and WEIER, T. E. (1967). *Pl. Physiol., Lancaster*, **42**, 736.

BARTELS, P. G. and WATSON, C. W. (1978). *Weed Sci.*, **26**, 198.

BARTELS, P. G. and WEIER, T. E. (1969). *Amer. J. Bot.*, **56**, 1.

BARUFFINI, A., BORGNA, P., CALDERARA, G., MAZZA, M. and GIALDI, F. (1970). *Farmaco (Ed. Sci.)*, **25**, 10.

BASLER, E. and HANSON, T. L. (1964). *Bot. Gaz.*, **125**, 50.

BASLER, E. and NAKAZAWA, K. (1961). *Bot. Gaz.*, **122**, 228.

BAUMANN, G. and GÜNTHER, G. (1976a). *Biochem. Physiol. Pflanzen*, **169**, 163.

BAUMANN, G. and GÜNTHER, G. (1976b). *Biochem. Physiol. Pflanzen.*, **169**, 337.

BAUMANN, I., BAUMANN, G. and GÜNTHER, G. (1977). *Biochem. Physiol. Pflanzen*, **171**, 157.

BAUR, J. R. (1979). *Weed Sci.*, **27**, 69.

BAUR, J. R. and BOWMAN, J. J. (1972). *Physiol. Pl.*, **27**, 354.

BAUR, J. R. and BOWMAN, J. J. (1973). *Physiol. Pl.*, **28**, 372.

BAXTER, R. (1967). *The Effect of 2,4-D upon Metabolism and Composition of Soybean Mitochondria*, Ph.D. Thesis, University of Illinois; *Diss. Abstr. Int. B.*, **28**, 431.

BAXTER, R., KNELL, V. C., SOMERVILLE, H. J., SWAIN, H. M. and WEEKS, D. P. (1973). *Nature New Biol.*, **243**, 139.

BEGONIA, G. B., MERCADO, B. L. and ROBLES, R. P. (1973). *Weed Abstr.*, **22**, 12, 316.

BELL, S. L., SCHWARZ, O. J. and HUGHES, K. W. (1976). *Can. J. Gen. Cytol.*, **18**, 93.

BELLOTTI, A., COGHI, E., BARUFFINI, A., PAGANI, G. and BORGNA, P. (1968). *Farmaco (Ed. Sci.)*, **23**, 591.

BERGMANNOVA, E. and TAIMR, L. (1980). *Weed Res.*, **20**, 243.

BESTE, C. E. (1971). *Diss. Abstr. Int. B.*, **32**, 771.

BESTE, C. E. and SCHREIBER, M. M. (1972a). *Weed Sci.*, **20**, 8.

BESTE, C. E. and SCHREIBER, M. M. (1972b). *Weed Sci.*, **20**, 4.

BISHOP, N. I. (1962). *Biochem. Biophys. Acta*, **57**, 186.

BLACK, C. C. (1965). *Science*, **149**, 62.

BLACK, C. C. and HUMPHREYS, T. E. (1962). *Pl. Physiol., Lancaster*, **37**, 66.

BLACK, C. C. and MEYERS, L. (1966). *Weeds*, **14**, 331.

BOGER, P. and KUNERT, K. J. (1978). *Z. Naturforsch.*, **33C**, 688.

BOGER, P. and SCHLUE, U. (1976). *Weed Res.*, **16**, 149.

BOLHA'R-NORDENKAMPF, H. R. (1975). *Biochem. Physiol., Pflanzen*, **167**, 41.

BOLTON, P. and HARWOOD, J. L. (1976). *Phytochem.*, **15**, 1507.

BONNER, J. (1949). *Amer. J. Bot.*, **36**, 323.

BORST, P. and SLATER, E. C. (1961). *Biochem. Biophys. Acta*, **48**, 362.

BOTTRILL, D. E. (1965). 'Studies on the uncoupling action of 2,4-dichlorophenoxy-acetic acid', *Diss. Abstr. Int. B*, **26**, 57.

BOTTRILL, D. E. and HANSON, J. B. (1969). *Aust. J. Biol. Sci.*, **22**, 847.

BOULWARE, M. A. and CAMPER, N. D. (1972). *Physiol. Pl.*, **26**, 313.

BOURKE, J. B., BUTTS, J. S. and FANG, S. C. (1962). *Pl. Physiol., Lancaster*, **37**, 233.

BOURKE, J. B., BUTTS, J. S. and FANG, S. C. (1964). *Weeds*, **12**, 272.

BRACHA, P., LUWISCH, M. and SHAVIT, N. (1971). *Proc. 2nd Int. IUPAC Congr. Pestic. Chem.* (Ed. A. S. Tahori), **5**, 141.

BRECKE, B. J. and DUKE, W. B. (1977). *Abstr. Weed Sci. Soc. Amer.*, 87.

BREWER, P. E., ARNTZEN, C. J. and SLIFE, F. W. (1979). *Weed Sci.*, **27**, 300.

BRIQUET, M. V. and WIAUX, A. L. (1967). *Meded. Rijksfac. Landb. Wet. Gent.*, **32**, 1040.

BRODY, T. M. (1952). *Proc. Soc. Exp. Biol. Med.*, **80**, 533.

BRUINSMA, J. (1965). *Residue Rev.*, **10**, 1.

BRUNETTI, N., PICCIURRO, G. and FERRANDI, L. (1972). *Agrochimica*, **16**, 33.

BÜCHEL, K. H. (1972). *Pestic. Sci.*, **3**, 89.

BÜCHEL, K. H. and DRABER, W. (1969). In *Progress in Photosynthesis Research* (Ed. H. Metzner), Int. Union Biol. Sci., Tübingen, p. 1777.

BUGG, W., RIECK, C. E., COHEN, W. S. and WHITMARSH, J. (1979). *Pl. Physiol., Lancaster*, **63**, (Suppl.) 41.

BURNS, E. R., BUCHANAN, G. A. and CARTER, M. C. (1971). *Pl. Physiol., Baltimore*, **47**, 144.

BUSH, P. B. and RIES, S. K. (1974). *Weed Sci.*, **22**, 227.

BUTLER, R. D. and SIMON, E. W. (1971). *Adv. Generontol. Res.*, **3**, 73.

BYSTROVA, E. I. (1974). *Fiz. Rast.*, **21**, 148.

CALDERBANK, A. (1968). *Adv. Pest Control Res.*, **8**, 127.

CALDERBANK, A. and CROWDY, S. H. (1962). *Rept. Progr. Appl. Chem.*, **47**, 536.

CAMPBELL, W. F., EVANS, J. O. and REED, S. C. (1973). *Proc. W. Soc. Weed Sci.*, **26**, 34.

CAMPBELL, W. F., EVANS, J. O. and REED, S. C. (1976). *Weed Sci.*, **24**, 22.

CARLSON, W. C., LIGNOWSKI, E. M. and HOPEN, H. J. (1975). *Weed Sci.*, **23**, 155.

CARTER, G. E., JR. and CAMPER, N. D. (1973). *Abstr. Weed Sci. Soc. Amer.*, 57.

CASELEY, J. (1970). *Pestic. Sci.*, **1**, 28.

CERVELLI, S., NANNIPIERI, P., GIOVANNINI, G. and PERNA, A. (1975). *Pestic. Biochem. Physiol*, **5**, 221.

CHANDLER, J. M., CROY, L. I. and SANTELMANN, P. W. (1972). *J. Agric. Fd. Chem.*, **20**, 661.

CHANG, C. T., MARSH, H. V. JR. and JENNINGS, P. H. (1975). *Pestic. Biochem. Physiol.*, **5**, 323.

CHANG, C. T. and SMITH, D. W. (1972a). *Abstr. Weed Sci. Soc. Amer.*, 64.

CHANG, C. T. and SMITH, D. W. (1972 b). *Weed Sci.*, **20**, 220.

CHEN, L. G., SWITZER, C. M. and FLETCHER, R. A. (1972). *Weed Sci.*, **20**, 53.

CHEN, T. M., SEAMAN, D. E. and ASHTON, F. M. (1968). *Weed Sci.*, **16**, 28.

CHERRY, J. H. (1976). In *Herbicides: Physiology, Biochemistry, Ecology* (Ed. L. J. Audus), Vol. 1, Academic Press, New York and London, p. 525.

CHESALIN, G. A. and TIMOFEEVA, A. A. (1971). *Trudy, Vses nauch. Inst. Udobr. Agropoch.*, **51**, 195.

CHESALIN, G. A. and TIMOFEEVA, A. A. (1973). In *The Use of the Stable Isotope* ^{15}N *in Agricultural Research* (Ed. P. M. Smirnov), Moscow, USSR, p. 254.

CHKANIKOV, D. I., MIKITYUK, P. D. and MAKEEV, A, (1973). *Fiz. Rast.*, **20**, 603.

CODD, G. A. and COSSAR, J. D. (1978). *Biochem. Biophys. Res. Commun.*, **83**, 342.

COLBY, S. R., WOJTASZEK, T. and WARREN, G. F. (1965). *Weeds*, **13**, 87.

COLE, D. J., DODGE, A. D. and CASELEY, J. C. (1979). *Pl. Physiol., Lancaster*, **63**, (5, Suppl.) 96.

COOKE, A. R. (1956). *Weeds*, **4**, 397.

COPPING, L. G. and DAVIS, D. E. (1972). *Weed Sci.*, **20**, 86.

CORBETT, J. R. (1974). *The Biochemical Mode of Action of Pesticides*, Academic Press, London, 330 pp.

COSS, R. A. and PICKETT-HEAPS, J. D. (1974). *J. Cell Biol.*, **63**, 84.

COUCH, R. W. and DAVIS, D. E. (1966). *Weeds*, **14**, 251.

CURREY, W. L. (1972). *Diss. Abstr. Int. B.*, **33**, 2428.

DALVI, R. R., SINGH, B. and SALUNKHE, D. K. (1972). *J. Agric. Fd. Chem.*, **20**, 1000.

DAS, K., SINGH, B. D., SINGH, R. B., SINGH, J. and SINGH, R. M. (1978). *Indian J. Exp. Biol.*, **16**, 446.

DAS, V. S. R. and SANTAKUMARI, M. (1975). *Proc. Indian Acad. Sci. B*, **82**, 108.

DAVENPORT, H. E. (1963). *Proc. Roy. Soc., London*, **157B**, 332.

DAVIES, P. J., DRENNAN, D. S. H., FRYER, J. D. and HOLLY, K. (1968). *Weed Res.*, **8**, 241.

DAVIS, C. and HARVEY, R. G. (1979). *Proc. N.E. Weed Sci. Soc.*, **33**, 112.

DAVIS, D. E. (1968). *Proc. 21st S. Weed Control Conf.*, 346.

DAVIS, D. E., PILLAI, P. and TRUELOVE, B. (1979). *Abstr. Weed Sci. Soc. Amer.*, 99.

DAVIS, D. G., HOERAUF, R. A., DUSBABEK, K. E. and DOUGALL, D. K. (1977). *Physiol. Pl.*, **40**, 15.

DAVIS, D. G. and SHIMABUKURO, R. H. (1973). *Abstr. Weed Sci. Soc. Amer.*, 71.

DEAL, L. M. and HESS, F. D. (1980). *Weed Sci.*, **28**, 168.

DECLEIRE, M. and DE CAT, W. (1978). *Biol. Pl.*, **20**, 431.

DECLEIRE, M., DE CAT, W. and BASTIN, R. (1974). *Biochem. Physiol. Pflanzen*, **165**, 175.

DECLEIRE, M., DE CAT, W. and BASTIN, R. (1976). *Biochem. Physiol. Pflanzen*, **169**, 409.

DECLEIRE, M. and ROEY, G. VAN (1979). *Biochem. Physiol. Pflanzen*, **174**, 240.

DECLEIRE, M., ROEY, G. VAN and BASTIN, R. (1974). *Biochem, Physiol. Pflanzen*, **166**, 411.

DEEVA, V. P., VOINILO, V. A. and EREMENKO, L. (1972). *Fiz. Biokhim. Kul'turn. Rast.*, **4**, 57.

DELCOURT, A. and DEYSSON, G. (1976). *Cytologia*, **41**, 75.

DEVLIN, R. M. and KARCZMARCZYK, S. J. (1975a). *Proc. N.E. Weed Sci. Soc.*, 181.

DEVLIN, R. M. and KARCZMARCZYK, S. J. (1975b). *Proc. N.E. Weed Sci. Soc.*, 161.

DEVLIN, R. M., KISIEL, M. J. and KARCZMARCZYK, S. J. (1976). *Weed Res.*, **16**, 125.

DEVLIN, R. M., KISIEL, M. J. and KOSTUSIAK, A. S. (1979). *Abstr. Weed Sci. Soc. Amer.*, 3.

DEVLIN, R. M., SARAS, C. N., KISIEL, M. J. and KOSTUSIAK, A. S. (1978). *Weed Sci.*, **26**, 432.

DE VRIES, M. L. (1963). *Weeds*, **11**, 220.

DHILLON, N. S. (1971). *Diss. Abstr. Int. B.*, **31**, 7137.

DHILLON, N. S. and ANDERSON, J. L. (1972). *Weed Res.*, **12**, 182.

DICKS, J. W. (1978). *Pestic. Sci.*, **9**, 59.

DINER, A. M., TRUELOVE, B. and DAVIS, D. E. (1978). *Proc. 31st Mtg. S. Weed Sci. Soc.*, 250.

DODDS, E. C. and GREVILLE, G. D. (1934). *Lancet*, **112**, 398.

DODGE, J. D. and LAWES, G. B. (1972). *Ann. Bot.*, **36**, 315.

DOROZHKINA, L. A., KALANIN, V. A., KUZ'MINSKAYA, V. A. and LEONOVA, I. N. (1973). *Hort. Abstr.*, **44**, 1575.

DOW, W. A. (1952). *The Effect of Growth Substances on the Oxidative Activity of Plant Mitochondria*, M.Sc. Thesis, University of Wisconsin.

DRABER, W., BÜCHEL, K. H., DICKORÉ, K., TREBST, A. and PISTORIUS, E. (1969). In *Progress in Photosynthesis Research* (Ed. H. Metzner), Int. Union Biol. Sci., Tübingen, p. 1789.

DRABER, W., BÜCHEL, K. H., TIMMLER, H. and TREBST, A. (1974). In *Mechanism of Pesticide Action* (Ed. G. K. Kohn), ACS Symp. Ser. 2, Washington Amer. Chem. Soc., p. 100.

DUKE, S. O. and HOAGLAND, R. E. (1978). *Pl. Sci. Letters*, **11**, 185.

DUKE, S. O. and HOAGLAND, R. E. (1979a). *Pl. Physiol., Lancaster*, **63**, (5, Suppl.) 106.

DUKE, S. O. and HOAGLAND, R. E. (1979b). *Abstr. Weed Sci. Soc. Amer.*, 96.

DUKE, S. O., HOAGLAND, R. E. and ELMORE, C. D. (1979). *Physiol. Pl.*, **46**, 307.

DUKE, W. B., SLIFE, F. W. and HANSON, J. B. (1967). *Abstr. Weed Sci. Soc. Amer.*, 50.

DUKE, W. B., SLIFE, F. W., HANSON, J. B. and BUTLER, H. S. (1975). *Weed Sci.*, **23**, 142.

DUNCAN, D. N., MEGGITT, W. F. and PENNER, D. (1978). *Abstr. Weed Sci. Soc. Amer.*, 84.

EHRENFEST, E. and RONZONI, E. (1933). *Proc. Soc. Exp. Biol. Med.*, **31**, 318.

ÉIDEL'NANT, N. M., ESIPOVA, I. V., VENCHIKOVA, T. A., USMANOVA, R. F. and DANIL'CHENKO, A. M. (1974). *Fiz. Rast.*, **21**, 794.

ELEK, É., BORBÉLY, F. and KECSKÉS, M. (1974). *Acta Agron. Acad. Sci. Hung.*, **23**, 456.

EXER, B. (1958). *Experientia*, **13**, 136.

EXER, B. (1961). *Weed Sci.*, **1**, 233.

EZERZHA, A. A. and KULIKOV, B. N. (1973). *Fiz. Biokhim. Kul'turn. Rast.*, **5**, 644.

EZERZHA, A. A., KULIKOV, B. N. and IBRAGIMOVA, L. Ya. (1975). *Fiz. Biokhim. Kul'turn. Rast.*, **7**, 404.

EZERZHA, A. A., KULIKOV, B. N., MOCHALKIN, A. I. and POPOV, L. N. (1971). *Fiz. Biokhim. Kul'turn. Rast.*, **3**, 140.

FARRINGTON, J. A. (1976). *Proc. 1976 Brit. Crop Protect. Conf. Weeds*, **1**, 225.

FEDTKE, C. (1972). *Pestic. Biochem. Physiol.*, **2**, 312.

FEDTKE, C. (1973). *Pestic. Sci.*, **4**, 653.

FEDTKE, C. (1974). *Ber. dt. Botan. Ges.*, **87**, 155.

FEDTKE, C. (1975). *8th Int. Pl. Protect. Cong. Chem. Control, Moscow, Part 1*, 232.

FEDTKE, C., DEISCHGRÄBER, C. and SCHNEPF, E. (1977). *Biochem. Physiol Pflanzen*, **171**, 307.

FEIERABEND, J. and SCHUBERT, B. (1978). *Pl. Physiol., Lancaster*, **61**, 1017.

FERRARI, T. E. and MORELAND, D. E. (1967). *Abstr. Mtg. Weed Sci. Soc. Amer.*, 57.

FERRARI, T. E. and MORELAND, D. E. (1969). *Pl. Physiol., Lancaster*, **44**, 429.

FINKE, R. L., WARNER, R. L. and MUZIK, T. J. (1977). *Weed Sci.*, **25**, 18.

FLORE, J. A. and BUKOVAC, M. J. (1974). *HortSci.*, **9**, 33.

FLORE, J. A. and BUKOVAC, M. J. (1975). *Abstr. 72nd Mtg. Amer. Soc. Hort.*

FLORE, J. A. and BUKOVAC, M. J. (1976). *Abstr. 73rd Mtg. Amer. Soc. Hort.*

FLOYD, R. (1972). *Pl. Physiol., Lancaster*, **49**, 455.

FOY, C. L. and PENNER, D. (1965). *Weeds*, **13**, 226.

FRIEND, J. and OLSSON, R. (1967). *Nature (London)*, **214**, 942.

FRIESKE, S. (1971). *Biol. Inst. Ochr. Roslin*, **48**, 19.

FROSCH, S., JABBEN, M., BERGFELD, R., KLEINIG, H. and MOHR, H. (1979). *Planta*, **145**, 497.

FUNDERBURK, H. H. and CARTER,M. C. (1965). *Proc. 18th S. Weed Control Conf.*, 607.

FUNDERBURK, H. H. and DAVIS, D. E. (1963). *Weeds*, **11**, 101.

GABBOTT, P. A. (1968). *Soc. Chem. Ind. Monograph Series*, No. 29, 335.

GABBOTT, P. A. (1969). In *Progress in Photosynthesis Research* (Ed. H. Metzner), Int. Union Biol. Sci., Tübingen, p. 1712.

GAST, A. and GROB, M. (1960). *Pest. Tech.*, **3**, 68.

GAUR, B. K. and BEEVERS, H. (1959). *Pl. Physiol., Lancaster*, **34**, 427.

GEIKE, F. (1972). *Z. Pflkrankh. Pflschutz.*, **79**, 677.

GEIKE, F. (1974). *Gesunde Pflanzen*, **26**, 112.

GEORGE, K. (1971). *Indian J. Exp. Biol.*, **9**, 481.

GIANNOPOLITIS, C. N. and AYERS, G. S. (1978). *Weed Sci.*, **26**, 440.

GIBBONS, L. K., KOLDENHOVEN, E. F., NETHERY, A. A., MONTGOMERY, R. E. and PURCELL, W. P. (1976). *J. Agric. Fd. Chem.*, **24**, 203.

GNANARETHINAM, J. L. (1976). *Weed Res.*, **16**, 255.

GOELTENBOTH, F. (1978). *Proc. 5th EWRS Int. Symp. Aquatic Weeds, Amsterdam, 1978*, 383.

GOOD, N .E. (1961). *Pl. Physiol, Lancaster*, **36**, 788.

GOSS, J. R., RICHARD, E. P., ARNTZEN, C. J. and SLIFE, F. W. (1978). *Abstr. Weed Sci. Soc. Amer.*, 74.

GRUENHAGEN, R. D. and MORELAND, D. E. (1971). *Weed Sci.*, **19**, 319.

GRUZDEV, L. G. (1979). *Fiz. Rast.*, **26**, 153.

GRUZDEV, L. G., POSHITNAYA, L. V. and GRUZDEV, G. S. (1979). *Izv. Tim. Sel'skok. Akad.*, No. 4, 105.

GRZESIUK, S., JANUSZEWICZ, E. and SÓJKA, E. (1973). *Bull l'Acad. Polon. Sci. Ser. Sci. Biol.*, **21**, 439.

GRZESIUK, S., LOGIN, A, and NOWAK, J. (1973). *Bull. l'Acad. Polon. Sci. Ser. Biol.*, **21**, 689.

GRZESIUK, S., PALIWODA, Z. and SÓJKA, E. (1973). *Bull. l'Acad. Polon. Sci. Ser. Sci. Biol.*, **21**, 369.

GYSIN, H. and KNÜSLI, E. (1960). *Advances in Pest Control Research* (Ed. R. L. Metcalf), Interscience Publishers, New York, p. 289.

GZIK, A. and GRÄSER, H. (1972). *Biochem. Physiol. Pflanzen*, **163**, 458.

HADERLIE, L. C. (1976). *Diss. Abstr. Int. B.*, **36**, 4253.

HADERLIE, L. C., WIDHOLM, J. M. and SLIFE, F. W. (1974). *Proc. N. Central Weed Control Conf.*, **29**, 92.

HALL, W. C., JOHNSON, S. P. and LEINWEBER, C. L. (1954). *Bull. Texas Agric. Exp. Sta.*, **789**, 1.

HANSCH, C. (1969). In *Progress in Photosynthesis Research* (Ed. H. Metzner), Vol. III, Int. Union Biol. Sci., Tübingen, p. 1685.

HANSCH, C. and DEUTSCH, E. W. (1966). *Biochem. Biophys. Acta*, **112**, 381.

HARPER, D. B. and HARVEY, B. M. R. (1978). *Pl. Cell Environ.*, **1**, 211.

HARRIS, N. and DODGE, A. D. (1972a). *Planta*, **104**, 201.

HARRIS, N. and DODGE, A. D. (1972b). *Planta*, **104**, 210.

HARVEY, B. M. R., CHANG, F. Y. and FLETCHER, R. A. (1975). *Can. J. Bot.*, **53**, 225.

HASSAN, H. M. and FRIDOVICH, J. (1978). *J. Biol. Chem.*, **253**, 8143.

HATZIOS, K. K. and PENNER, D. (1978). *Proc. N. Central Weed Control Conf.*, **33**, 48.

HATZIOS, K. K. and PENNER, D. (1979). *Abstr. Weed Sci. Soc. Amer.*, 104.

HATZIOS, K. K., PENNER, D. and BELL, D. (1979). *Pl. Physiol., Lancaster*, **63**, (Suppl.) 41.

HAUNOLD, E. and ZSOLDOS, F. (1976). *Bodenkultur*, **27**, 331.

HAUSKA, G., TREBST, A., KOTTER, C. and SCHULZ, H. (1975). *Z. Naturforsch. C.*, **30**, 505.

HENDRIX, D. L. and MUENCH, S. R. (1969). *Pl. Physiol., Lancaster, Abstr.*, **44**, 5–26.

HERRETT, R. A. and BERTHOLD, R. V. (1965). *Science, N.Y.*, **149**, 191.

HILL, E. R., PUTALA, E. C. and VENGRIS, J. (1968). *Weeds*, **16**, 377.

HILTON, J. L., SCHAREN, A. L., ST. JOHN, J. B., MORELAND, D. E. and NORRIS, K. H. (1969). *Weed Sci.*, **17**, 541.

HIRANPRADIT, H., FOY, C. L. and SHEAR, G. M. (1972a). *Agron. J.*, **64**, 267.

HIRANPRADIT, H., FOY, C. L. and SHEAR, G. M. (1972b). *Agron. J.*, **64**, 274.

HOAGLAND, R. E. and DUKE, S. O. (1978). *Abstr. 175th Amer. Chem. Soc. Nat. Mtg. 1978, Pestic.*, 42.

HOAGLAND, R. E. and DUKE, S. O. (1979). *Abstr. Weed Sci. Soc. Amer.*, 96.

HOAGLAND, R. E., DUKE, S. O. and ELMORE, C. D. (1978). *Pl. Sc. Letters*, **13**, 291.

HOAGLAND, R. E., DUKE, S. O. and ELMORE, C. D. (1979). *Physiol. Pl.*, **46**, 357.

HOAGLAND, R. E. and PAUL, R. N. (1978a). *Abstr. Weed Sci. Soc. Amer.*, 78.

HOAGLAND, R. E. and PAUL, R. N. (1978b). *Proc. 31st Ann. Mtg. S. Weed Sci. Soc.*, 284.

HOAGLAND, R. E. and PAUL, R. N. (1978c). *Pl. Physiol., Lancaster*, **61**, (4, Suppl.) 42.

HOAGLAND, R. E. and PAUL, R. N. (1979). *Abstr. Weed Sci. Soc. Amer.*, 7.

HOFFMANN, C. E. (1971). *Second Int. Conf. Pestic. Chem.*

HOFSTRA, G. and SWITZER, C. M. (1968). *Weed Sci.*, **16**, 23.

HOLM, R. E. and STALLARD, D. E. (1974). *Weed Sci.*, **22**, 10.

HOOGSTRATEN, S. D. VAN. (1973). *Diss. Abstr. Int. B.*, **34**, 564.

HOPKINS, T. R., NEIGHBORS, R. P. and PHILLIPS, L. V. (1967). *J. Agric. Fd. Chem.*, **15**, 501.

HUMPHREYS, T. E. and DUGGER, W. M. (1957a). *Pl. Physiol., Lancaster*, **32**, 136.

HUMPHREYS, T. E. and DUGGER, W. M. (1957b). *Pl. Physiol., Lancaster*, **32**, 530.

HUMPHREYS, T. E. and DUGGER, W. M. (1958). *Pl. Physiol., Lancaster*, **33**, 112.

HUMPHREYS, T. E. and DUGGER, W. M. (1959). *Pl. Physio., Lancaster*, **34**, 580.

HUNYADI, K. and BOGDAN, I. (1976). *Novényvédelem*, **12**, 58.

HYAKUTAKE, H. and ISHIZUKA, K. (1976). *Proc. 5th Asian-Pacific Weed Sci. Conf., Tokyo, Japan*, 192.

ILIEV, L. and GRÄSER, H. (1975). *The Effect of Atrazine on the Incorporation of ^{32}P into the r-RNA and TB-RNA*, Inst. Fiz. Rast. M. Popov, BAN, Sofia 13, Bulgaria.

INGLE, M. and RODGERS, B. J. (1961). *Weeds*, **9**, 264.

INOUE, Y., ISHIZUKA, K. and MITSUI, S. (1967). *Agric. Biol. Chem.*, **31**, 422.

IZAWA, S. (1968). In *Comparative Biochemistry and Biophysics of Photosynthesis.* (Eds. K. Shibata, A. Takamya, A. T. Jagendorf and R. K. Fuller), Univ. of Tokyo Press, Tokyo and Univ. Park Press, State College, Penn., p. 140.

IZAWA, S. and GOOD, N. E. (1965). *Biochem. Biophys. Acta*, **102**, 20.

JACOBSON, A. B. and ROGERS, B. J. (1961). *Pl. Physiol., Lancaster, Abstr.*, **36**, XI.

JAGENDORF, A. T. and AVRON, M. (1959). *Arch. Biochem. Biophys.*, **80**, 246.

JAIN, M. L., KURTZ, E. B. and HAMILTON, K. C. (1966). *Weeds*, **14**, 259.

KAHLON, P. S. and BANERJEE, M. R. (1979). *Bull. Environ. Contam. Toxicol.*, **22**, 365.

KALININ, F. L., MUSIYAKA, V. K., STASEVSKAYA, I. P. and RYBAL'CHENKO, V. K. (1975). *Fiz. Biokhim. Kul'turn. Rast.*, **7**, 256.

KANDLER, O. (1958). *Physiol. Pl.*, **11**, 675.

KARUNEN, P. and WILKINSON, R. E. (1975). *Physiol. Pl.*, **35**, 228.

KAWAMURA, Y., MIYAKE, H. and MAEDA, E. (1976). *Proc. Crop Sci. Soc. Japan*, **45**, 538.

KERR, M. W. and WAIN, R. L. (1964). *Ann. Appl. Biol.*, **54**, 441.

KEY, J. L. and HANSON, J. B. (1961). *Pl. Physiol., Lancaster*, **36**, 145.

Key, J. L., Lin, C. Y., Gifford, E. M. and Dengler, R. (1966). *Bot. Gaz.*, **127**, 87.

Key, J. L. and Shannon, J. C. (1964). *Pl. Physiol., Lancaster*, **39**, 360.

Khakimov, Y. I. and Kvitko, K. V. (1978). *Genetika*, **14**, 1319.

Khristov, K. (1974). *Rast. Nauki*, **11**, 105.

Killion, D. D. and Frans, R. E. (1969). *Weed Sci.*, **17**, 468.

Kim, W. K. and Bidwell, R. G. S. (1967). *Can. J. Bot.*, **45**, 1751.

Kirkwood, R. C. (1976). In *Herbicides: Physiology, Biochemistry, Ecology* (Ed. L. J. Audus), Vol. 1, Academic Press, London and New York, p. 441.

Klepper, L. A. (1975). *Weed Sci.*, **23**, 188.

Kleudgen, H. K. (1978). *Pestic. Biochem. Physiol.*, **9**, 57.

Kleudgen, H. K. (1979). *Z. Naturforsch.*, **34**, 110.

Kobayashi, K. and Ishizuka, K. (1974). *Weed Sci.*, **22**, 131.

Kok, B., Rurainski, R. J. and Owens, O. V. J. (1965). *Biochem. Biophys. Acta*, **109**, 347.

Kolattukudy, P. E. (1968). *Science, N.Y.*, **159**, 498.

Krishchenko, V. P., Kovach, I. and Gruzdev, L. G. (1973). *Agrokhimiya*, **10**, 113.

Krishning, J. (1965). *Phytopath. Z.*, **53**, 65.

Kulandaivelu, G. and Gnanam, A. (1975). *Physiol. Pl.*, **33**, 234.

Kunert, K. J. and Böger, P. (1979). *Pl. Physiol., Lancaster*, **63**, (Suppl.) 42.

Ladonin, V. F. (1966). *Vest. sel'-khoz. Nauki Mosk.*, **11**, 137.

Ladonin, V. F. (1967a). *Agrokhimiya*, **2**, 85.

Ladonin, V. F. (1967b). *Khim. sel'-khoz.*, **5**, 36.

Ladonin, V. F. (1972). *Vest. sel'-khoz. Nauki Mosk.*, **11**, 19.

Ladonin, V. F. and Beketova, L. I. (1972). *Dokl. Vses. Akad. sel'-khoz. Nauk*, No. 5, 18.

Ladonin, V. F., Khachatryan, S. M. and Gal'per-Blichenko, E. M. (1973). *Agrokhimiya*, **10**, 103.

Ladonin, V. F., Samoilov, L. N., Posmitnaya, L. V., Gruzden, L. G. and Dyukina, N. N. (1974). *Agrokhimiya*, **11**, 120.

Ladonin, V. F. and Svittser, K. M. (1967). *Soviet Pl. Physiol.*, **14**, 853.

Lalova, M. (1973). *Rast. Nauki*, **10**, 151.

Lapado, B. E. and Mercado, B. L. (1973). *Weed Abstr.*, **22**, 316.

Lay, M. M. and Ilnicki, R. D. (1971). *Proc. N.E. Weed Sci. Soc.*, **25**, 141.

Lay, M. M. and Ilnicki, R. D. (1972a). *Proc. N.E. Weed Sci. Soc.*, **26**, 80.

Lay, M. M. and Ilnicki, R. D. (1972b). *Proc. N.E. Weed Sci. Soc.*, **26**, 85.

Leavitt, J. R. C., Duncan, D. N., Penner, D. and Meggitt, W. F. (1978). *Pl. Physiol., Lancaster*, **61**, 1034.

Leavitt, J. R. C. and Penner, D. (1978). *Abstr. Weed Sci. Soc. Amer.*, 79.

Leavitt, J. R. C. and Penner, D. (1979). *Weed Sci.*, **27**, 47.

Lehmann, K., Turowski, W. and Frieske, S. (1971). *Biol. Inst. Ochr. Roslin*, **50**, 149.

Lignowski, E. M. and Scott, E. G. (1971). *Pl. Cell Physiol.*, **12**, 701.

Lignowski, E. M. and Scott, E. G. (1972). *Weed Sci.*, **20**, 267.

Lin, M. and Anderson, J. L. (1967). *Proc. 21st W. Weed Control*, 41.

LIN, P. P. C. (1978). *Pl. Physiol., Lancaster*, **61**, (4, Suppl.) 7.

LINCK, A. J. (1976). In *Herbicides: Physiology, Biochemistry, Ecology* (Ed. L. J. Audus), Vol. 1, Academic Press, New York and London, p. 83.

LITVINOV, I. A. (1972). *Trudy, Khar'k. Sel'skok. Inst.*, No. 172, 167.

LOLAS, P. C. (1980). *Weed Res.*, **20**, 239.

LOTLIKAR, P. D. (1960). 'The effects of herbicides on oxidative phosphorylation in mitochondria from cabbage (*Brassica oleracea*)', *Diss. Abstr. Int. B.*, **21**, 446.

LOTLIKAR, P. D., REMMERT, L. F. and FREED, V. H. (1968). *Weed Sci.*, **16**, 161.

LUZHNOVA, M. I. and SHEKHTMAN, L. M. (1975). *Pestic. Biochem. Physiol.*, **5**, 205.

LUZHNOVA, M. I., SHEKHTMAN, L. M. and SHARKOV, W. I. (1972). *Symp. Wirkungsmechanismen von Herbiziden und Synthetischen Wachstumregulatoren, Wissen*schaftliche Beiträge der Martin-Luther-Universität, Halle-Wittenberg, No. 10, 153.

MCDANIEL, J. L. and FRANS, R. E. (1969). *Weed Sci.*, **17**, 192.

MCWHORTER, C. G. (1959). *Some Effects of 3-Amino-1,2,4-triazole on the Respiratory Activities of* Zea mays *L.*, Ph.D. Thesis, Louisiana State University (*Diss. Abstr. Int. B*, **19**, 1929).

MCWHORTER, C. G. and PORTER, W. K. (1960). *Weeds*, **8**, 29.

MALAKANDAIAH, N. and FANG, S. C. (1978). *Pestic. Biochem. Physiol.*, **9**, 33.

MALHOTRA, S. S. and HANSON, J. B. (1966). *Pl. Physiol., Lancaster, Abstr.*, **41**, VI.

MALHOTRA, S. S. and HANSON, J. B. (1970). *Weed Sci.*, **18**, 1.

MANN, J. D., COTA-ROBLES, E., YUNK, K. H., PU, M. and HAID, H. (1967). *Biokhim. Biophys. Acta*, **138**, 133.

MANN, J. D., JORDAN, L. S. and DAY, B. E. (1965a). *Weeds*, **13**, 63.

MANN, J. D., JORDAN, L. S. and DAY, B. E. (1965b). *Pl. Physiol., Lancaster*, **40**, 840.

MANN, J. D. and PU, M. (1968). *Weed Sci.*, **16**, 197.

MARSH, H. V., BATES, J. and DOWNS, S. (1976). *Pl. Physiol., Lancaster*, **57**, (5, Suppl.) 61.

MASHTAKOV, S. M., DEEVA, V. P. and VOLYNETS, A. P. (1967). In *Physiological Effects of Herbicides on Varieties of Crop Plants* (Ed. S. M. Mashtakov), Nauka, Tekhnika, Minsk.

MATLIB, M. A. (1970). *Effect of Selected Phenoxy Acid Herbicides on the Oxidative and Phosphorylative Activities of Mitochondria from* Vicia faba, Ph.D. Thesis, University of Strathclyde, Glasgow.

MATLIB, M. A., KIRKWOOD, R. C. and PATTERSON, J. D. E. (1971). *Weed Res.*, **11**, 190.

MATLIB, M. A., KIRKWOOD, R. C. and SMITH, J. E. (1972). *J. Exp. Bot.*, **23**, 886.

MEES, G. C. (1960). *Ann. Appl. Biol.*, **48**, 601.

MEIKLE, R. W. (1970). *Weed Sci.*, **18**, 475.

MERBACH, W. and SCHILLING, G. (1977). *Biochem. Physiol. Pflanzen*, **171**, 171.

MERCADO, B. L. and BALTAZAR, A. M. (1973). *Weed Abstr.*, **22**, 12, 316.

MERCADO, B. L. and TALATALA, R. L. (1973). *Weed Abstr.*, **22**, 12, 316.

MEREDITH, F. I. and LANGDALE, G. (1978). *Ann. Mtg. Amer. Soc. Pl. Physiol.* with *Amer. Inst. Biol. Sci.*, New Orleans.

MEREZHINSKII, YU. G. and LAPINA, T. V. (1974). *Fiz. Biokhim. Kul'turn. Rast.*, **6**, 596.

MEREZHINSKII, YU. G., MITROFANOV, B. A. and IVANISCHEV, Y. N. (1975). *Fiz. Biokhim. Kul'turn. Rast.*, **7**, 345.

METCALF, E. C. and COLLIN, H. A. (1978). *New Phytol.*, **81**, 243.

MIKHNO, A., KALININ, F. L. and LESNEVICH, L. A. (1975). *Ukr. Bot. Zh.*, **32**, 489.

MIKHNO, A. N., KALININ, F. L. and LESNEVICH, L. A. (1978). *Bot. Zh.*, **63**, 1632.

MIKHNO, A. N., MUSIYAKA, Y. K. and KALININ, F. L. (1972). *Tsitologiya i Genetika*, **6**, 516.

MIKITYUK, O. D., PETELINA, G. G. and CHKANIKOV, D. I. (1977). *Fiz. Rast.*, **24**, 1242.

MILLER, C. S. and HALL, W. C. (1957). *Weeds*, **5**, 218.

MINE, A. and MATSUNAKA, S. (1975). *Pestic. Biochem. Physiol.*, **5**, 444.

MISRA, L. P., KAPOOR, L. D. and CHOUDHRI, R. S. (1974). *Photosynthetica*, **8**, 302.

MONACO, T. J. (1968). *The Partitioning and Distribution of Simazine and Propanil in Spinach Chloroplasts*, Ph.D. Thesis, North Carolina State University, Raleigh.

MONOD, J., CHANGEUX, J. P. and JACOB, F. (1963). *J. Molec. Biol.*, **6**, 306.

MORELAND, D. E. (1967). *Ann. Rev. Pl. Physiol.*, **18**, 365.

MORELAND, D. E. (1969). In *Progress in Photosynthesis Research* (Ed. H. Metzner), Int. Union Biol. Sci., Tübingen, p. 1737.

MORELAND, D. E. (1974). *Abstr. Weed Sci. Soc. Amer.*, 76.

MORELAND, D. E. (1980). *Ann. Rev. Pl. Physiol.*, **31**, 597–638.

MORELAND, D. E. and BLACKMON, W. J. (1968). *Abstr. Papers 156th. Mtg. Amer. Chem. Soc.*, AGFD-74.

MORELAND, D. E. and BLACKMON, W. J. (1970). *Weed Sci.*, **18**, 419.

MORELAND, D. E., BLACKMON, W. J., TODD, H. G. and FARMER, F. S. (1970). *Weed Sci.*, **18**, 636.

MORELAND, D. E., FARMER, F. S. and HUSSEY, G. G. (1972a). *Pestic. Biochem. Physiol.*, **2**, 342.

MORELAND, D. E., FARMER, F. S. and HUSSEY, G. G. (1972b). *Pestic. Biochem. Physiol.*, **2**, 354.

MORELAND, D. E., GENTNER, W. A., HILTON, J. L. and HILL, K. L. (1959), *Pl. Physiol., Lancaster*, **34**, 432.

MORELAND, D. E. and HILL, K. L. (1959). *J. Agric. Fd. Chem.*, **7**, 832.

MORELAND, D. E. and HILL, K. L. (1962). *Weeds*, **10**, 229.

MORELAND, D. E. and HILL, K. L. (1963). *Weeds*, **11**, 55.

MORELAND, D. E., HILL, K. L. and HILTON, J. L. (1958). *WSSA Abstr.*, 40.

MORELAND, D. E. and HILTON, J. L. (1976). In *Herbicides: Physiology, Biochemistry, Ecology* (Ed. L. J. Audus), Vol. 1, Academic Press, New York and London, p. 493.

MORELAND, D. E., HUSSEY, G. G. and FARMER, F. S. (1974). *Pestic. Biochem. Physiol.*, **4**, 356.

MORELAND, D. E., MALHOTRA, S. S., GRUENHAGEN, R. D. and SHOKRAII, E. H. (1969). *Weed Sci.*, **17**, 556.

MORIMOTO, M., SHIMIZU, N. and UEKI, K. (1972). *Weed Res., Japan.*, No. 14, 40.

MUR, A. R. DE., SWADER, J. A. and YOUNGNER, V. B. (1972). *Pestic. Biochem. Physiol.*, **2**, 337.

MUSCHINEK, G., GARAB, G. I. and MUSTÁRDY, L. A. (1979). *Weed Res.*, **19**, 101.

MUSIYAKA, V. K. and KALININ, F. L. (1972). *Fiz. Biokim. Kul'turn. Rast.*, **4**, 477.

MUSIYAKA, V. K. and KALININ, F. L. (1975). *Fiz. Biokim. Kul'turn. Rast.*, **7**, 486.

MUSIYAKA, V. K. and KALININ, F. L. (1979). *Fiz. Biokim. Kul'turn. Rast.*, **11**, 241.

MUSIYAKA, V. K., MIKHNO, A. N., KALININ, F. L. and MARTYN, G. I. (1972). *Ukr. Bot. Zh.*, **29**, 158.

MUZYCHENKO, M. D. and BORODYCHEVA, Z. N. (1976). *Agronomiya*, No. 3, 11.

NAKAYAMA, H. (1974). *Bull. Hokuriku Agric. Exp. Sta.*, No. 16, 1.

NARSAIAH, D. B. and HARVEY, R. G. (1977). *Weed Sci.*, **25**, 197.

NASYROVA, T., ARIPOVA, F. and BABAEV, A. (1974). *Dokl. Akad. Nauk Uzbek. SSR*, **31**, 60.

NASYROVA, T., MIRKASIMOVA, KH. and PAZILOVA, S. (1968). *Uzbek. Biol. Zh.*, **12**, 23.

NEGI, N. S., FUNDERBURK, H. H., SCHULTZ, D. P. and DAVIS, D. E. (1967). *WSSA Abstr.*, 58.

NEGI, N. S., FUNDERBURK, H. H., SCHULTZ, D. P. and DAVIS, D. E. (1968). *Weed Sci.*, **16**, 83.

NEVZOROVA, L. N. and VOEVODIN, A. V. (1977). *Trudy Vses. Nauchno-Issled., Inst. Zash. Rast*, No. 54, 74.

NISHIDA, K. (1968). *Comparative Biochemistry and Biophysics of Photosynthesis* (Eds. K. Shibata, A. Takamya, A. T. Jagendorf and R. C. Fuller), University of Tokyo Press, p. 97.

NISHIMURA, K., KAWATA, T., ASADA, K. and NAKAJIMA, M. (1975). *Agric. Biol. Chem.*, **39**, 867.

NISHIMURA, M. and TAKAMIYA, A. (1966). *Biochem. Biophys. Acta*, **120**, 45.

O'BRIEN, M. C. and PRENDEVILLE, G. N. (1979). *Weed Res.*, **19**, 331.

OKUBO, C. K. and SWITZER, C. M. (1965). *Pl. Physiol., Lancaster, Abstr.*, **40**, XIV.

OLECH, K. (1966). *Annls. Univ. Mariae Curie-Sktodowska (E)*, **21**, 289.

OLOFINOBA, M. O. and FAWOLE, M. O. (1976). *Turrialba*, **26**, 167.

OORSCHOT, J. L. P. VAN (1964). *Proc. 7th Brit. Weed Control Conf.* , 321.

OORSCHOT, J. L. P. VAN (1965). *Weed Res.*, **5**, 84.

OORSCHOT, J. L. P. VAN and BELKSMA, M. (1961). *Weed Res.*, **1**, 245.

OORSCHOT, J. L. P. VAN and LEEUWEN, P. H. VAN (1979). *Weed Res*, **19**, 63.

OVERBEEK, J. L. P. VAN (1962). *Weeds*, **10**, 170.

OYOLU, C. and HUFFAKER, R. C. (1964). *Crop. Sci.*, **4**, 95.

PALLETT, K. E. and DODGE, A. D. (1976). *Proc. 1976 Brit. Crop. Prot Conf. – Weeds, London*, **1**, 235.

PALMER, R. D. and ALLEN, W. S. (1962). *Proc. 15th S. Weed Control Conf.*, 271.

PARKER, M. V. (1965). *Biochem. J.*, **97**, 658.

PARKS, J. P., TRUELOVE, B. and BUCHANAN, G. A. (1971). *Proc. 24th Ann. Mtg. S. Weed Sci. Soc.*, 354.

PARKS, J. P., TRUELOVE, B. and BUCHANAN, G. A. (1972). *Weed Sci.*, **20**, 89.

PATON, D. and SMITH, J. E. (1965). *Weed Res.*, **5**, 75.

PATON, D. and SMITH, J. E. (1967). *Can. J. Biochem.*, **45**, 1891.

PAUL, W., REUTHER, R., RÖHLER, E., RÖHLER, I., ILIEW, L., STÖLZEL, G. and GRASER, H. (1973). *Wissenschaft. Beit. der Martin-Luther-Universitat, Halle-Wittenberg*, **10**, 86.

PENNER, D. (1968). *Weed Sci.*, **16**, 519.

PENNER, D. (1970). *Weed Sci.*, **18**, 360.

PENNER, D. and EARLY, R. W. (1972). *Pl. Physiol., Lancaster*, **49**, (Suppl.) 29.

PENNER, D. and RIVERA, C. M. (1977). *Proc. N. Central. Weed Control Conf.*, **32**, 115.

PEREIRA, J. F., SPLITTSTOESSER, W. E. and HOPEN, H. J. (1971). *Weed Sci.*, **19**, 662.

PETERSON, C. A. (1977). *Diss. Abstr. Int. B.*, **37**, 4244.

PETKOVA, R. A. (1972). *Compte. Rend. de l'Academie Bulgare des Sciences*, **25**, 985.

PFISTER, K. and ARNTZEN, C. J. (1979). *Z. Naturforsch.*

PFISTER, K., RADOSEVICH, S. R. and ARNTZEN, C. J. (1979a). *Abstr. Weed Sci. Soc. Amer.*, 113.

PFISTER, K., RADOSEVICH, S. R. and ARNTZEN, C. J. (1979b). *Pl. Physiol., Lancaster*, **64**, 995.

PILLAI, C. G. P. and DAVIS, D. E. (1974). *Abstr. Weed Sci. Soc. Amer.*, 80.

PILLAI, C. G. P. and DAVIS, D. E. (1975). *Proc. 28th Ann. Mtg. S. Weed Sci. Soc.*, 308.

PILLAI, C. G. P., DAVIS, D. E. and TRUELOVE, B. (1978). *Proc. 31st Mtg. S. Weed Sci. Soc.*, 278.

PILLAI, C. G. P., DAVIS, D. E., TRUELOVE, B. and THOMPSON, O. C. (1973). *Proc. 26th Ann. Mtg. S. Weed Sci. Soc.*, 414.

PINCHOT, G. B. (1963). *Fedn. Proc. Fedn. Amer. Soc. Exp. Biol.*, **22**, 1076.

PINCHOT, G. B. (1967). *J. Biol. Chem.*, **242**, 4577.

PLOSZYŃSKI, M. (1972). *Pam. Pulawski.*, No. 51, 51.

PLOSZYŃSKI, M. and ROLA, J. (1974). *Pam. Pulawski.*, No. 60, 105.

PLOSZYŃSKI, M. and ZURAWSKI, H. (1971). *Acta Agrobotanica*, **24**, 205.

POCHINOK, Kh. N. (1971). *Fiz. Biokhim. Kul'turn. Rast.*, **3**, 538.

PORTER, E. M. and BARTELS, P. G. (1977). *Weed Sci.*, **25**, 60.

POTTER, J. R. (1977). *Weed Sci.*, **25**, 241.

POTTER, J. R. and WERGIN, W. P. (1974). *Pl. Physiol., Lancaster*, **53**, (Suppl.) 5.

POTTER, J. R. and WERGIN, W. P. (1975). *Pestic. Biochem. Physiol.*, **5**, 458.

PULVER, E. L. (1972). *Diss. Abstr. Int. B.*, **33**, 1881.

PULVER, E. L. and RIES, S. K. (1973). *Weed Sci.*, **21**, 233.

PUSZTAI, T. and VEGH, A. (1978). *Acta Bot. Acad. Sci. Hung.* , **24**, 327.

PUTALA, E. C. (1967). *A Study of the Ultrastructure and Histochemistry of the Apoplastidic Effect of 3-Amino-1,2,4-triazole on the Shoot Apex of Tomato,* Ph. D. Thesis, Univ. Calif., Berkeley, 51 pp.

PYFROM, H. T., APPOEMAN, D. and HEIM, W. G. (1957). *Pl. Physiol., Lancaster,* **32**, 674.

PYNE, W. J., SZABO, S. S. and HOLM, R. E. (1979). *J. Agric. Fd. Chem.*, **27**, 537.

QUIMBY, P. C. (1967). *Studies Relating to the Selectivity of Dicamba for Wild Buckwheat (*Polygonum convolvulus *L.) vs. Selkirk Wheat (*Triticum aestivum *L.) and a Possible Mode of Action*, Ph.D. Thesis, N. Dakota State University, 87 pp.

RACKER, E. (1961). *Adv. Enzymol.*, **23**, 323.

RADOSEVICH, S. R. (1976). *Proc. W. Weed Sci. Soc.*, **29**, 80.

RADOSEVICH, S. R. and DEVILLIERS, O. T. (1976). *Weed Sci.*, **24**, 229.

RADOSEVICH, S. R., STEINBACK, R. E. and ARNTZEN, C. (1979). *Weed Sci.*, **27**, 216.

RADTSEV, V. S., RADTSEVA, G. E., ZAVEZENOVA, N. B. and CHISTOVA, N. P. (1975). In *Materialy 10go Mezhdunarodnogo Simpoziuma Stran-Chlenov SEV, Pushchino, SSSR*, **1**, 41.

RAFII, Z. E., ASHTON, F. M. and GLENN, R. K. (1979). *Weed Sci.*, **27**, 422.

RAJU, M. V. S. and GROVER, R. (1976). *Weed Sci.*, **24**, 175.

RAO, V. S. (1974). *Diss. Abstr. Int. B.*, **35**, 641.

RAO, V. S. and DUKE, W. B. (1974). *Abstr. Weed Sci. Soc. Amer.*, 87.

RAO, V. S. and DUKE, W. B. (1976). *Weed Sci.*, **24**, 616.

REIMER, S., LINK, K. and TREBST, A. (1979). *Z. Naturforsch., C.*, **34**, 419.

REJOWSKI, A., TLUCZKIEWICZ, J. and LOGIN, A. (1973). *Rocz. Nauk. Roln., A*, **9**, 25.

RENSBURG, A. J. J. VAN and VILLIERS, O. T. DE (1979). *Agroplantae*, **11**, 3.

RENSEN, J. J. S. VAN (1974). *Proc. 3rd Int. Cong. Photosynthesis, Rehovot, Israel*, 683.

RENSEN, J. J. S. VAN (1975). *Physiol. Pl.*, **1**, 42.

RENSEN, J. J. S. VAN (1979). *Pl. Physiol., Lancaster*, **63**, (Suppl.) 42.

RENSEN, J. J. S. VAN, VET, W. VAN DER and VLIET, W. P. A. VAN (1977). *Photochem. Photobiol.*, **25**, 579.

RENSEN, J. J. S. VAN, WONG, D. and GOVINDJEE (1978). *Z. Naturforsch. C. Biosciences*, **33**, 413.

RETZLAFF, G. and HAMM, R. (1976). *Weed Res.*, **16**, 263.

RICHARDSON, J. T. and FRANS, R. E. (1975). *Proc. 28th Ann. Mtg. S. Weed Sci. Soc.*, 296.

RIES, S. K. and GAST, A. (1965). *Weeds*, **13**, 272.

RIES, S. K., LARSEN, R. P. and KENWORTHY, A. L. (1963). *Weeds*, **11**, 270.

RIES, S. K., PULVER, E. L. and BUSH, P. B. (1974). *3rd IUPAC Symp., Helsinki*, 244.

RIZK, T. Y. (1973). *Egypt. J. Bot.*, **16**, 281.

ROBERTSON, M. M. and KIRKWOOD, R. C. (1970). *Weed Res.*, **10**, 94.

ROBLES, R. P. and MERCADO, B. L. (1973). *Weed Abstr.*, **22**, 12, 316.

ROHWER, F. and FLÜCKIGER, W. (1979). *Angew. Bot.*, **53**, 59.

ROISCH, V. and LINGENS, F. (1974). *Angew. Chem. Int.*, **13**, 400.

ROSS, M. A. (1966). *Abstr. Weed Sci. Soc. Amer.*, 50.

ROSS, M. A. and SALISBURY, F. B. (1960). *WSSA Abstr.*, 43.

ROST, T. L. and BAYER, D. E. (1974). *Abstr. Weed Sci. Soc. Amer.*, 89.

ROTTMAN, G. A., TWEEDY, J. A. and KAPUSTA, G. (1974). *Agron. J.*, **66**, 701.

RUSNESS, D. G. and STILL, G. G. (1974). *Pestic. Biochem. Physiol.*, **4**, 24.

RUSSELL, J. (1957). *Can. J. Bot.*, **35**, 409.

SANTAKUMARI, M. and DAS, V. S. R. (1978). *Pestic. Biochem. Physiol.*, **9**, 119.

SARKISSIAN, I. V. and MCDANIEL, R. G. (1966). *Biochem. Biophys. Acta*, **128**, 413.

SCHMIDT, R. R., DRABER, W., EVE, L. and TIMMLER, H. (1975). *Pestic. Sci.*, **6**, 239.

SCHNEIDER, I. and GÜNTHER, G. (1974a). *Wiss. Zeit. Pädag. Hoch. Potsdam.*, **18**, 15.

SCHNEIDER, I. and GÜNTHER, G. (1974b). *Biol. Zent.*, **93**, 459.

SCHULTZ, H. (1969). In *Progress in Photosynthesis Research* (Ed. H. Metzner), Int. Union Biol. Sci., Tübingen, p. 1737.

SCHWEIZER, C. J. and RIES, S. K. (1969). *Science*, **165**, 73.

SELL, H. M., LUECKE, R. W., TAYLOR, B. M. and HAMNER, C. L. (1949). *Pl. Physiol., Lancaster*, **24**, 295.

SEN, S. (1975). *Naturwissenschaften*, **62**, 184.

SEN, S. (1978). *Experientia*, **34**, 724.

SHANER, D. L. and LYON, J. L. (1979a). *Abstr. Weed Sci. Soc. Amer.*, 97.

SHANER, D. L. and LYON, J. L. (1979b). *Pl. Sci. Letters*, **15**, 83.

SHANNON, J. C. (1963). *The Effect of 2,4-Dichlorophenoxyacetic Acid on the Growth and Ribonuclease Content of Seedling Tissues*, Ph.D. Diss., University of Illinois, Urbana.

SHARAKY, M. M. and EL-GANDOUR, M. A. (1975). *Ann. Agric. Sci. Moshtohor*, **3**, 85.

SHARMA, M. P. and VANDEN BORN, W. H. (1970). *Weed Sci.*, **18**, 57.

SHAUKAT, S. S. (1975). *Pakist. J. Bot.*, **7**, 57.

SHAUKAT, S. S., MOORE, H. G. and LOVELL, P. H. (1975). *Physiol. Pl.*, **33**, 295.

SHAYBANY, B. and ANDERSON, J. L. (1972). *Weed Sci.*, **12**, 164.

SHIMABUKURO, R. H. and SWANSON, H. R. (1968). *Abstr. Papers 155th Nat. Mtg. Amer. Chem. Soc.*, A24.

SHIRAKAWA, V. and UEKI, K. (1974). *J. Japan Soc. Hortic. Sci.*, **43**, 55.

SHIVE, C. J. and HANSEN, H. L. (1958). *Hormolog.*, **2**, 9.

SHOKRAII, E. H. and MORELAND, D. E. (1966). *WSSA Abstr.*, 46.

SIERRA, J. N. and VEGA, M. R. (1967). *Philipp. Agric.*, **51**, 438.

SIKKA, H. C. and DAVIS, D. E. (1969). *Weed Sci.*, **17**, 122.

SINGH, B. D., CAMPBELL, W. F. and SALUNKHE, D. K. (1972a). *Amer. J. Bot.*, **59**, 568.

SINGH, B. D. and HARVEY, B. L. (1975). *Experientia*, **31**, 785.

SINGH, B. D., SALUNKE, D. K. and LIPTON, S. H. (1972b). *HortSci.*, **47**, 441.

SINGH, B. D., VADHAWA, O. P. and SALUNKHE, D. K. (1972c). *HortSci.*, **7**, 196.

SIVTSEV, M. V. and KUZNETSOVA, E. A. (1973a). *Fiz. Biokhim. Kul'turn. Rast.*, **5**, 163.

SIVTSEV, M. V. and KUZNETSOVA, E. A. (1973b). *Agrokhimiya*, **10**, 134.

SLATER, E. C. (1963). In *Metabolic Inhibitors* (Eds. R. M. Hochster and J. H. Quastel), Vol. 2, Academic Press, New York and London.

SLATER, E. C. (1970). *Methods Enzymol.*, **10**, 48.

SLATER, J. W. (1968). *Weed Res.*, **8**, 149.

SLOVTSOV, R. I., ARABI, A. K. and DOROZHKINA, L. (1976). *Khim. Sel'. Khoz.*, **14**, 55.

SMITH, J. E., PATON, D. and ROBERTSON, M. M. (1966). *Proc. 8th Brit. Weed Control Conf.*, **1**, 279.

SMITH, J. E. and SHENNAN, J. L. (1966). *J. Gen. Microbiol.*, **42**, 293.

SMITH, L. W., PETERSON, R. L. and HORTON, R. F. (1971). *Weed Sci.*, **19**, 174.

SMITH, W. F. and ILNICKI, R. D. (1972). *Proc. N.E. Weed Sci. Soc., New York*, **36**, 79.

SOLECKA, M., PROFIC, H. and MILLKAN, D. F. (1969). *J. Amer. Soc. Hort. Sci.*, **93**, 55.

SOUZA MACHADO, V., ARNTZEN, C. J., BANDEEN, J. D. and STEPHENSON, G. R. (1978a). *Weed Sci.*, **26**, 318.

SOUZA MACHADO, V., BANDEEN, J. D., STEPHENSON, G. R. and ARNTZEN, C. J. (1978b). *Abstr. Weed Sci. Soc. Amer.*, 69.

SPIKES, J. B. (1956). *Pl. Physiol., Lancaster, Abstr.*, **31**, XXXII.

SPILSBURY, R. D. (1972). *Diss. Abstr. Int. B.*, **33**, 2443.

SRINIVASAN, P. S. and SAKHARAM RAO, J. (1973). *Madras Agric. J.*, **60**, 1756.

STENLID, G. and SADDIK, K. (1962). *Physiol. Pl.*, **15**, 369.

STERRETT, R. B. and FRETZ, T. A. (1974). *HortSci.*, **9**, 33.

STERRETT, R. B. and FRETZ, T. A. (1975). *HortSci.*, **10**, 161.

STILL, G. G., DAVIS, D. G. and ZANDER, G. L. (1970). *Pl. Physiol., Baltimore*, **46**, 307.

STILL, G. G., RUSNESS, D. G. and MANSAGER, E. R. (1974). *Abstr. Papers 167th Nat. Mtg. Amer. Chem. Soc. Pestic.*, 40.

STIRBAN, M. and SOREANU, I. (1975). *Contributii Botanice, Gradina Botanica, Universitatea 'Babes-Bolyai' din Clij-Napoca*, p. 189.

ST. JOHN, J. B. (1976). *Pl. Physiol., Lancaster*, **57**, 38.

ST. JOHN, J. B. and HILTON, J. L. (1973). *Weed Sci.*, **21**, 477.

ST. JOHN, J. B. and HILTON, J. L. (1976). *Pl. Growth Reg. Bull.*, **3**, 38.

STONER, C. D., HODGES, T. K. and HANSON, J. B. (1964). *Nature (London)*, **203**, 258.

STONOV, L. D., LUZHNOVA, M. I., BAKUMENKO, I. and SHEKHTMAN, L. M. (1974). *3rd IUPAC Symp., Helsinki*, 205.

STRANGER, C. E. (1972). *Diss. Abstr. Int. B.*, **32**, 3726.

STREETER, J. C. (1972). *Agron. J.*, **64**, 315.

STRUCKMEYER, B. E., BINNING, L. K. and HARVEY, R. G. (1976). *Weed Sci.*, **24**, 366.

SUMIDA, S. and UEDA, M. (1973). *Pl. Cell Physiol.*, **14**, 781.

SUSHEELA DEEVI, L. and PERUR, N. G. (1978). *Mysore J. Agric. Sci.*, **12**, 94.

SWADER, J. A. (1978a). *Abstr. Weed Sci. Soc. Amer.*, 49.

SWADER, J. A. (1978b). *Pestic. Biochem. Physiol.*, **9**, 140.

SWADER, J. A. and HOWE, C. M. (1978). *Pl. Physiol., Lancaster*, **61**, (Suppl.) 75.

SWADER, J. A. and HOWE, C. M. (1979). *Weed Sci.*, **27**, 232.

SWITZER, C. M. (1957). *Pl. Physiol. Lancaster*, **32**, 42.

TAKEUCHI, Y., KONNAI, M. and TAKEMATSU, T. (1972). *Weed Res., Japan*, **13**, 37.

TALBERT, R. E. (1967). *Abstr. Mtg. Weed Sci. Soc. Amer.*, 50.

TERRILLON, G. and PAYNOT, M. (1973). *Compt. Rend., D.*, **277**, 2489.

TETLEY, R. M., IKUMA, H. and HAYASHI, F. (1975). *Pl. Physiol., Lancaster*, **56**, 62.

THANGARAT, M. and SAKHARAM RAO, J. (1973). *Madras Agric. J.*, **60**, 1739.

THOMPSON, O. C., TRUELOVE, B. and DAVIS, D. E. (1969). *J. Agric. Fd. Chem.*, **17**, 997.

THOMPSON, O. C., TRUELOVE, B. and DAVIS, D. E. (1970). *Proc. Weed Sci. Soc. Amer.*, 318.

TIMOFEEV, M. M. (1978). *Fiz. Biokhim. Kul'turn. Rast.*, **10**, 171.

TLUCZKIEWICZ, J. (1974). *Bull. l'Acad. Polonaise des Sciences, Sci. Biol.*, **22**, 639.

TONECKI, J. (1975). *Acta Soc. Bot. Pol.*, **44**, 42.

TREBST, A. and DRABER, W. (1978). *IUPAC Symp. Zurich, Switzerland*, 223.

TREBST, A. and DRABER, W. (1979). *Advances in Pesticide Science*, Parts 1–3, Pergamon Press, Oxford, p. 223.

TREBST, A. and HARTH, E. (1974). *Z. Naturforsch. C*, **29**, 232.

TREBST, A. and WIETOSKA, H. (1975). *Z. Naturforsch. C*, **30**, 497.

TRUELOVE, B. and DAVIS, D. E. (1969). *Abstr. Mtg. Weed Sci. Soc. Amer.*, 179.

TWEEDY, J. A. and RIES, S. K. (1967). *Pl. Physiol., Lancaster*, **42**, 280.

TYMONKO, J. M. (1979). *Diss. Abstr. Int. B*, **39**, 3624.

TYMONKO, J. M. and FOY, C. L. (1978a). *Abstr. Weed Sci. Soc. Amer.*, 70.

TYMONKO, J. M. and FOY, C. L. (1978b). *Pl. Physiol., Lancaster*, **61**, (4, Suppl.) 41.

UPADHYAYA, M. K. and NOODEN, L. D. (1976). *Pl. Physiol., Lancaster*, **57**, (4, Suppl.) 63.

UPADHYAYA, M. K. and NOODEN, L. D. (1978). *Pl. Physiol., Lancaster*, **61**, (4, Suppl.) 56.

VAISBERG, A. J. and SCHIFF, J. A. (1976). *Pl. Physiol., Lancaster*, **57**, 260.

VANSTONE, D. E. and STOBBE, E. H. (1979). *Weed Sci.*, **27**, 88.

VASYUTA, A. N., POPOVA, L. P., SKOBLIN, A. P. and GOLOVCHANSKAYA, L. I. (1974). *Dokl. Vses Akad. Sel'skok. Nauk. Lenina.*, **8**, 13.

VAVERKOVA, S. and VANCOVA, A. (1973). *Biol. Czech.*, **28**, 563.

VEERASEKARAN, P., KIRKWOOD, R. C. and FLETCHER, W. W. (1977). *Weed Res.*, **17**, 85.

VERLOOP, A. (1972). *Residue Rev.*, **43**, 55.

VOEVODIN, A. V., NEVZOROVA, L. I. and KAZARINA, E. M. (1977). *Trudy Vses. Nauchno-Issled. Inst. Zash. Rast.*, **54**, 32.

VOINILO, V. A., DEEVA, V. P. and MASHTAKOV, S. M. (1967). *Dokl. Akad. Nauk Belorussk. SSR*, **11**, 638.

VOINILO, V. A., DEEVA, V. P. and MASHTAKOV, S. M. (1968). *Dokl. Akad. Nauk Belorussk. SSR*, **12**, 460.

WAIN, R. L. (1964). *Proc. 7th Brit. Weed Control Conf.*, **1**, 1.

WAKAMORI, S. (1973). *Agric. Biol. Chem.*, **37**, 307.

WANG, B., GROOMS, S. and FRANS, R. E. (1974). *Weed Sci.*, **22**, 64.

WATANABI, H. and KIUCHI, T. (1975). *Soil Sci. Pl. Nutr.*, **21**, 151.

WEDDING, R. T. and BLACK, M. K. (1962). *Pl. Physiol., Lancaster*, **37**, 364.

WEEKS, D. P. and BAXTER, R. (1972). *Biochemistry*, **11**, 3060.

WEIER, T. E. and IMAM, A. A. (1965). *Amer. J. Bot.*, **52**, 631.

WEINBACH, E. C. and GARBUS, J. (1964). *Science, N.Y.*, **145**, 824.

WEINBACH, E. C. and GARBUS, J. (1965). *J. Biol. Chem.*, **240**, 1811.

WEINBACH, M. B. and CASTELFRANCO, P. A. (1975). *Weed Sci.*, **23**, 185.

WELLER, L. E., LUECKE, R. W., HAMMER, C. L. and SELL, H. M. (1950). *Pl. Physiol., Lancaster*, **25**, 289.

WERGIN, W. P. and POTTER, J. P. (1974). *Abstr. Weed Sci. Soc. Amer.*, 125.

WERGIN, W. P. and POTTER, J. R. (1975). *Pestic. Biochem. Physiol.*, **5**, 265.

WESSELS, J. S. C. (1959). *Biochem. Biophys. Acta*, **36**, 264.

WESSELS, J. S. C. and VAN DER VEEN (1956). *Biochem. Biophys. Acta*, **19**, 548.

WEST, L. D., MUZIK, T. J. and WITTERS, R. E. (1976). *Weed Sci.*, **24**, 68.

WEST, S. H., HANSON, J. B. and KEY, J. L. (1960). *Weeds*, **8**, 333.

WHITE, J. A. and HEMPHILL, D. D. (1972). *Weed Sci.*, **20**, 478.

WILKINSON, R. E. (1974). *Pl. Physiol., Lancaster*, **53**, 269.

WILKINSON, R. E. (1978). *Pestic. Biochem. Physiol.*, **8**, 208.

WILKINSON, R. E. and SMITH, A. E. (1973). *Proc. 26th Mtg. S. Weed Sci. Soc.*, 415.

WILKINSON, R. E. and SMITH, A. E. (1974). *Abstr. Weed Sci. Soc. Amer.*, 76.

WILKINSON, R. E. and SMITH, A. E. (1975a). *Weed Sci.*, **23**, 90.

WILKINSON, R. E. and SMITH, A. E. (1975b). *Weed Sci.*, **23**, 100.

WILKINSON, R. E. and SMITH, A. E. (1976a). *Weed Sci.*, **24**, 235.

WILKINSON, R. E. and SMITH, A. E. (1976b). *Phytochemistry*, **15**, 841.

WINKLE, D. VAN and EWING, E. E. (1974). *Pl. Physiol., Lancaster*, **53**, (Suppl.) 8.

WIT, J. G. and VAN GENDEREN, H. (1966). *Biochem. J.*, **101**, 707.

WOJTASZEK, T. (1966). *Weeds*, **14**, 125.

WOJTASZEK, T., CHERRY, J. H. and WARREN, G. F. (1966). *Pl. Physiol., Lancaster*, **41**, 34.

WORT, D. J. (1964). In *The Physiology and Biochemistry of Herbicides* (Ed. L. J. Audus), Academic Press, New York and London, p. 291.

WORT, D. J. and SHRIMPTON, M. (1958). *Res. Rep. Nat. Weed Committee, West Sect. Canad. Dept. Agric.*, 124 (*cit.* Wort, D. J., 1964).

WU, M. T., SINGH, B. and SALUNKHE, D. K. (1972a). *Pl. Physiol., Lancaster*, **48**, 517.

WU, M. T., SINGH, B. and SALUNKHE, D. K. (1972b). *J. Exp. Bot.*, **23**, 793.

WU, M. T., SINGH, B. and SALUNKHE, D. K. (1972c). *Experientia.*, **28**, 1002.

WYSE, D. L., MEGGITT, W. F. and PENNER, D. (1974a). *Abstr. Weed Sci. Soc. Amer.*, 77.

WYSE, D. L., MEGGITT, W. F. and PENNER, D. (1974b). *Abstr. Weed Sci. Soc. Amer.*, 78.

YORK, A. C. and ARNTZEN, C. J. (1979). *Abstr. Weed Sci. Soc. Amer.*, 103.

YORK, A. C., ARNTZEN, C. J. and SLIFE, F. W. (1978). *Abstr. Weed Sci. Soc. Amer.*, 76.

YUKINAGA, H., IDE, K. and ITO, K. (1973). *Weed Res. (Japan)*, **15**, 34.

ZHIRMUNSKAYA, N. M. (1966). *Khim. Sel. Khoz.*, **4**, 46.

ZSOLDOS, F., KARVALY, B., TOTH, I. and ERDEI, L. (1978). *Physiol. Pl.*, **44**, 395.

ZWEIG, G. (1969). *Residue Rev.*, **25**, 69.

ZWEIG, G. and AVRON, M. (1965). *Biochem. Biophys. Res. Commun.*, **19**, 397.

ZWEIG, G., HITT, J. E. and MCMAHON, R. (1968). *Weed Sci.*, **16**, 69.

ZWEIG, G. N., SHAVIT, N. and AVRON, M. (1965). *Biochem. Biophys. Acta*, **109**, 332.

CHAPTER SIX

Metabolism of Herbicides

The degradation of herbicides by plants is an important mechanism of detoxification generally minimising the possibility of transmission of the compound through food chains and acting as an important basis for selective toxicity. Reviews on the metabolism of herbicide compounds include those on organic pesticides by Casida and Lykken (1969), herbicides in higher plants, soil and animal systems, edited by Kearney and Kaufman (1976), herbicides in higher plants by Ashton and Crafts (1973) and Naylor (1976); selective detoxification and activation mechanisms have been considered by Wain and Smith (1976).

Degradation mechanisms in plants

Herbicide degradation in higher plants may result from a wide range of chemical reactions, most of which are catalysed by specific enzymes though a few appear to be non-enzymatic in nature. The various types of reactions have been reviewed by Ashton and Crafts (1973) and include oxidation, decarboxylation, deamination, dehalogenation, dethioation, dealkylation, dealkyloxylation, dealkylthiolation, hydrolysis, hydroxylation and conjugation mechanisms.

OXIDATION

Three different types of oxidation reactions (α-, β- and ω-oxidation) have been reported to occur in the phenoxyalkanoic acids, involving oxidation at three different sites on the side chain. The β-oxidation of the ω-phenoxyalkanoic acids has been investigated by several workers, including Synerholm and Zimmerman (1947) (fig. 6.1). They postulated that only the 2,4-dichlorophenoxy acids with an even number of carbon atoms in the side chain were active when they were metabolised by β-oxidation to 2,4-D; those with an odd number of carbon atoms degraded to the inactive 2,4-dichlorophenol. However, Fawcett *et al.* (1954) found that degradation of 10-phenoxy-n-

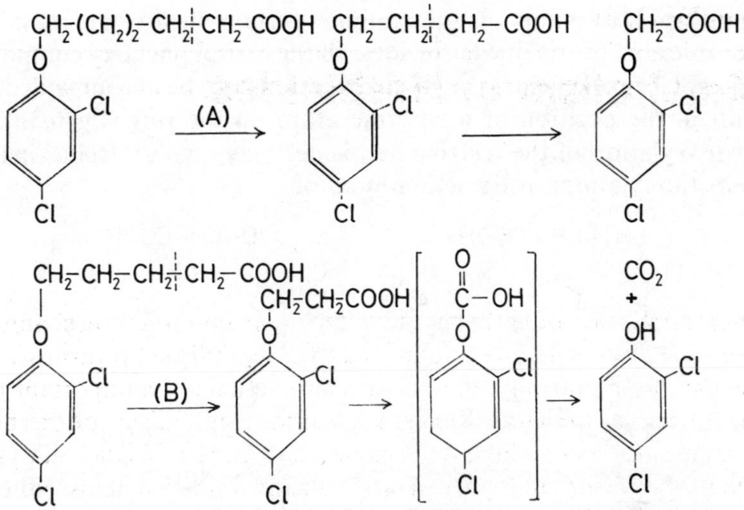

Fig. 6.1 β-Oxidation of 2,4-dichlorophenoxyalkanoic acids in plants showing (A) the formation of 2,4-D from acids with an even number of carbons in the side chain and (B) the formation of 2,4-dichlorophenol from acids with an odd number of carbons in the side chain (after Synerholm and Zimmerman, 1947, reprinted from Loos, 1976).

decanoic acid resulted in large amounts of phenol being produced and they suggested that this results from degradation involving ω-oxidation (fig. 6.2).

DECARBOXYLATION

Certain herbicides of the phenoxyalkanoic acid, benzoic acid and heterocyclic types may undergo decarboxylation by a reaction which requires a molecule of water and thus the reaction could be considered to be a hydrolysis mechanism.

HYDROXYLATION

Ring hydroxylation has been demonstrated to occur following

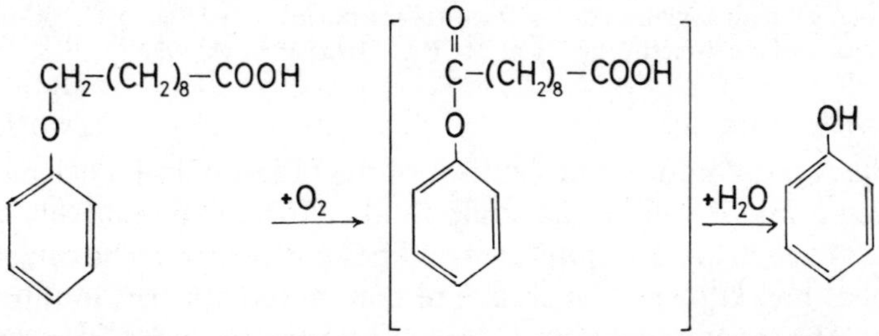

Fig. 6.2 ω-Oxidation of 10-phenoxy-n-decanoic acid in flax (after Fawcett *et al.*, 1954, reprinted from Loos, 1976).

application to plants of, for example, certain phenoxyalkanoic acid, benzoic acid or triazine herbicides. In the case of phenoxy compounds but not benzoic acids, ring hydroxylation may be accompanied by a shift in the position of a chlorine atom on the ring (fig. 6.3). Ring hydroxylation of the triazine herbicides may involve dechlorination, demethoxylation, or demethylthioation.

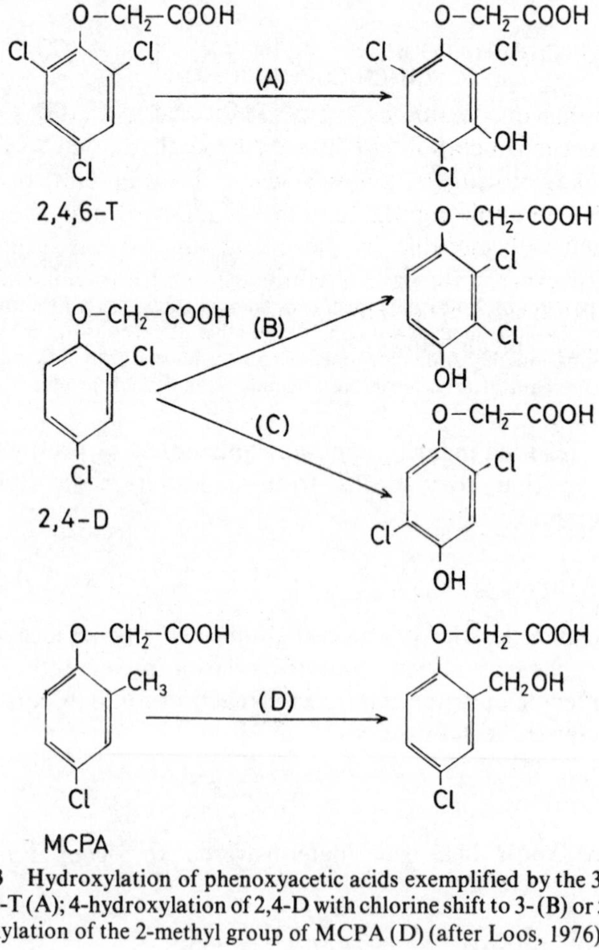

2,4,6-T

(A)

2,4-D

(B)

(C)

MCPA

(D)

Fig. 6.3 Hydroxylation of phenoxyacetic acids exemplified by the 3-hydroxylation of 2,4,6-T (A); 4-hydroxylation of 2,4-D with chlorine shift to 3- (B) or 5- (C) positions; hydroxylation of the 2-methyl group of MCPA (D) (after Loos, 1976).

HYDROLYSIS

Hydrolysis is a common mechanism of degradation in higher plants, involving a major split in the compound to form two relatively large fragments which are non-phytotoxic. These, in turn, may be subject to additional breakdown. The classes of compound affected in this way includes the phenoxyalkanoic acids, carbamates, thiocarbamates, ureas and triazines.

DEALKYLATION

Substitution of alkyl (or alkyloxy) groups may be an important mechanism of detoxification and act as a basis for selectivity. Dealkylation and dealkyloxylation have been reported to occur in several groups of herbicides including the dinitroanilines, carbamates, thiocarbamates, ureas and triazines.

CONJUGATION

There are numerous examples in the literature of the conjugation of herbicides or their metabolites (often hydroxy derivatives), generally involving linkage to sugars, amino acids, or less frequently protein or lignin (fig. 6.4). Simple conjugates of herbicides with sugars or amino acids are generally soluble in the extracting solvent and readily detected. However, in the case of conjugation with protein or lignin, treatment of the insoluble residue by mild hydrolysis may be necessary to release the herbicide molecule.

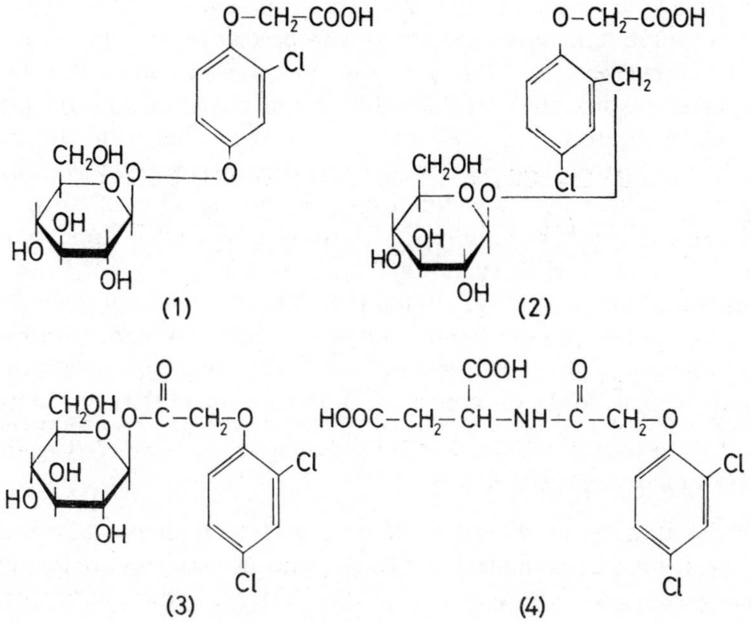

O–CH₂–COOH
Cl
CH₂OH
O
O
OH
HO
OH
(1)

O–CH₂–COOH
CH₂
CH₂OH
O O
Cl
OH
HO
OH
(2)

O
‖
CH₂OH O–C–CH₂–O
O
Cl
OH
HO
OH
Cl
(3)

COOH O
| ‖
HOOC–CH₂–CH–NH–C–CH₂–O
Cl
Cl
(4)

Fig. 6.4 The conjugated forms of phenoxyacetic acids detected in plants including the β-D-glucoside of the 4-hydroxy derivative of 2-CPA (1), the β-D-glucoside of 4-chloro-2-hydroxymethylphenoxyacetic acid (2), the β-D glucose ester of 2,4-D (3) and 2,4-dichlorophenoxyacetylaspartic acid (4) (after Loos, 1976).

RING CLEAVAGE

Generally the various substitutions on the ring are cleaved off and the ring itself persists in the plant as a non-toxic compound. However, in certain cases, the splitting of the aromatic or heterocyclic ring structure of the herbicide may proceed very slowly in higher plants; generally this is indicated by the minute levels of loss of $^{14}CO_2$ from ^{14}C-ring-labelled compounds.

The metabolism of soil- or foliage-applied herbicides

HALOALKANOIC ACIDS

In general, **dalapon** and **TCA** are relatively stable in higher plants and animals, but rapidly degraded in soils (Foy, 1969). Short-term studies with ^{14}C-dalapon or ^{36}Cl-dalapon revealed no breakdown products in a number of species (Ashton and Crafts, 1973); in long-term studies (9–10 weeks) however, low levels of non-extractable radioactive residues were detected in cotton (Foy, 1961; Smith and Dyer, 1961). The metabolic pathway of degradation of dalapon has still to be elucidated, since the intermediates are formed so slowly that they do not accumulate in sufficient amount to be readily detected. It seems likely, however, that the pathway involves dechlorination with concurrent hydroxylation followed by dehydroxylation and concurrent oxidation resulting in the removal of the other chlorine atom producing pyruvic acid. This would be further degraded by the normal processes of respiratory metabolism (Ashton and Crafts, 1973)

The third member of this group, **chlorfenprop-methyl**, appears to be readily metabolised in susceptible and resistant plants of *Avena* sp. Fedtke and Schmidt (1977) found that it is hydrolysed completely as soon as it penetrates the leaf in both sensitive and resistant plants with the production of the herbicidally active chlorfenprop. Selectivity is apparently due to the differences in concentration at the site of action.

PHENOXYALKANOIC ACIDS

Early studies in the 1950s and 1960s on the fate of phenoxyacetic acid herbicides in plants suggest that three basic mechanisms are involved: (1) degradation of the acetic acid side chain, (2) hydroxylation of the aromatic ring, (3) conjugation with a plant constituent (Naylor, 1976). These conclusions have been confirmed by subsequent work and the observations extended to other species.

Side-chain degradation appears to be widespread in plants, but may be of major importance in only a few species in which it is sufficiently rapid and extensive to influence selectivity. Such species include red currant (Luckwill and Lloyd-Jones, 1960a), Cox's orange pippin and McIntosh apples (Luckwill and Lloyd-Jones, 1960b; Edgerton and Hoffman, 1961), strawberry and lilac (Luckwill and Lloyd-Jones, 1960b). These workers treated the test plants with ^{14}C-side-chain-labelled **2,4-D** and measured the rate and extent of $^{14}CO_2$ release. However, a proportion of the ^{14}C is not released but rather is incorporated into plant products including acids, sugar, dextrins, starch, protein, pectin and cell wall compounds (Weintraub *et al.*, 1956; Leafe, 1962).

A second major route involves ring hydroxylation, generally in the 4-position if that is unsubstituted (e.g. **2,6-CPA**) or possibly at the 3-position if the 4-position is already substituted (e.g. **2,4,6-T**) (Thomas *et al.*, 1964). Hydroxylation in the 4-position of phenoxyacetic acids has also been reported to occur in wheat pea tissue (Fawcett *et al.*, 1959), roots of oats, barley and corn (Wilcox *et al.*, 1963), and a range of seven weed species (Fleeker and Steen, 1971). However, in these weed species the lack of correlation between the amount of 2,4-D hydroxylated and the amount of herbicide tolerated suggests that hydroxylation *per se* does not account for variation in their sensitivity to 2,4-D.

Esterification of glucose with 2,4-D has been found to take place in wheat coleoptile cylinders (Klambt, 1961) and with 2,4-D and 4-CPA in oat mesocotyls (Thomas *et al.*, 1964). Conjugation of 2,4-D with aspartic acid has been reported in peas (Andreae and Good, 1957), red and black currants (Luckwill and Lloyd-Jones, 1960b), wheat (Klambt, 1961), cucumber (Slife *et al.*, 1962) and soybean callus (Feung *et al.*, 1973, 1974). Feung *et al.*, also found that glutamic acid, alanine, valine, leucine, phenylalanine and trytophan conjugated with 2,4-D.

During the last decade a number of papers have been published concerning analytical methods for detection of phenoxyalkanoic acids, including 2,4-D (Frankoski and Siggia, 1972; McLeod and Wales, 1972; Munro, 1972; Yip, 1972; Hammond, 1973; Hargreaves and Rapkins, 1976; Hatjula and Raisanen, 1976; Hazemoto, Kamo and Kobatake, 1976; Roder and Laass, 1976; Tuinstra, Roos and Bronsgeest, 1976), 2,4,5-T (Chow *et al.*, 1971; Munro, 1972; Hammond, 1973; Hargreaves and Rapkins, 1976; Hatjula and Raisanen, 1976), **MCPA** and other phenoxy herbicides (Agemian and Chau, 1976; Hatjula and Raisanen, 1976; Roder and Laass, 1976; Tuinstra *et al.*, 1976). Most methods involve separation by g.l.c. though high performance liquid chromatography (h.p.l.c.) has been successfully used for a number of compounds (Tuinstra *et al.*, 1976).

Evidence of amino acid or sugar conjugates and hydroxylated metabolites of 2,4-D has been obtained by a number of workers. The metabolism of 1-^{14}C-labelled 2,4-D has been investigated using callus tissue culture of soybean (Feung *et al.*, 1972), carrot, jackbean (*Canavalia ensiformis*), sunflower and maize (Feung *et al.*, 1974) and rice (Feung *et al.*, 1973).

The two ring-hydroxylated aglycones, 4-hydroxy-2,3-dichlorophenoxyacetic acid and 4-hydroxy-2,5-dichlorophenoxyacetic acid are major metabolites in all callus tissues except rice where they represent only 0.4% of the applied dose. Glutamic acid and aspartic acid conjugates are also important metabolites in all species except rice, in which the major metabolite appears to be a sugar ester of 2,4-D.

Chkanikov *et al.* (1977), using bean (*Phaseolus vulgaris*) and soybean found that the major metabolites are 4-O-β-D-glucosides of 4-hydroxy-2,5-dichloro- and 4-hydroxy-2,3-dichlorophenoxyacetic acids with, in addition, considerable accumulation of *N*-(2,4-dichlorophenoxyacetyl)-L-aspartic acid and *N*-(2,4-dichlorophenoxyacetyl)-L-glutamic acids. Using cereals, these workers found evidence of conjugation of 2,4-D with glucose, while in strawberry the glycoside of 2,4-dichlorophenol (DCP) prevails.

Hydroxylation of ^{14}C-labelled 'Hoe' 23 408 OH (**diclofop-methyl**) in wheat has also been reported, the main metabolites being 2-(4-[2,4-dichlorophenoxy]phenoxy)propionic acid and the corresponding 5-hydroxy derivative; two possible isomers are the 3-hydroxy and 6-hydroxyphenoxy derivatives (Gorbach, Kuenzler and Asshauer, 1977).

Degradation of the side chain may involve a stepwise reduction in length involving decarboxylation or cleavage into a 2-carbon fragment (fig. 6.5). Examples of both types of mechanism can be found in the literature. Decarboxylation of 2,4-D has been reported to occur in pea and maize (Ugulava and Ugrekhelidze, 1973) and of 2,4,5-T in woody plants (Brady, 1973). On the other hand, removal of the acetate moiety resulted from application of 2,4-D to excised leaves of *Ribes sativum* (Fleeker, 1973) and resistant grass species (Hagin and Linscott, 1972). In the latter case, however, the major detoxification mechanism involved the formation of 3-(2,4-dichlorophenoxy)propionic acid.

Examination of the fate of 2,4-D (^{14}C-labelled in the ring, carboxyl or methylene groups) in strawberry revealed that decarboxylation of 2,4-D occurs at almost the same rate as the formation of dichlorophenol (DCP) suggesting cleavage of the side-chain of 2,4-D into a 2-carbon fragment (Grigor'eva *et al.*, 1975). DCP is immediately converted to the glycoside which may persist for several months during

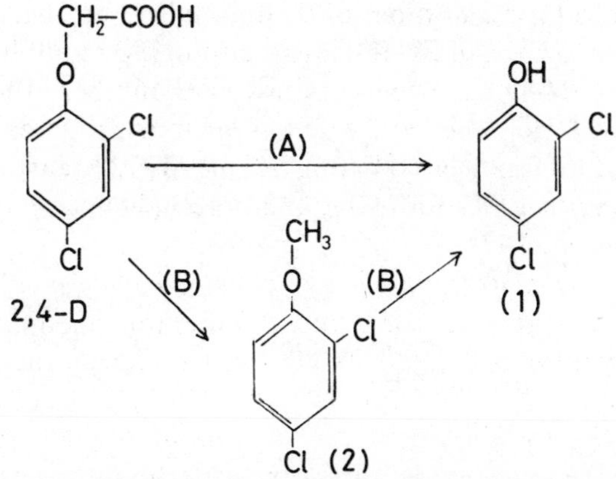

Fig. 6.5 The degradation of 2,4-D in plants to 2,4-dichlorophenol (1) by hypothetical pathways involving either (A) cleavage at the ether linkage, or (B) the formation of the intermediary 2,4-dichloroanisole (2) (after Loos, 1976).

which time strawberry plants develop normally suggesting that the glycoside can be regarded as a product of detoxification. Differential destruction of the side-chain might reasonably be expected to be an important basis of selectivity, but there was no evidence of a relationship between plant resistance to 2,4-D and the rate of decarboxylation in eighteen species differing in sensitivity to this herbicide (Dubovoi, Chkanikov and Makeev, 1973). Summaries of the known and postulated steps in the degradation of 2,4-D and MCPA have been presented by Pillmoor and Gaunt (1981) and Williams (1976) (figs. 6.6 and 6.7 respectively).

Side-chain degradation of **2,4-DB** by β-oxidation has been reported by Wathana and Corbin (1972) who examined the metabolism of 2,4-DB in soybean and cocklebur (*Xanthium sp.*). Both species β-oxidised 2,4-DB to 2,4-D at each stage of growth examined, though degradation was less rapid in soybean. A second transformation pathway involved elongation of the fatty acid side chain with the formation of 10-(2,4-dichlorophenoxy)decanoic acid which was found in the cuticle of treated leaves. However, β-oxidation of phenoxybutyric acids does not necessarily result in the formation of the respective acetic acid derivatives. Smith and Oswald (1978) examining the detoxification of ^{14}C-2,4-DB by clover cell suspensions found that a water-soluble intermediate was formed which was not the acetic acid homologue.

The rate of herbicide metabolism, however, appears to vary according to plant age, region and species. For example, Fykse (1976) examined the metabolism of **MCPA** in *Sonchus arvensis* and found that degradation occurred more rapidly in older plants and in primary

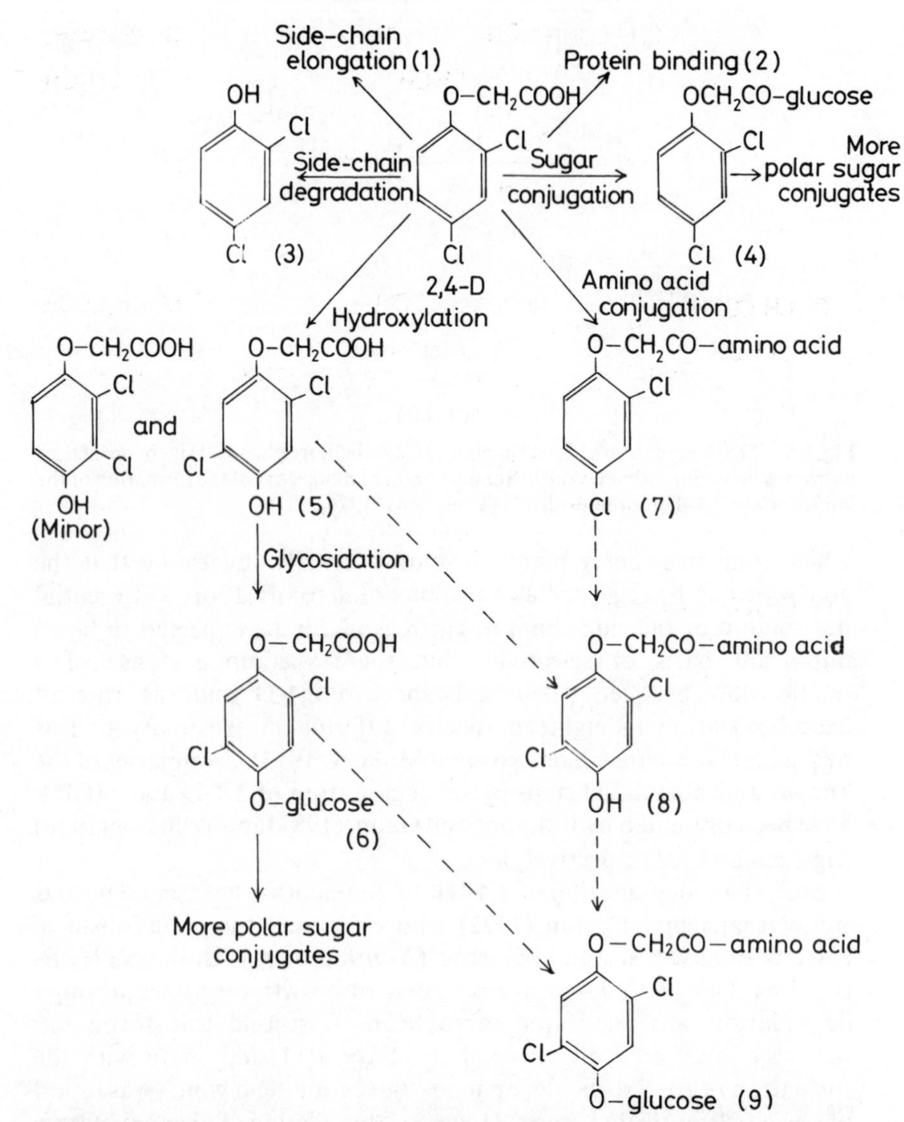

Fig. 6.6 Metabolism of 2,4-D (after Pillmoor and Gaunt, 1981). References to the reactions and pathways are as follows: (1) Linscott *et al.* (1968), (2) Rakitin *et al.* (1966); Dexter *et al.* (1971) (3) Chkanikov *et al.* (1965) (4) Klambt (1961) (5) Thomas *et al.* (1963); (6) Thomas *et al.* (1964); (7) Andreae and Good (1957); Feung *et al.* (1972) (8) Mumma and Hamilton (1975) (9) Chkanikov *et al.* (1977). Routes yet to be substantiated are indicated - - - - ->.

compared with secondary roots and leaves. Nazarova *et al.* (1978) also found that the rate of degradation (hydroxylation) of 2,4-D increased in progressively older plants of barley and suggested that this may explain its increased resistance at tillering compared with younger

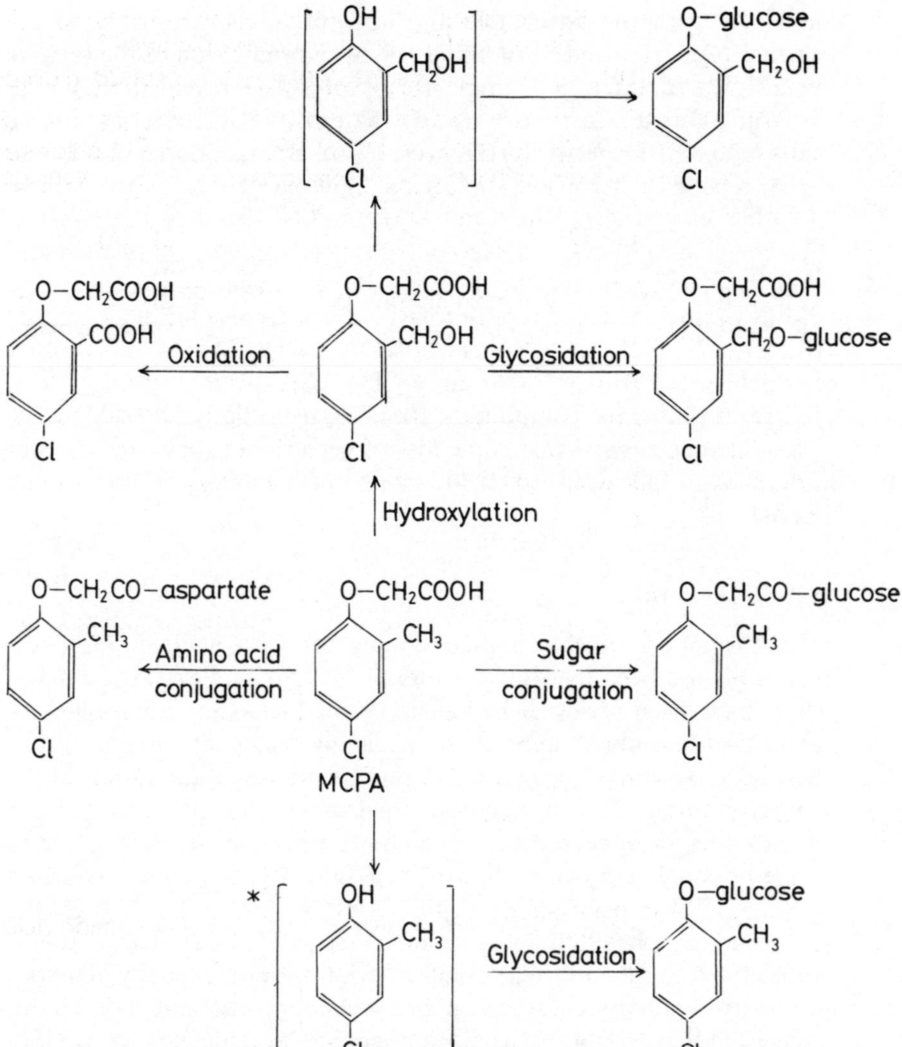

Fig. 6.7. Metabolism of MCPA (after Williams, 1976, reprinted from Pillmoor & Gaunt, 1981) (*this compound was not isolated from plant tissue in free form).

stages. Differential absorption and metabolism are involved in the selective activity of 2,4-D in common milkweed (*Asclepias syriaca*) and hemp dogbane (*Apocynum cannabinum*) (Wyrill and Burnside, 1976).

Other forms of loss of herbicide from the plant may involve volatilisation and photochemical degradation. Volatilisation of formulated butyl esters of 2,4-D from leaf (or Pyrex glass) surfaces increased in proportion to the available surface area/applied dose ratio and inversely with the adsorptive and absorptive characteristics of the surface (Que Hee and Sutherland, 1975). Examination of the relative volatilities of ester and amine forms of 2,4-D revealed that high volatile esters are lost more readily than low volatile esters or amine salts (440:33:1 respectively) (Grover, 1976). Degradation of 2,4-D and other phenoxy herbicides involving light has been observed by a number of workers. Klotz and Duysen (1972) studied the effect of 2,4-D on the growth and metabolism of *Chlorella pyrenoidosa* and their results suggest that light causes the cells to become insensitive to 2,4-D or to inactivate it. Photolysis of aqueous solutions of 2,4-D (Silber, Previtali and Silbera 1976), 2,4-D, MCPA and other herbicides (Gryzlova, 1976) or 4-CPA (Crosby and Wong, 1973) indicates that these compounds are at least partially photodegraded These results suggest that some loss of herbicidal activity must occur following application to field crops, particularly in hot, sunny climates.

AROMATIC ACIDS

The formation of **chloramben** conjugates with endogenous plant materials has been reported for a range of crop and weed species and these have been reviewed by Frear (1976). The major metabolite, *N*-glucosylchloramben, appears to be a glycoside of the unchanged herbicide (Colby, 1965), there being no evidence of alteration of the ring structure or decarboxylation in plants. In addition, certain minor metabolites have been detected in chloramben-treated plants, one of these possibly being an isomer of the *N*-glucosyl conjugate (Swanson *et al.*, 1966). Another less polar minor metabolite designated chloramben-X has been reported in a number of plant species (Stoller, 1968, 1969; Stoller and Wax, 1968). Variation in the capacity of tissues to form *N*-glucosylchloramben has been reported and it has been suggested that resistance to chloramben may be associated with ability of resistant species to sustain higher internal concentrations of the free herbicide and conjugate-absorbed chloramben more rapidly than susceptible species (Stoller, 1969).

Extensive metabolism of **dicamba** has been reported to occur in several grass species including wheat (Quimby and Nalewaja, 1961; Broadhurst, Montgomery and Freed, 1966; Chang and Vanden Born, 1971a), barley (Chang and Vanden Born, 1971a; Ray and Wilcox, 1967), blue grass (*Poa pratensis*) (Chang and Vanden Born, 1971a), maize (Ray and Wilcox, 1967) and Johnson grass (*Sorghum halepense*) (Hull and Weisenberg), 1967). In contrast only limited breakdown was evident in bean (Hull and Weisenberg, 1967), purple nutsedge (*Cyperus rotundus*) (Magalhaes, Ashton and Foy, 1968; Ray and Wilcox, 1969), Tartary buckwheat (*Fagopyrum tataricum*) (Quimby and Nalewaja, 1961; Chang and Vanden Born, 1971a; Chang and Vanden Born, 1971b), wild mustard (*Sinapsis arvensis*) and Canada thistle (*Cirsium arvense*) (Chang and Vanden Born, 1968).

The pathways of metabolism of dicamba have been reviewed by Frear (1976). The major pathway appears to involve ring hydroxylation at the 5-position with subsequent formation of polar acid-labile conjugates. The absence of appreciable quantities of the free phenol (5-hydroxy-2-methoxy-3,6-dichlorobenzoic acid) indicates that the conjugation step may occur more rapidly than the initial ring-hydroxylation step. Minor pathways of degradation include demethoxylation of dicamba and possibly 5-hydroxy-dicamba to yield 3,6-dichlorosalicylic acid (DCSA) and 3,6-dichlorogentisic acid (DCGA) which may also become conjugated to form glycosides (Broadhurst *et al.*, 1966; Chang and Vanden Born, 1971b; Ray and Wilcox, 1967); dicamba may also be directly conjugated in several species (Broadhurst *et al.*, 1966; Chang and Vanden Born, 1971b).

Dicamba was metabolised in Russian knapweed (*Centaurea repens*) to form 3,6-dichlorosalicylic acid, salicylic acid, benzoic acid and an unidentified metabolite (Berezovskii and Nikhotin, 1971). In the same species Krumzdorov (1975a) found 3,6-dichlorosalicylic acid, *p*-aminobenzoic and benzoic acid even 3 years after application of dicamba (2 kg/ha); these metabolites were also detected in maize (Krumzdorov, 1974) and proso millet (Krumzdorov, 1975b). While dicamba disappeared from the roots of *C. repens* after 12 months, 3,6-dichlorosalicylic and benzoic acids persisted throughout the 3 years of observation (Krumzdorov, 1976). In maize, root-absorbed 'VEL-4207' (a slow-release derivative of dicamba) and dicamba yielded 2,5-dichlorosalicylic as a major metabolite (Harger, Rieck and Standiford, 1975). Field application of dicamba in spring to living tissue of a quasi-dominant rhizome of bracken (*Pteridium aquilinum* var. *pubescens*) resulted in the formation of 5-hydroxy-3,6-dichloro-*o*-anisic and 3,6-dichlorogentisic acid within 10 days (Robocker and Zamora, 1976).

AMIDES

Diphenamid undergoes a stepwise dealkylation in plants, the process involving the formation of glucosides of intermediate methylol derivatives (Still and Herrett, 1976). Conjugates may be formed with glucose, gentitobiose and amino acids, such as alanine or aspartic acid. The role of dealkylation and selective toxicity is, however, somewhat controversial. Initially it was believed that resistance was attributable to rapid dealkylation (Lemin, 1966), and then later that dealkylation to monomethyldiphenamid was required for herbicidal activity (Holly, 1968). More recent evidence supports the former view that monomethyldiphenamid is less toxic than diphenamid and that dealkylated 2,2-diphenylacetamide is not toxic (Genther, 1969; Schultz and Tweedy, 1971).

Fumigation of tomato plants with low levels of ozone (30 p.p.m.) has little effect on the metabolism of diphenamid to water-soluble conjugates, but the proportion of specific conjugates is markedly altered (Hodgson *et al.*, 1973). Twenty-four hours after treatment, the predominant conjugate formed in non-fumigated and fumigated tomato plants were the β-glucoside (MDAG) and the β-gentio-bioside (MDAGB) respectively of *N*-hydroxymethyl-*N*-methyl-2,2-diphenyl-acetamide. The ratios of MDAG:MDAGB in non-fumigated and ozone-fumigated tissue were 8.2:1 and 0.6:1 respectively. The rate of metabolism was subsequently found to be rather more rapid in fumigated plants of tomato (Hodgson, Dusbadek and Hoffer, 1974) and pepper (*Capsicum* sp.) (Hodgson and Hoffer, 1974). In soybean, in addition to the glucose and gentiobiose conjugates of the *N*-hydroxymethyl-*N*-methyl herbicide derivative, evidence was obtained for the formation of a glucose-malonyl ester of diphenamid (Hoffer and Hodgson, 1978). Rice and Putnam (1980) also detected *N*-methyl-2,2-diphenylacetamide and possibly a glucose conjugate in tomato plants treated with diphenamid.

The removal of ^{14}C-diphenamid from water and its degradation to various metabolites by aquatic plants has been investigated by Bingham and Shaver (1977). *Myriophyllum brasiliense* and *Eichhornia crassipes* removed larger quantities of diphenamid than did algae and *Potamogeton diversifolius*, though algae and *E. crassipes* degraded diphenamid further to unknown products. *N*-Methyl-2,2-diphenyl-acetamide and 2,2-diphenylacetic acid were identified as metabolites.

NITRILES

The major pathway of metabolism of **dichlobenil** in bean, rice and wheat appears to be ring hydroxylation followed by phenol conjugation and incorporation into insoluble plant residues (Verloop and Nimmo, 1969, 1970; Verloop, 1972) (fig. 6.8). Ring hydroxylation results in the formation of 3-hydroxy- and 4-hydroxy-2,6-dichloro-benzonitrile, the former being predominant in bean, rice and wheat seedlings; both have been isolated from plants as phenols and acid-labile conjugates believed to be glycosides. The major soil degradation product, 2,6-dichlorobenzamide can also become hydroxylated and

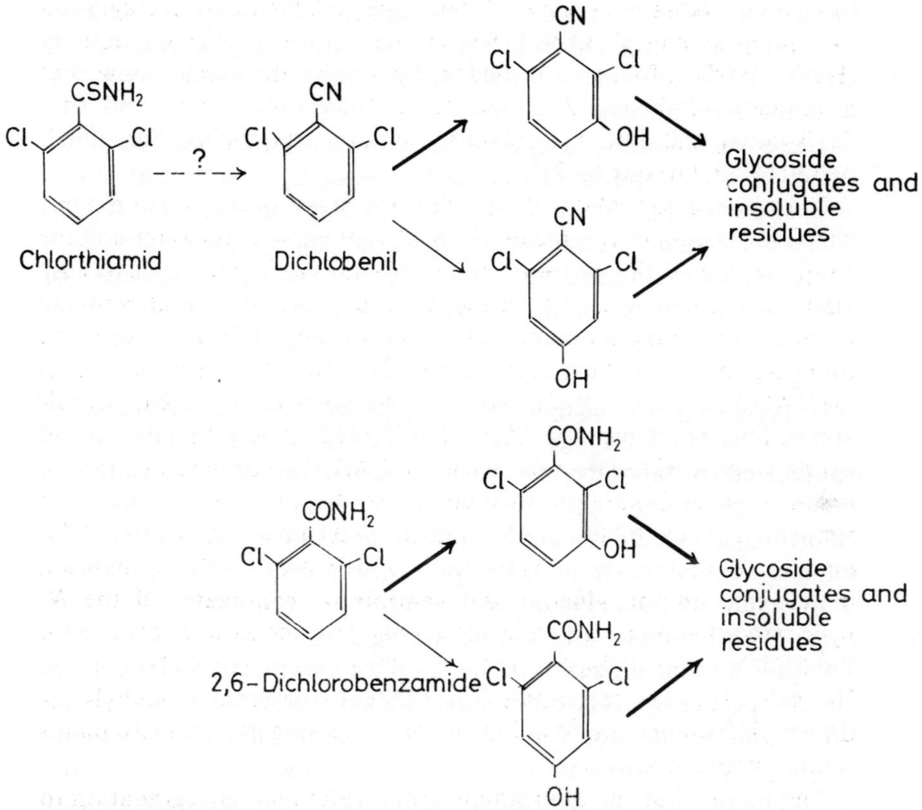

Fig. 6.8 Degradation pathways of chlorthiamid, dichlobenil and 2,6-dichlorobenzamide in plants (after Verloop, 1972 and Beynon and Wright, 1972, reprinted Frear, 1976).

conjugated in plants, but at a slower rate than dichlobenil (Verloop, 1972). Differential rates of metabolism have been observed and appear to be an inherent property of each species and independent of the stage of development (Verloop, 1972). Hydrolysis of dichlobenil to 2,6-dichlorobenzamide (2,6-BAM) or to 2,6-dichlorobenzoic acid appears to be a minor mechanism of degradation and normally only trace quantities of these are present in treated plants (Verloop and Nimmo, 1969, 1970).

The uptake and metabolism of soil-incorporated **chlorthiamid** in higher plants has been studied by Beynon and Wright (1968) (fig. 6.8). It appears that most, if not all, of the applied chlorthiamid is degraded in the soil to dichlobenil and possibly to 2,6-dichlorobenzamide prior to plant uptake. Thus plants grown in chlorthiamid-treated soils may contain dichlobenil, its 3-hydroxy and 4-hydroxy metabolites and their conjugates (mainly *O*-glucosides or *O*-glycosides). In addition to these compounds both free and conjugated 3-hydroxy-2,6-dichlorobenzamide were isolated from apple or wheat plants treated with soil-applied chlorthiamid or apple leaves treated directly with 2,6-dichlorobenzamide.

These metabolites have also been detected in wheat seedlings (Verloop, 1972) and in kale (*Brassica oleracea*) (Leach *et al.*, 1971) after short-term (1–5 day) root and soil uptake studies with ^{14}C-2,6-dichlorobenzamide. Long-term (14 weeks) soil uptake studies with this compound revealed that the main metabolites in the leaves of apple are 3-hydroxy- and 4-hydroxy-2,6-dichlorobenzamide, about 85% of these metabolites being conjugated as glycosides (Leach *et al.*, 1971).

In reviewing the degradation of dichlobenil and chlorthiamid in plants, Frear (1976) suggests that further work is required to identify conjugated metabolites, establish precursor–product relationships between metabolites and insoluble residues, determine the extent of chlorthiamid metabolism and hydrolysis of dichlobenil and characterise the enzyme systems responsible for hydroxylation and conjugation.

In reviewing the degradation of **ioxynil** and **bromoxynil** in plants, Frear (1976) points out that there is little detailed information on the isolation and identification of metabolites and the pathways remain largely speculative. However, work by Zaki *et al.* (1967), Davies *et al.* (1968) and Schafer and Chilcote (1970) suggests that the unhindered nitrile groups of bromoxynil and ioxynil are slowly hydrolysed within the plant to form the benzamide and benzoic acid derivatives, with possible further degradation through decarboxylation, dehalogenation and conjugation of the benzyl moiety or the aromatic ring (Frear, 1976).

ANILIDES

The degradation of **propanil** in higher plants has been studied more intensively than any other anilide herbicide; the subject has been included in reviews by Casida and Lykken (1969), Matsunaka (1969), Ashton and Crafts (1973), Still and Herrett (1976).

The initial reaction in plant metabolism is generally considered to involve hydrolysis of the amide bond with the formation of 3,4-dichloroaniline (McRae, Yih and Wilson, 1964; Adachi, Tonegawa and Uejima, 1966; Ishizuka and Mitsui, 1966; Still and Kuzirian, 1967; Yih, McRae and Wilson, 1968a, b). Still (1968a), examining the fate of ^{14}C-labelled propanil (C-1 or C-3 labelled) in pea (susceptible) or rice (tolerant), found that intact propionic acid was cleaved from the propanil and subsequently degraded by β-oxidation to $^{14}CO_2$.

3,4-Dichloroacetanilide was believed to be an intermediate between propanil and 3,4-dichloroacetaniline (Yih *et al.*, 1968a), but it appears to be a transient intermediate in rice which is not normally isolated (Ashton and Crafts, 1973). The formation of 3,4-dichloroaniline is catalysed by a macromolecule (arylacylamidase) (Still and Kuzirian, 1967) which has been partially purified and characterised by Frear and Still (1968).

3,4-Dichloroaniline is converted to three complexes, one being *N*-(3,4-dichlorophenyl) glucosylamine, the others having a reducing sugar component (Still, 1968b). An enzyme which catalyses the biosynthesis of *N*-glucosylarylamines has been isolated from soybean (Frear, 1968). Yih *et al.* (1968a) found evidence of transient propanil metabolites which contained 3,4-dichloroaniline conjugates with glucose, xylose and fructose; complexing with cell wall components such as cellulose, hemicellulose and lignin also occurs (Yih *et al.*, 1968a; Still and Mansager, 1971).

There is some evidence that differential rates of metabolism of propanil may account for its selective activity. For example, propanil is degraded in rice (resistant) more rapidly than in most weed species. The rate of inactivation by homogenates of rice was ten times more rapid than by those of barnyard grass (*Echinochloa crus-galli*) (Adachi *et al.*, 1966). This is consistent with the report of Frear and Still (1968) who found an arylacylamidase enzyme in rice capable of hydrolysing the amide bond of propanil; the leaves of rice contained sixty times more enzyme units than those of barnyard grass. Variation in the levels of arylacylamidases in range of agricultural and horticultural crop species has been reported by Hoagland, Graf and Handel (1974).

The metabolism of propanil by an arylacylamidase enzyme (arylacylamine amidohydrolase EC 3.5.1.a) which is present in rice leaves has been studied by Tsai (1974). He purified and characterised the enzyme and outlined a pathway for the metabolism of propanil in rice. Ray and Still (1975) have also studied the metabolism of propanil in this species and compared propanil amidase activities in rice plant and callus issues.

It is interesting to note that propanil injury to rice was increased by organophosphate and carbamate insecticides (Bowling and Hodgins, 1966) and this is due to their inhibition of acylamidase activity. It is concluded that differential detoxication of propanil by arylacylamidase is a major factor determining the selectivity of this herbicide (Still and Herrett, 1976).

The rate of metabolism of propanil is also governed by environmental factors such as temperature and day length (Hodgson,

1971). Absorption and metabolism are most rapid under high temperature (37° C) and long day (16 h) conditions, greater quantities of 3,4-dichloroaniline and *N*-(3,4-dichlorophenyl)glucosylamine being recovered.

The activity and selectivity of **benzoylprop-ethyl** in cereals and in *Avena fatua* have been investigated by workers at Shell Research, Sittingbourne. Jeffcoat and Harris (1973) found that toxicity of foliar-applied benzoylprop-ethyl to wild oat (*Avena fatua*) is dependent on its metabolism to benzoylprop which is biologically active. Selectivity is dependent on the rate of de-esterification and subsequent detoxication of the acid to inactive conjugates. De-esterification is most rapid in oat and least rapid in wheat. In the latter the rate of detoxication prevents the accumulation of acid to phytotoxic levels, while in oat, although the rate of detoxication is greater, it is insufficient to prevent the occurrence of phytotoxic levels of benzoylprop.

The degradation of foliage-applied benzoylprop-ethyl in cereal seedlings has revealed that hydrolysis occurs to give the des-ethyl analogue which is conjugated with plant sugars (Beynon, Roberts and Wright, 1947a). A minor route of degradation involves debenzoylation of benzoylprop-ethyl to give metabolites which are also complexed or conjugated; there was no evidence for the formation of 3,4-dichloroaniline. After treatment of wheat plants at the late-tillering to early-shoot stages, the major residues consisted of benzoylprop-ethyl and benzoylprop in free and conjugated forms (Beynon, Roberts and Wright, 1974b).

Comparison of the activity of benzoylprop-ethyl, **flamprop-isopropyl** and **flamprop-methyl** in oats (cv. Mostyn), wild oats (*Avena fatua*) and wheat revealed that the flamprop-methyl was superior in activity against oats and wild oats with tolerance in wheat (Jeffcoat, Harris and Thomas, 1977). Like benzoylprop-ethyl, selectivity of flamprop-methyl is dependent on the rate of degradation and the subsequent detoxication of the biologically active acid to inactive conjugates. The rate of degradation is comparable to that of benzoylprop-ethyl but the greater activity of the acid accounts for the improved performance of flamprop-methyl.

Conjugation metabolites of **propachlor** have also been reported to occur in onions and in organic soils following soil application of this herbicide (Frank *et al.*, 1977). Conjugated *N*-isopropylacetanilide was detected in onions, the levels increasing with increasing rate of application of the herbicide and at the later timing of treatments. In the soil, conjugated metabolite was found to be present even 2 years after application.

NITROPHENOLS

The metabolism of phenolic herbicides in plants has received relatively little investigation, though studies involving t.l.c. (Gruebner, 1971; Mosinska and Kotarski, 1972) or oscillographic polarography (Supin, Vaintraub and Makarova, 1971) has been used to detect nitrophenol residues. Bandal and Casida (1972) compared the degradation of ^{14}C-**dinoseb** and the related acaricide dinobuton (2-s-butyl-4,6-dinitrophenyl isopropyl carbonate) and found their metabolism to be similar. They found that metabolism of ^{14}C-dinoseb resulted in a rapid loss of radioactivity from the organosoluble fraction with an increase occurring in the aqueous and organoinsoluble fractions. The metabolites formed were not identified but were assumed to involve modification of the s-butyl and/or nitro groups and the formation of conjugates. In plants and animals, dinobuton is hydrolysed to form dinoseb and knowledge of probable metabolite formation of the latter is based on studies carried out with ^{14}C-ring-labelled dinobuton. In addition to dinoseb, t.l.c. of ether extracts of water-soluble materials treated with β-glucosidase revealed the presence of 4-amino-6-nitro-2-s-butylphenol, 6-amino-4-nitro-2-s-butylphenol and other unidentified products. It is presumed that similar analysis of water-soluble extracts of plants treated with dinoseb would give similar results.

Neither dealkylation nor oxidation of the s-butyl group of either dinoseb or dinobuton was found to occur in plants treated with these compounds (Bandal and Casida, 1972).

It seems that phenolic pesticides are readily conjugated to form water-soluble metabolites and reduction of the nitro groups with further conjugation of the amines might be expected (Kaufman, 1976).

NITROPHENYL ETHERS

Investigation of the basis of selectivity of **nitrofen** in wild turnip (*Brassica campestris*), redroot pigweed (*Amaranthus retroflexus*) and green foxtail (*Setaria viridis*) has been carried out by Hawton and Stobbe (1971a, b). Several metabolites were produced, at least two of which were believed to be lipid–nitrofen conjugates or nitrofen polymers; there was also evidence of the products of cleavage at the ether linkage (Hawton and Stobbe, 1971b). However, Matsunaka (1976) points out that identification of the derivatives is not positive since they failed to co-chromatograph the unknowns with authentic materials at the same time.

The requirements of light for nitrofen activity and its role in diphenyl ether metabolism were first demonstrated by Hawton and

Stobbe (1971b). Supin, Vaintraub and Makarova (1971) confirmed that nitrofen required light to inhibit the growth of rice seedlings. Limited metabolism of ^{14}C-nitrofen was observed in excised shoots of rice in light or dark or root-treated intact plants of sunflower or soybean. No cleavage of the diphenyl ether linkage was evident however, and light activation of nitrofen did not appear to be related to metabolism of nitrofen.

The fate of **fluorodifen** in peanut (*Arachis hypogea*) seedlings has been examined by Eastin (1971). Major degradation products included *p*-nitrophenol, two unknown compounds (unknowns I and II), 2-aminofluorodifen, 2-amino-4-trifluoromethylphenol and a number of trace compounds and unknowns. *p*-Nitrophenol was also the major metabolite detected in maize (*Zea mays*) and soybean (*Glycine max*) by Geissbühler and others (cit. Matsunaka, 1976); in addition a polar compound was identified tentatively as the β-D-glucoside of 4-nitrophenol.

Eastin (1971) suggested that unknowns I and II may be conjugates of *p*-nitrophenol and 2-amino-4-trifluoromethylphenol, respectively. He postulated a major pathway of degradation of fluorodifen in peanut involving reduction to 2-aminofluorodifen which is cleaved at the ether linkage to yield *p*-nitrophenol and 2-amino-4-trifluoromethylphenol, and that these were then conjugated to plant material. Eastin (1972) also reported one major photolysis product of fluorodifen (probably a polymer of fluorodifen) and low levels of several minor metabolites.

Rogers (1969) treated soybeans and cucumber (*Cucumis sativus*) with fluorodifen and detected *p*-2-diaminofluorodifen in both species. He proposed that metabolism in soybean involved cleavage of the ether linkage and believed that reduction of the nitro substituents on the ring was of minor importance (Rogers, 1971).

Matsunaka (1976), in reviewing the metabolism of diphenyl ethers believed that Roger's results do not deny Eastin's observations and he proposed a pathway of fluorodifen degradation in higher plants adapted from Eastin (1971).

Locke and Baron (1972) applied ^{14}C-labelled 'Preforan' (fluorodifen) to tobacco cells in suspension culture. ^{14}C-1-labelled fluorodifen produced conjugates of 4-nitrophenol (possibly glucosides or conjugates with amino acids or protein) together with certain unidentified acidic conjugates. Metabolites resulting from treatment with $^{14}CF_3$-labelled fluorodifen probably represent ^{14}C incorporation into natural products, following oxidation and cleavage of the $^{14}CF_3$ group from the parent compound.

Further evidence of fluorodifen conjugation is reported by

Shimabukuro *et al.* (1973) who found that the herbicide is cleaved in groundnut to produce the metabolite *S*-(2-nitro-4-trifluoromethyl-phenyl)glutathione. A 'soluble' glutathione *S*-transferase which catalyses the cleavage of fluorodifen has been isolated from the epicotyl tissue of pea seedlings and partially characterised (Frear and Swanson, 1973).

Bifenox appears to have been little investigated though Eastin (1976) has reported a t.l.c. method which separates the herbicide from four of its degradation products.

Chlomethoxynil is rapidly absorbed and degraded in rice and barnyard millet (*Echinochloa crus-galli* var. *frumentacea*) to form three major metabolites (Niki, Kuwatsuka and Yokomichi, 1976). Metabolic products appear to be present largely as conjugates of amino and desmethyl derivatives of chlomethoxynil, possibly reflecting conjugation with saccharides, amino acids, lipids, lignin and other substances.

NITROANILINES

The literature concerning the detection of dinitroanilines has largely been concerned with the use of g.l.c. techniques for the separation of the two isomeric herbicides **trifluralin** and **benfluralin** (Koons and Day, 1972; Hambleton, 1973; Hall and Mallen, 1976; Downer, Hall and Mallen, 1976). In reviewing the chemical and physical properties of the dinitroaniline herbicides, Helling (1976) observed that the degradation processes (in soil) include oxidation, *N*-dealkylation, cyclisation and nitro group reduction; certain of these mechanisms have been reported to occur in plants. Losses due to volatilisation or photodecomposition have also been reported for trifluralin and **nitralin** (Rieck *et al.*, 1971); trifluralin vapour undergoes oxidative breakdown in the atmosphere to dealkylated residues (Crosby, Leffingwell and Moilanen, 1975).

Trifluralin is apparently extensively metabolised in certain plant species. For example, application of ^{14}C-n-propyl- or ^{14}C-trifluoro-methyl-labelled trifluralin to soybean (*Glycine max*) and cotton (*Gossypium hirsutum*) resulted in extensive metabolism but little production of $^{14}CO_2$ (Probst *et al.*, 1967). The ^{14}C was distributed in various plant components including lipids, glycosides, hydrolysis products, proteins and cellular fractions. Examination of the glycoside fraction and hydrolytic products revealed no major metabolites in soybean plants or cotton seeds. In comparison trifluralin is relatively persistent in carrots (*Daucus carota*), this being the major portion ($>$80%) of the residual material in the roots (Golab *et al.*, 1967) (fig.

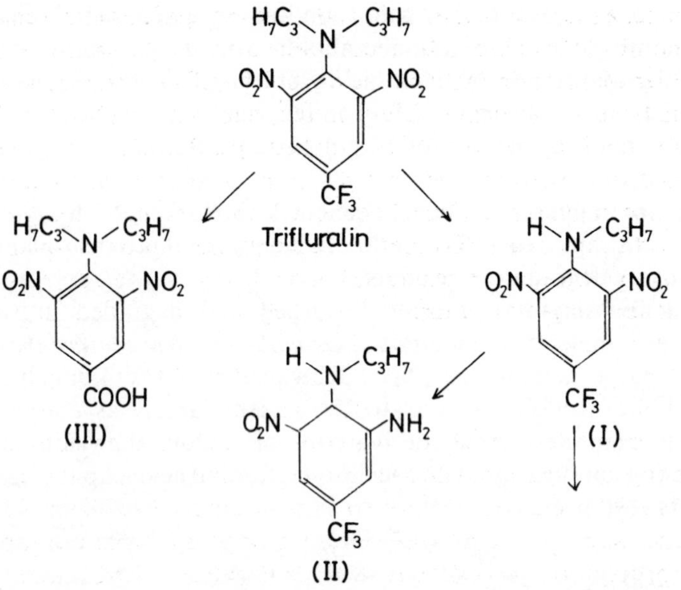

Fig. 6.9 Degradation of trifluralin in carrot root, the major metabolite being α,α,α-trifluoro-2,6-dinitro-*N*-(n-propyl)-*p*-toluidine (I) with trace amounts of α,α,α-trifluoro-5-nitro-*N*-propyltoluene-3,4-diamine (II) and 4-(dipropylamino)-3,5-dinitrobenzoic acid (III) (after Probst and Tepe, 1969).

6.9). The major metabolite was α,α,α-trifluoro-2,6-dinitro-*N*-(n-propyl)-*p*-toluidine (I), which appears to form as a result of dealkylation; two minor metabolites (II and III) are also formed by processes involving reduction and dealkylation in one case and oxidation in the other. Two routes of degradation have been proposed which are believed to operate simultaneously (Golab *et al.*, 1967, 1970). Dealkylated and reduced derivatives of trifluralin have also been detected in extracts from peanut (*Arachis hypogea*) and sweet potato (*Ipomoea batatas*) treated with ^{14}C-trifluralin (Biswas and Hamilton, 1969). In passing, it should be mentioned that trifluralin may contain *N*-nitroso-di-n-propylamine (NDPA) as a contaminant. A study of the uptake and transformation of root-applied ^{14}C-NDPA in soybean revealed that the transformation products are extremely water soluble, but they have yet to be identified (Berard and Rainey, 1979).

The fate of soil-applied ^{14}C-'Benefin' (benfluralin) in peanuts and alfalfa has been investigated by Golab *et al.* (1970). Perhaps significantly, the derivative compounds detected in the plants were identical to those found in the soil. After 4 months the polar compounds identified were α,α,α-trifluoro-5-nitrotoluene-3,4-diamine-2,6-dinitro-α,α,α-trifluoro-*p*-cresol and α,α,α-trifluorotoluene-3,4,5-

triamine; a major portion of the radioactivity was unextractable. After 7½ months 65% of the radioactivity was extractable, most of it being in the form of polar derivatives. However it is still uncertain whether seed plants have a significant ability to degrade benefin (Naylor, 1976).

Soil application of $^{14}CF_3$-labelled **profluralin** to groundnuts resulted in concentrations of 5.8 p.p.m. w after 5 weeks and only 2 p.p.m. w at maturity (Wright, Rieck and Harger, 1974). Ninety per cent of the absorbed ^{14}C compound partitioned into the polar or non-extractable phases, fifteen metabolites being identified as non-polar and at least eight as polar.

CARBAMATES

The literature on the metabolism of carbamate herbicides has been reviewed by Schlagbauer and Schlagbauer (1972a, b), Ashton and Crafts (1973) and Naylor (1976). The analysis of carbamate residues includes such techniques as cellulose column chromatography (for chlorpropham) (Ercegovich and Witkonton, 1972) and h.p.l.c. (Sparacino and Hines, 1976; Lawrence 1976).

Propham and **chlorpropham** are widely used as pre-emergence herbicides in broad-leaved crops and there is evidence that they may undergo degradation in tolerant species. Prendeville *et al.* (1968) and James and Prendeville (1969) found evidence of water-soluble β-glucosides of a modified form of chlorpropham in two crop and two weed species. The 2-propyl ester portion of the molecule is apparently modified by all four species, though there is no evidence of cleavage of the carbamate bond.

Still and Mansager (1972a, 1973a) treated soybean with ^{14}C-chlorpropham and reported that polar derivatives and insoluble residues are rapidly produced in the roots. The dominant polar metabolite is the *O*-glucoside of isopropyl-5-chloro-hydroxycarbanilate, but it is rapidly converted to insoluble residues. In the shoot, isopropyl-5-chloro-2-hydroxycarbanilate has been detected and it may be the precursor of other polar compounds. Subsequently Still and Mansager (1973a) investigated the metabolism of ^{14}C-chlorpropham in cucumber plants and found that it is rapidly converted to isopropyl-4-hydroxy-3-carbanilate which is then conjugated with unknown plant materials. In soybean ^{14}C-chlorpropham is hydroxylated by root and shoot tissues to form 5-chloro-2-hydroxycarbanilate; direct acetylation and characterisation also showed the presence of *O*-glucoside of this aglycone (Still and Mansager, 1973b).

Root treatment of lucerne with ^{14}C-phenyl-labelled chlorpropham resulted in the formation of two aryl hydroxylated metabolites conjugated with carbohydrates in addition to the herbicide (Still and Mansager, 1974). Similar treatment of lucerne with ^{14}C-phenyl-labelled propham resulted in the production of glucoside conjugates of 2-hydroxy- and 4-hydroxypropham, the latter accounting for 46% of the ^{14}C in the shoots (Still and Mansager, 1975). In oat, chlorpropham is similarly degraded to the 2-hydroxy and 4-hydroxy metabolites but the latter is converted to an *S*-cysteinyl-hydroxychlorpropham thioether conjugate (Rusness and Still, 1975).

The metabolism of **phenmedipham** in several susceptible and resistant species has been studied by Bischof *et al.* (1970) and Kassebeer (1971). Sensitivity was related to the rate of uptake and ability to metabolise the herbicide. In beet and *Amaranthus retroflexus*, both relatively tolerant species, phenmedipham was converted to a metabolite, tentatively identified as a hydroxylated derivative, which may form complexes with plant constituents. In contrast, 'Swep' appears to form complexes of lignin without first undergoing hydrolysis (Chin *et al.*, 1964).

Barban is a carbamate of high selectivity which is chiefly used for the control of wild oats (*Avenu fatua*) in a number of crops including wheat, barley, flax, peas, sugar-beet and soybeans, which have a high degree of tolerance. The metabolism of barban has been studied by Riden and Hopkins (1961, 1962) who found that it is rapidly metabolised in wheat and wild oats to an unknown water-soluble compound (X) which releases 3-chloroaniline on hydrolysis with alkali. This 3-chloroaniline moiety is complexed into several water-soluble derivatives. Jacobsohn and Anderson (1972) found that compound X is not phytotoxic and postulated that its build-up may reduce the rate of metabolism of barban. Lamoureaux *et al.* (1971) reported the formation of a barban/glutathione conjugate.

Still and Mansager (1972b), applied ^{14}C-phenyl-, ^{14}C-carbonyl- and ^{14}C-1-butynyl-labelled barban to soybeans and found that the barban molecule was not cleaved. However, their results indicate that barban ^{14}C-metabolites are formed by the alteration of the 4-chloro-2-butynyl side chain and not by hydroxylation of the aromatic nucleus.

THIOCARBAMATES

The degradation of thiocarbamate herbicides has been recently reviewed by Ashton and Crafts (1973) and by Fang (1976).

EPTC has been found to be more rapidly degraded in the seedlings of resistant as compared to susceptible crops species. ^{35}S from ^{35}S-labelled EPTC was incorporated into cysteic acid, cystine, methionine, methionine sulphone, and two unidentified compounds (Fang and Yu,

1959). Similarly Nalewaja, Behrens and Schmid (1964) found that ^{14}C from ethyl-1-^{14}C-EPTC was incorporated into fructose, glucose and several amino acids including aspartic acid, asparagine, glutamic acid, glutamine, serine, threonine and alanine. They presented evidence which suggests that cleavage occurs between the sulphur atom and ethyl group. Treatment of maize plants with ^{14}C-carbonyl-labelled EPTC resulted in the loss of 21% of the absorbed activity as $^{14}CO_2$; only 80% of the radioactivity remained as EPTC, 66% being metabolised to water-soluble products. The latter was identified as a glutathione conjugate and its breakdown products (Carringer and Rieck, 1976).

The metabolism of *S*-(1-^{14}C-propyl) **vernolate** in soybean (*Glycine* max) has been examined by Bourke and Fang (1968). $^{14}CO_2$ production increased with age but was reduced by pretreatment with vernolate suggesting that inhibition of enzyme synthesis may be a primary site of action of thiocarbamate herbicides in germinating seedlings. There was evidence of four labelled metabolites, one being tentatively identified as citric acid.

The rate of degradation of 1-^{14}C-propyl **pebulate** has been compared in germinating seeds of mung bean (*Phaseolus aureus*) and wheat (*Triticum aestivum*) (Fang and George, 1962). The degradation rate was higher in the mung bean seeds which are more resistant than in wheat (susceptible). While the rate of breakdown of pebulate in mung bean was unaffected by concentrations below 10 p.p.m., the rate of degradation in wheat declined as the concentration increased, suggesting that a detoxication system was absent in this species. As in EPTC and vernolate, oxidation of pebulate to CO_2 is rapid, 18–40% of the absorbed radioactivity being evolved from root and shoot tissues. This is consistent with the findings of Long, Thompson and Rieck (1974) that ^{14}C-pebulate is not recovered in detectable amounts in leaves or roots of tobacco seedlings after a 32 h incubation period, suggesting rapid incorporation into acetyl-CoA.

The degradation of root-applied 1-^{14}C-ethyl **molinate** or 2-^{14}C-azepine molinate in rice also results in $^{14}CO_2$ production (Gray, 1969). Paper chromatography of the cell sap from rice shoots harvested 3 days after application of ^{14}C-chain-labelled molinate revealed the presence of several amino acids (asparagine, glycine, threonine, alanine, tryptophan, phenylalanine and isoleucine) and organic acids including lactic and glycolic acids; cellulose and protein fractions were also labelled. Treatment of rice with 2-^{14}C-azepine molinate also led to the formation of organic acid and amino acid metabolites.

The fate of ring-labelled **cycloate** in sugar-beet has been examined by Antognini, Gray and Menn (1970). The ^{14}C label was incorporated into CO_2 and plant constituents including glycine, asparagine, proline, alanine, valine, leucine, phenylalanine, sucrose, glucose and cellulose. Extensive degradation of 2-^{14}C-CDEC (2-chloroallyl diethyldithiocarbamate) in cabbage has also been reported with the formation of lactic acid as a major metabolite (Jaworski, 1964).

UREAS

Examination of the degradation of urea herbicides in plants has ranged from a colorimetric method for diuron residues in sorghum and lucerne (Gupta and Dutta, 1974), to column and t.l.c. of siduron in Kentucky blue grass (*Poa pratensis*) (Jordan *et al.*, 1975), to g.l.c. (electron capture and electrolytic conductivity) and h.p.l.c. of linuron and other herbicides in crop plants (Lawrence, 1976) or weeds (Glad, Popoff and Theander, 1977).

A number of degradative reactions have been demonstrated for phenyl ureas, namely *N*-demethylation, *N*-demethoxylation, ring hydroxylation, oxidation of ring substituents and aniline formation. These so-called phase I reactions introduce or lead to relatively polar and biochemically reactive groups (e.g. —NH_2, —OH and —COOH groups) (Parke and Williams, 1969), Phase II reactions involve binding to plant (or animal or microbial metabolites) and include acetylation of anilines, glucoside/glucuronide formation and binding to nitrogen constituents. The nature of these degradative reactions has been reviewed by Voss and Geissbühler (1971), Ashton and Crafts (1973), Geissbühler, Martin and Voss (1976).

N-Dealkylation is the major initial pathway of urea herbicide metabolism which has been demonstrated for a number of dialkyl- and alkoxyalkylphenylureas in soil, microbial media, plants and mammals. In plants these reactions have been observed for **fenuron**, **monuron** (fig. 6.10), **diuron**, **fluometuron**, **chlortoluron**, **chloroxuron**, **metoxuron**, **monolinuron**, **metobromuron**, **linuron** and **chlorbromuron** in a variety of crop species and have been reviewed by Geissbühler *et al.* (1976). The mechanism of *N*-demethoxylation is still unknown though there is some evidence that the corresponding hydroxyl-amides are formed as transient intermediates (Anon, 1972 *cit.* Geissbühler *et al.*, 1976; Börner, Burgemeister and Schroeder, 1969; Voss and Geissbühler, 1966). In maize leaves, sequential demethylation is the major pathway of transformation of monuron (Lee and Fang, 1973). Resistance of wheat as compared to barley seedlings to **tebuthiuron** appears to be due mainly to differential rates of *N*-demethylation (Lee and Ishizuka, 1976). This was not the case, however, in sugar-cane, where susceptible and resistant cultivars both metabolised diuron to the monomethyl

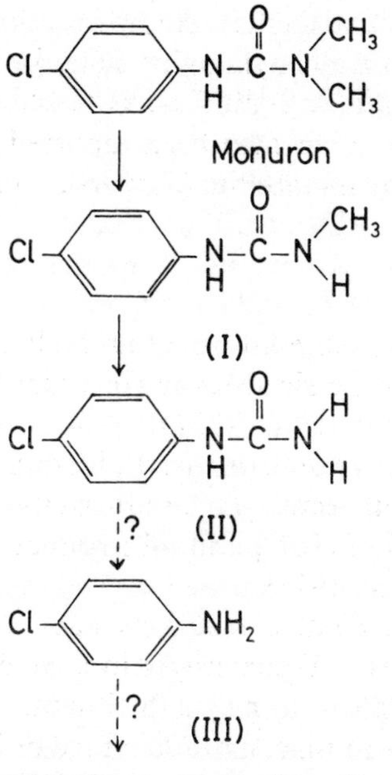

$$Cl-C_6H_4-NH-CO-N(CH_3)_2$$

Monuron

$$\downarrow$$

$$Cl-C_6H_4-NH-CO-NH(CH_3)$$

(I)

$$\downarrow$$

$$Cl-C_6H_4-NH-CO-NH_2$$

(II)

$$\downarrow ?$$

$$Cl-C_6H_4-NH_2$$

$$\downarrow ?$$

(III)

Fig. 6.10 A proposed pathway for the metabolism of monuron in resistant plants showing the products of *N*-demethylation (I and II) and deamination (III) (after Frear, Swanson and Tanaka, 1972, reprinted from Naylor, 1976).

derivatives and its conjugates (Liu, Shimabukuro and Nalewaja, 1978). Resistance of wheat and *Galium aparine* to **methabenzthiazuron** (2×10^{-4} M) depended on their ability to rapidly hydroxylate the *N*-methyl group (Collet, 1973). A subsequent *in vitro* study using a range of species again revealed urea chain degradation of methabenzthiazuron to form the 1-hydroxymethyl derivative; the loss of this hydroxymethyl group and condensation with glucose may result in further metabolites (Pont *et al.*, 1974).

There is some evidence that ring hydroxylation of phenylureas occurs in plants. For example, there are indications that fluometuron is hydroxylated in the 6-position in cotton tissue (Anon, 1972 *cit.* Geissbühler *et al.*, 1976) and monuron ring-hydroxylated in the 2-position with the subsequent formation of polypeptide complexes and β-D-glucosides.

Oxidation of ring substituents has been demonstrated to take place in chlortoluron in wheat. The major portion of the hydroxymethyl metabolite (3-[3-chloro-4 hydroxymethylphenyl]-1,1-dimethylurea) was initially trapped as water-soluble conjugate(s) and later oxidised to the corresponding carboxy derivative; *N*-demethylation did not proceed beyond the *N*-monomethyl stage (Anon, 1972, cit. Geissbühler *et al.*, 1976).

It has been postulated that stepwise *N*-demethylation and/or *N*-demethyloxylation may be followed by hydrolysis of the dealkylated ureas to the corresponding anilines. However the amounts of free anilines which have been detected in treated plants have been consistently small (Geissbühler *et al.*, 1963; Rogers and Funderburk, 1967; Onley, Yip and Aldridge, 1968; Smith and Sheets, 1967; Börner *et al.*, 1969; Kuratle, Rahn and Woodmansee, 1969; Nashed and Ilnicki, 1970; Nashed, Katz and Ilnicki, 1970), and there is no evidence that further rapid transformation of the aromatic amine masks the rate of aniline formation (Geissbühler *et al.*, 1976).

In addition to these mechanisms, there is evidence that phenylureas and their *N*-dealkylated and/or hydroxylated metabolites are conjugated to carbohydrate and peptide material in plants. For example, Jordan *et al.* (1975) found evidence that 2-hydroxy and 4-hydroxy derivatives of **siduron** were subsequently conjugated in Kentucky blue grass (*Poa pratensis*) cv. Merion. Evidence suggesting the complexing to low molecular protein or peptide has been reported for ^{14}C-carbonyl-labelled monuron applied to bean leaves (Fang *et al.*, 1955) and excised bean and maize leaves (Lee and Fang, 1973). The latter estimate the molecular weight of the monuron complex to be greater than 5000. The sorption of phenylureas to plant protein fractions has also been found to take place following the metabolism of linuron and chlorbromuron in maize and cucumber (Nashed and Ilnicki, 1970; Nashed, Katz and Ilnicki, 1970). Sorption of phenylureas to bovine serum albumin (BSA) is postulated to involve bond formation between either or both the amide hydrogen and the carbonyl oxygen of the herbicides with free amino groups of the protein (Camper and Moreland, 1971). The extent to which such sorption phenomena might account for the protein/polypeptide complexes occurring in plant material is however, uncertain (Geissbühler *et al.*, 1976).

In addition to *N*-dealkylated metabolites, much more polar degradation products of carbonyl and ^{14}C-ring-labelled phenylureas have been detected (Voss and Geissbühler, 1966; Swanson and Swanson, 1968). These represent glucosidic conjugates of hydroxymethyl intermediates formed during *N*-dealkylation (Geissbühler and Voss, 1971; Frear and Swanson, 1971). Similarly, β-D-glucosides of monuron and fluometuron have been detected in excised leaves of cotton treated with ^{14}C-trifluoromethyl fluometuron and with ^{14}C-monuron (Frear and Swanson, 1972); a β-glucoside metabolite has also been detected in wheat seedlings treated with ^{14}C-phenyl **buturon** (Anon, 1976; Schuphan and Ebing, 1977).

The biochemical and enzymatic mechanisms involved in the

formation of these metabolites have been reviewed by Geissbühler *et al.* (1976).

TRIAZINES

Examination of the metabolites formed by degradation of s-triazine herbicides has been carried out using a range of techniques including alkali flame and electrolytic conductivity detectors (Greenhalgh and Cochrane, 1972; Cochrane, Wilson and Greenhalgh, 1973; Purkayatha and Cochrane, 1973) and chromatographic techniques (Beynon, 1972) (t.l.c., Heede and Heyndricks, 1976; partition and column chromatography, Suzuki, Nagayoshi and Kashiwa, 1976; and g.l.c., Hormann and Eberle, 1972; Schroeder *et al.*, 1972; Hofberg and Murphy, 1973; Purkayatha and Cochrane, 1973; Hofberg, Heinricks and Gentry, 1976; Josepovits, Dobrovolszky and Jeney, 1976).

The mechanisms of degradation and the nature of the metabolites formed have been reviewed by Matsunaka (1969), Ashton and Crafts (1973), Naylor (1976) and Esser *et al.* (1976)

The first evidence of dechlorination of chlorotriazines accompanied by substitution of a hydroxyl group in the 2-position was obtained for **simazine** and **atrazine** with *in vitro* systems of maize (Roth, 1957; Gysin and Knüsli; 1960; Castelfranco, Foy and Deutsch, 1961; Freed, Montgomery and Kief, 1961) (fig. 6.11). In resistant maize or wheat, an active compound capable of inactivating simazine and atrazine has been isolated which has the same properties as a cyclic hydroxamate (Roth and Knüsli, 1961; Castelfranco and Brown, 1962; Anderson, 1964; Foy, 1967). Its structure has been shown to be 2,4-dihydroxy-7-methoxy-1,4-benzoxazine-3-one (benzoxazinone) (Wahlroos and Virtanen, 1959; Hamilton and Moreland, 1962; Hamilton, 1964). Hydroxylation appears to occur predominantly in roots (Shimabukuro *et al.*, 1970; Thompson *et al.*, 1970) and is attributed to the catalytic action of benzoxazinones (Willard and Penner, 1976).

A variety of *in vivo* studies have revealed that chlortriazines are degraded to hydroxytriazines, this hydrolytic capacity being evident to varying degree in a variety of plants (Brown, Furmidge and Grayson, 1972; Beynon, Stoydin and Wright, 1972a, b; Gzik and Graser, 1972; Esser *et al.*, 1976). However, such hydroxylation does not ensure resistance to these herbicides (Hamilton, 1964; Shimabukuro, 1967). In the case of methoxy- and methylthiotriazine compounds, hydrolysis appears to be a less effective mechanism than in the case of chlorotriazines.

Detoxication of s-triazines may also take place through the production of peptide conjugates (Shimabukuro and Swanson, 1969;

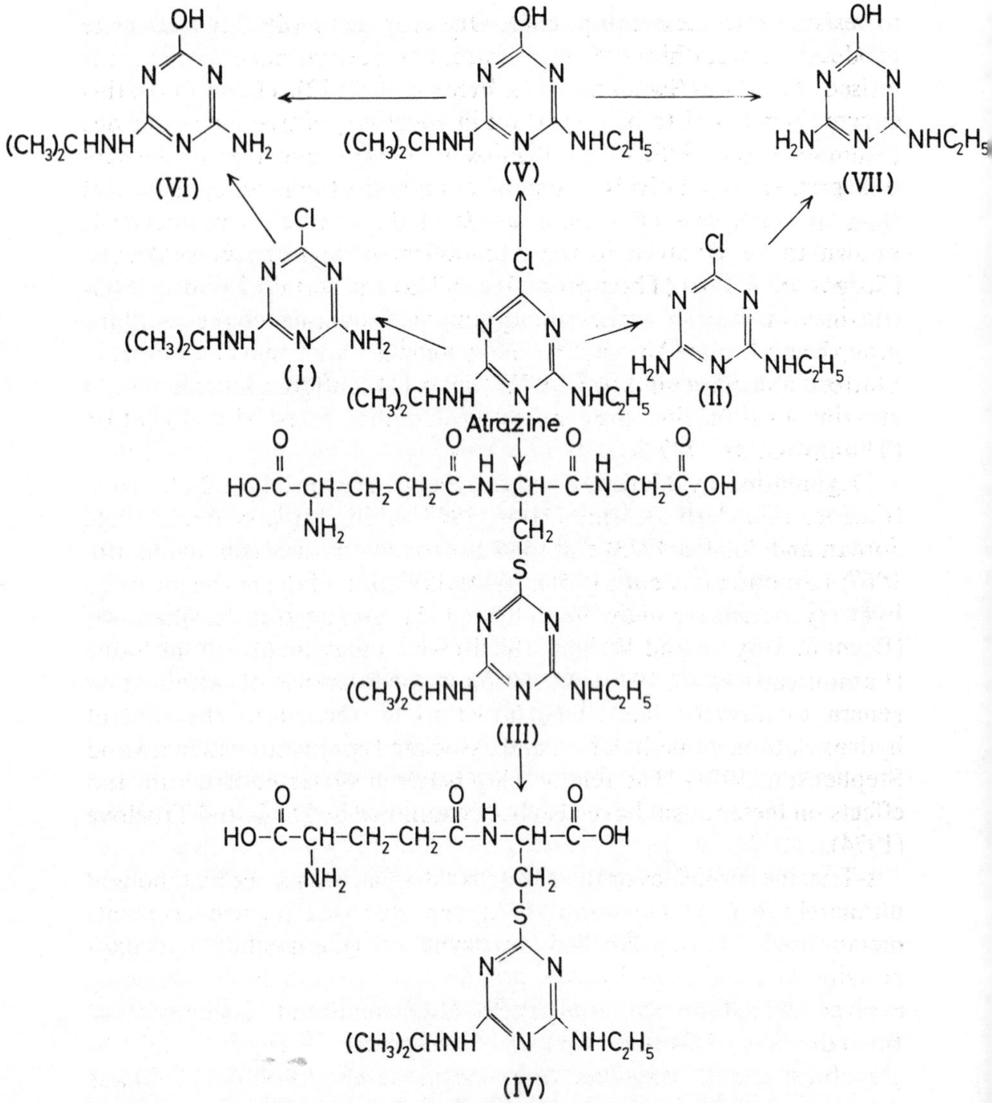

Fig. 6.11 Proposed pathways of degradation of atrazine involving the formation of hydroxy derivatives, viz. hydroxyatrazine (V), 2-hydroxy-4-amino-6-isopropyl-amino-s-triazine (VI) and 2-hydroxy-4-amino-6-ethylamino-s-triazine (VII) (after Lamoureaux *et al.*, 1970, adapted from Naylor, 1976).

Shimabukuro, Swanson and Walsh, 1970; Thompson, Slife and Butler, 1970; Shimabukuro *et al.*, 1971) which have been detected in cultivated sorghum (*Sorghum bicolor*) (Lamoureaux *et al.*, 1970), *Setaria* and *Panicum* species (Thompson, Jr., 1972b); in the latter, production of peptide conjugates of atrazine and simazine was related to resistance to these compounds. One enzyme involved in conjugate production, glutathione-*S*-transferase, has been purified and characterised (Frear and Swanson, 1970; Frear *et al.*, 1972). The action of this enzyme is believed to be important in selectivity of triazine herbicides (Shimabukuro, 1975, 1976). Robinson (1974) found that its activity was greater in relatively resistant crabgrass (*Digitaria sanguinalis*) than in witchgrass (*Panicum capillare*). In addition glutathione is known to be involved in the metabolism of atrazine in wild cane (*Sorghum bicolor*) (Thompson, Jr., 1972a) and certain 2-choro-1,3,5-triazines in tolerant maize, sorghum and sugar-cane, the 2-chloro group being displaced by glutathione or α-glutamylcysteine (Lamoureaux, Stafford and Shimabukuro, 1972) (fig. 6.11). Differential response to atrazine or simazine may reflect differential rates of conjugation (Thompson, Jr., 1972a).

Degradation by dealkylation has also been reported for 2-chloro-s-triazines (Funderburk and Davis, 1963; Shimabukuro *et al.*, 1966; Jordan and Jolliffe, 1973) and for 2-hydroxy-s-triazines (Shimabukuro, 1967; Lamoureaux *et al.*, 1971). *N*-Dealkylation of the herbicide or its hydroxy derivative may be followed by conjugation mechanisms (Beynon, Stoydin and Wright, 1972b) which may involve glutathione (Lamoureaux *et al.*, 1973). Variation in the tolerance of certain grass genera to atrazine has been shown to be related to the rate of hydroxylation, demethylation and associated conjugation (Jensen and Stephenson, 1974). The relationship between s-triazine structure and effects on metabolism have also been examined by Davis and Truelove (1974).

s-Triazine herbicides or their hydroxy derivatives may be metabolised ultimately to CO_2. Goswami (1972) reported that sugar-cane plants metabolised ^{14}C-ring-labelled **ametryn**, ^{14}C-ring-labelled hydroxyatrazine and ^{14}C-ethyl hydroxyatrazine to $^{14}CO_2$, most $^{14}CO_2$ being evolved from the soil and roots (Goswami and Green, 1974). Breakdown of ^{35}S-**prometryne** to ^{35}S-sulphate, ^{35}S-cysteine and ^{35}S-glucobrassicin in fodder kale (*Brassica oleracea* var. *acephala* DC) was reported by Kudelova and Bergmannova (1975), intervarietal differences in the rate of metabolism being noted.

PYRIDINES

The bipyridylium herbicides **diquat** and **paraquat** are not metabolically degraded in plants though extensive photochemical degradation of the herbicide on the leaf surface may take place (Calderbank and Slade, 1976). Under United Kingdom conditions it appears that up to two-thirds of the paraquat applied to plants may be degraded with the formation of 4-carboxy-l-methylpyridinium ions and methylamine, most of this degradation occurring after the plant has been killed.

Picloram also is relatively resistant to biochemical degradation, very slow rates of metabolism being reported to occur in a number of species (Foy, 1976).

PYRIMIDINES (URACILS)

The degradation pathways involved in metabolism of substituted uracil herbicides have been reviewed by Gardiner (1976).

Bromacil is degraded by hydroxylation in plants. Treatment of young orange seedlings with 2-^{14}C-bromacil resulted in only a small proportion of the applied dose ($<5\%$) being absorbed. Extracts of the roots, stems and leaves contained three ^{14}C-labelled materials in the ratio 10:5:1, identified by their R_f values as bromacil, the 6-hydroxymethyl metabolite and a minor unidentified metabolite (Gardiner *et al.*, 1969). Similarly, hydroxy metabolites were detected in extracts from bromacil-treated orchard grass (*Dactylis glomerata*) and Kentucky blue grass (*Poa pratensis*), particularly in the latter which is relatively tolerant (Shriver, 1972; Shriver and Bingham, 1973). The metabolites detected included 5-bromo-3-(2-hydroxy-1-methylpropyl)-6-methyluracil and traces of 3-s-butyl-6-methyluracil and 5-bromo-3-s-butyl-6-methyluracil.

Evidence of metabolism of foliage- and root-applied 2-^{14}C-**terbacil** in ivy leaf morning glory (*Ipomoea hederacea*) (susceptible) and in peppermint (*Mentha* sp.) (tolerant) has also been reported by Barrentine and Warren (1970). There was evidence of a higher rate of degradation in peppermint than in morning glory; the metabolites were not identified. Similarly treatment of bean (susceptible) and citrus (resistant) seedlings with 2-^{14}C-terbacil by nutrient solution techniques resulted in greater metabolism of the herbicide in the resistant species (Herholdt, 1969). In a study in which orange seedlings were cultured in aqueous solutions of 2-^{14}C-terbacil, the herbicide was metabolised to form 3-t-butyl-5-chloro-6-hydroxymethyluracil which was conjugated to form a β-glucoside (Jordan *et al.*, 1975).

HETEROCYCLIC NITROGEN COMPOUNDS (UNCLASSIFIED)

The literature associated with the degradation of **aminotriazole** has been reviewed by Carter (1976). In higher plants the major metabolic alteration involves the formation of conjugates with plant constituents and this is generally regarded as a detoxication mechanism. Ring cleavage has also been reported (Freed, Montgomery and Kief, 1961;

Lund-Høie, 1970) though there is no conclusive evidence of rapid and extensive ring cleavage (Carter, 1976).

The first comprehensive studies of aminotriazole metabolism were carried out by Racusen (1958) who reported the formation of two metabolites ('X' and 'Y'); neither was as phytotoxic as aminotriazole and neither was a simple amide or aminoglucoside. One of these is a condensation product between aminotriazole and serine (3-amino-1,24-triazolylalanine) (3-ATAL) and this appears to represent a detoxication mechanism since the metabolite is less toxic than aminotriazole. It is significant that ammonium thiocyanate which synergises the action of this herbicide, inhibits the formation of 3-ATAL (Carter, 1965; Smith, Bayer and Foy, 1969). The other previously referred to as Y, compound 2, unknown I, etc., appears not to be the same compound in all studies; a third metabolite has been reported by certain workers.

Conjugation of aminotriazole with glucose has been suggested by several workers, including Fredricks and Gentile (1966), though Carter (1976) suggests that on current evidence, such a compound is simply an 'interesting artefact'.

There is also evidence that aminotriazole is metabolised in *Cirsium arvense* and pea (*Pisum sativum*) to three major products (unknown II, 1a and 1b) (Smith and Chang, 1973). Unknown II was found to be identical to β-(3-amino-1,2,4-triazolyl-1)-α-alanine while unknown 1a appeared to be a metabolic product of unknown II, and unknown 1b derived from unknown 1a.

While t.l.c. has been used to separate metabolites in the older studies, a technique involving g.l.c. has more recently been used in the analysis of aminotriazole–simazine formulations (Barrette and Scheuneman, 1973).

Conjugation of **bentazone** with natural plant components also appears to be a detoxication mechanism which may be important in selectivity. Mine (1974) found that the inhibition and recovery of photosynthesis in rice (resistant) and *Cyperus rotundus* (susceptible) corresponded to the rate of bentazone metabolism. Differential metabolism appears to be the mechanism of bentazone selectivity, the major water-soluble metabolite in rice apparently being a conjugate of some plant component.

ORGANOARSENIC COMPOUNDS

The fate of **MAA**, **MSMA** and **DSMA** has been examined in a range of plant species including purple nutsedge (*Cyperus rotundus*) (Duble, Holt and McBee, 1968), coastal Bermuda grass (*Cynodon dactylon*)

(Duble, Holt and McBee, 1968), Johnson grass (*Sorghum halepense*) (Sckerl and Frans, 1969), beans (*Phaseolus vulgaris* L. Black Valentine) (Sachs and Michael, 1971) and cotton (Keeley and Thullen, 1971). With the exception of cotton which apparently formed no complex during the 72 h treatment, one or more metabolites have been detected and a histidine complex has been proposed, though not proven. Determination of MAA by g.l.c. has been reported by Johnson, Gerhardt and Aue (1972). Limited studies of the degradation of GA in plants suggests that no degradation occurs (Woolson, 1976).

ORGANOPHOSPHORUS COMPOUNDS

Examination of **glyphosate**-treated sugar-cane and its processed products revealed no residues of the herbicide or its metabolites 40 days after treatment (Ching *et al.*, 1976). However, only limited metabolism of glyphosate in plants has been reported (Putnam, 1976; Wyrill and Burnside, 1976), aminomethylphosphonic acid (AMP) being the major metabolite isolated from both soil and plants (Putnam, 1976; Rueppel *et al.*, 1977; Sandberg, Meggett and Penner, 1980). Rueppel, Suba and Marvel (1976) have reported a method of derivatisation of aminoalkylphosphonic acids for characterisation by g.l.c.–mass spectrometry.

References

ADACHI, M., TONEGAWA, K. and UEJIMA, T. (1966). *Pestic. Technol.*, **14**, 19.

AGEMIAN, H. and CHAU, A. S. Y. (1976). *Analyst*, **101**, 732.

ANDERSON, R. N. (1964). *Weeds*, **12**, 60.

ANDREAE, W. A. and GOOD, N. E. (1957). *Pl. Physiol., Lancaster*, **32**, 566.

ANON (1976). *Jahresbericht 1975*, 137 pp.

ANTOGNINI, J., GRAY, R. A. and MENN, J. J. (1970). *Proc. Selective Weed Control in Beetcrops, 2nd Int.*, **1**, 293.

ASHTON, F. M. and CRAFTS, A. S. (1973). *Mode of Action of Herbicides*, Wiley–Interscience, 504 pp.

BANDAL, S. K. and CASIDA, J. E. (1972). *J. Agric. Fd. Chem.*, **20**, 1235.

BARRENTINE, J. L. and WARREN, G. F. (1970). *Weed Sci.*, **18**, 373.

BARRETTE, J. P. and SCHEUNEMAN, E. (1973). *J. Agric. Fd. Chem.*, **21**, 142.

BERARD, D. F. and RAINEY, D. P. (1979). *Bull. Environ. Contam. Toxicol.*, **23**, 136.

BEREZOVSKII, M. YA. and NIKHOTIN, V. V. (1971). *Dok. TSKRA*, 174.

BEYNON, K. I. (1972). *Pestic. Sci.*, **3**, 389.

BEYNON, K. I., ROBERTS, T. R. and WRIGHT, A. N. (1974a). *Pestic. Biochem. Physiol.*, **4**, 98.

BEYNON, K. I., ROBERTS, T. R. and WRIGHT, A. N. (1974b). *Pestic. Sci.*, **5**, 429.

BEYNON, K. I., STOYDIN, G. and WRIGHT, A. N. (1972a). *Pestic. Sci.*, **3**, 293.

BEYNON, K. I., STOYDIN, G. and WRIGHT, A. N. (1972b). *Pestic. Sci.*, **3**, 379.

BEYNON, K. I. and WRIGHT, A. N. (1968). *J. Sci. Fd. Agric.*, **19**, 727.

BEYNON, K. I. and WRIGHT, A. N. (1972). *Residue Rev.*, **43**, 23.

BINGHAM, S. W. and SHAVER, R. L. (1977). *Pestic. Biochem. Physiol.*, **7**, 8.

BISCHOF, VON F., KOCH, W., JAJUMDAR, J. C. and SCHWERDTLE, F. (1970). *Z. Pflkrankh. Pflpath. Pflschutz. Sonderh.*, 95.

BISWAS, P. K. and HAMILTON, W., Jr. (1969). *Weed Sci.*, **17**, 206.

BÖRNER, H., BURGEMEISTER, H. and SCHROEDER, M. (1969). *Z. Pflanzenkrankh. Plfanzenschutz*, **76**, 385.

BOURKE, J. B. and FANG, S. C. (1968). *Weed Sci.*, **16**, 290.

BOWLING, C. C. and HODGINS, H. R. (1966). *Weeds*, **14**, 94.

BRADY, H. A. (1973). *Proc. 26th. Ann. Mtg. S. Weed Sci. Soc.*, 282.

BROADHURST, N. A., MONTGOMERY, M. L. and FREED, V. H. (1966). *J. Agric. Fd. Chem.*, **14**, 585.

BROWN, N. P. H., FURMIDGE, C. G. L. and GRAYSON, B. T. (1972). *Pestic. Sci.*, **3**, 669.

CALDERBANK, A. and SLADE, P. (1976). In *Herbicides: Chemistry, Degradation and Mode of Action* (Eds. P. C. Kearney and D. D. Kaufman), Marcel Dekker, Inc., New York, p. 511.

CAMPER, N. D. and MORELAND, D. E.(1971). *Weed Sci.*, **19**, 269.

CARRINGER, R. D. and RIECK, C. E. (1976). *Proc. 29th Ann. Mtg. S. Weed Sci. Soc.*, 399.

CARTER, M. C. (1965). *Physiol. Pl.*, **18**, 1054.

CARTER, M. C. (1976). In *Herbicides: Chemistry, Degradation and Mode of Action* (Eds. P. C. Kearney and D. D. Kaufman), Vol. 1, Marcel Dekker, Inc., New York, p. 377.

CASIDA, J. E. and LYKKEN, L. (1969). *Ann. Rev. Pl. Physiol.*, **20**, 607.

CASTELFRANCO, P. and BROWN, M. S. (1962). *Weeds*, **10**, 131.

CASTELFRANCO, P. FOY, C. L. DEUTSCH, D. B. (1961). *Weeds*, **9**, 580.

CHANG, F. Y. and VANDEN BORN, W. H. (1968). *Weed Sci.*, **16**, 176.

CHANG, F. Y. and VANDEN BORN, W. H. (1971a). *Weed Sci.*, **19**, 107.

CHANG, F. Y. and VANDEN BORN, W. H. (1971b). *Weed Sci.*, **19**, 113.

CHIN, W. T., STANOVICK, R. P., CULLEN, T. E. and HOLSING, G. C. (1964). *Weeds*, **12**, 201.

CHING, W., NOMURA, N., YAUGER, W., UYEHARA, G. and HILTON, H. W. (1976). *1975 Ann. Rept. Exp. Sta. Hawaiian Sugar Planters Assoc.*, 38.

CHKANIKOV, D. I., MAKEYEV, A. M., PAVLOVA, N. N., GRYGORYEVA, L. Y., DUBOVOI, V. P. and KLIMOV, O. V. (1977). *Arch. Environ. Contam. Toxicol.*, **5**, 97.

CHKANIKOV, D. I. PAVLOVA, N. N. and GORTSUSKII, D. F. (1965). *Khim. v. Sel'sk. Khoz.* **3**, 56.

CHOW, C., MONTGOMERY, M. L. and TE, C. YU. (1971). *Bull. Environ. Contam. Toxicol.*, **6**, 576.

COCHRANE, W. P., WILSON, B. P. and GREENHALGH, R. (1973). *J. Chromatog.*, **75**, 207.

COLBY, S. R. (1965). *Science*, **150**, 619.

COLLET, G. F. (1973). *Compt. Rend.*, **276**, 1425.

CROSBY, D. G., LEFFINGWELL, J. T. and MOILANEN, K. W. (1975). *Environ. Quality and Safety*, **4**, 175.

CROSBY, D. G. and WONG, A.S. (1973). *J. Agric. Fd. Chem.*, **21**, 1049.

DAVIES, P. J., DRENNAN, D. S. H., FRYER, J. D. and HOLLY, K. (1968). *Weed Res.*, **8**, 241.

DAVIS, D. E. and TRUELOVE, B. (1974). *Abstr. 1974 Mtg. Weed Sci. Soc. Amer.*, 79.

DEXTER, A. G., SLIFE, F. W. and BUTLER, H. S. (1971). *Weed Sci.*, **19**, 721.

DOWNER, G. B., HALL, M. and MALLEN, D. N. B. (1976). *J. Agric. Fd. Chem.*, **24**, 1223.

DUBLE, R. L., HOLT, E. C. and MCBEE, G. (1968). *Weed Sci.*, **16**, 421.

DUBOVOI, V. P., CHKANIKOV, D. I. and MAKEEV, A. (1973). *Fiz. Rast.*, **20**, 1261.

EASTIN, E. F. (1971). *Weed Sci.*, **19**, 261.

EASTIN, E. F. (1972). *Weed Res.*, **12**, 75.

EASTIN, E. F. (1976). *J. Chromatog.*, **124**, 422.

EDGERTON, L. J. and HOFFMAN, M. B. (1961). *Science*, **134**, 341.

ERCEGOVICH, C. D. and WITKONTON, S. (1972). *J. Agric. Fd. Chem.*, **20**, 344.

ESSER, H. O., DUPUS, G. EBERT, E. and VOGEL, C. (1976). In *Herbicides: Chemistry, Degradation and Mode of Action* (Eds. P. C. Kearney and D. D. Kaufman), Vol. 1, Marcel Dekker, Inc., New York, p. 129.

FANG, S. C. (1976). In *Herbicides: Chemistry, Degradation and Mode of Action* (Eds. P. C. Kearney and D. D. Kaufman), Vol. 1, Marcel Dekker, Inc.,New York, p. 323.

FANG, S. C., FREED, V. H., JOHNSON, R. H. and COFFEE, D. R. (1955). *J. Agric. Fd. Chem.*, **3**, 400.

FANG, S. C. and GEORGE, M.(1962). *Pl. Physiol., Lancaster*, **37**, XXVI (Suppl.).

FANG, S. C. and YU, T. C. (1959). *West. Weed Control Conf. Res. Prog. Rept.*, 91.

FAWCETT, C. H., INGRAM, J. M. A. and WAIN, R. L. (1954). *Proc. Roy. Soc. (London)*, **B142**, 60.

FAWCETT, C. H., PASCAL, R. M., PYBUS, M. B., TAYLOR, H. F., WAIN, R. L. and WIGHTMAN, F. (1959). *Proc. Roy. Soc. (London)*, **B150**, 95.

FEDTKE, C. and SCHMIDT, R. R. (1977). *Weed Res.*, **17**, 233.

FEUNG, C. S., HAMILTON, R. H. and MUMMA, R. O. (1973), *J. Agric. Fd. Chem.*, **21**, 637.

FEUNG, C. S., HAMILTON, R. H., WITHAM, F. H. and MUMMA, R. O. (1972). *Pl. Physiol., Lancaster*, **50**, 80.

FEUNG, C. S., MUMMA, R. O. and HAMILTON, R. H. (1974). *Pl. Physiol., Lancaster*, **53** (Suppl.) 41.

FLEEKER, J. R. (1973). *Phytochemistry*, **12**, 757.

FLEEKER, J. and STEEN, R. (1971). *Weed Sci.*, **19**, 507.

FOY, C. L. (1961). *Pl. Physiol., Lancaster*, **36**, 698.

FOY, C. L. (1967). *Pl. Physiol., Lancaster*, **42**, 51 (Suppl.).

FOY, C. L. (1969). In *Degradation of Herbicides* (Eds. P. C. Kearney and D. D. Kaufman), Marcel Dekker, Inc., New York, p. 207.

FOY, C. L. (1976). In *Herbicides: Chemistry, Degradation and Mode of Action* (Eds. P. C. Kearney and D. D. Kaufman), Marcel Dekker, Inc., New York, p. 788.

FRANK, R., SIRONS, G. J., PAIK, N. J. and VALK, M. (1977). *Can. J. Pl. Sci.*, **57**, 473.

FRANKOSKI, S. P. and SIGGIA, S. (1972). *Anal. Chem.*, **44**, 507.

FREAR, D. S. (1968). *Phytochemistry*, **7**, 381.

FREAR, D. S. (1976). In *Herbicides: Chemistry, Degradation and Mode of Action* (Eds. P. C. Kearney and D. D. Kaufman), Marcel Dekker, Inc., New York, p. 541.

FREAR, D. S. and STILL, G. G. (1968). *Phytochemistry*, **7**, 913.

FREAR, D. S. and SWANSON, H. R. (1970). *Phytochemistry*, **9**, 2123.

FREAR, D. S. and SWANSON, H. R. (1971). *Abstr. 162nd Mtg. Amer. Chem. Soc. Washington D.C.*

FREAR, D. S. and SWANSON, H. R. (1972). *Phytochemistry*, **11**, 1919.

FREAR, D. S. and SWANSON, H. R. (1973). *Pestic. Biochem. Physiol.*, **3**, 473.

FREAR, D. S., SWANSON, H. R. and TANAKA, F. S. (1972). *Recent Adv. Phytochem.*, **5**, 225.

FREDRICKS, J. F. and GENTILE, A. C. (1966). *Phytochemistry*, **4**, 851.

FREED, Y. H., MONTGOMERY, M. and KIEF, M. (1961). *Proc. N.E. Weed Control Conf.*, **15**, 6.

FUNDERBURK, H. H. and DAVIS, D. E. (1963). *Weeds*, **11**, 101.

FYKSE, H. (1976). *Weed Res.*, **16**, 309.

GARDINER, J. A. (1976). In *Herbicides: Chemistry, Degradation and Mode of Action* (Eds. P. C. Kearney and D. D. Kaufman), Vol. 1, Marcel Dekker, Inc., New York, p. 293.

GARDINER, J. A., RHODES, R. C., ADAMS, J. B. Jr. and SOBOCZENSKI, E. J. (1969). *J. Agric. Fed. Chem.*, **17**, 980.

GEISSBÜHLER, H., HASELBACH, C. and AEBI, H. (1963). *Weed Res.*, **3**, 140.

GEISSBÜHLER, H., MARTIN, H. and VOSS, G. (1976). In *Herbicides: Chemistry, Degradation and Mode of Action* (Eds. P. C. Kearney and D. D. Kaufman), Vol. 1, Marcel Dekker, Inc., New York, p. 209.

GEISSBÜHLER, H. and VOSS, G. (1971). In *Pesticide Terminal Residues* (Ed. A. S. Tahori), Suppl. *Pure Appl. Chem.*, Butterworths, London, p. 305.

GENTHER, W. A. (1969). *Weed Sci.*, **17**, 284.

GLAD, G., POPOFF, T. and THEANDER, O. (1977). In *Weeds and Weed Control, 18th Swed. Weed Conf., Uppsala*, Pt. 1, 1.

GOLAB, T., DAY, E. W. J. and PROBST, G. W. (1967). *153rd Mtg. Amer. Chem. Soc., Miami*, 45.

GOLAB, T., HERBERG, R. J., GRAMLICH, J. V., RAUN, A. P. and PROBST, G. W. (1970). *J. Agric. Fd. Chem.*, **18**, 838.

GORBACH, S. G., KUENZLER, K. and ASSHAUER, J. (1977). *J. Agric. Fd. Chem.*, **25**, 507.

GOSWAMI, K. P. (1972). *Diss. Abstr. Int. B.*, **33**, 2436.

GOSWAMI, K. P. and GREEN, R. E. (1974). *J. Agric. Fd. Chem.*, **22**, 340.

GRAY, R. A. (1969). *Weed Sci. Soc. Amer. Abstr.*, 174.

GREENHALGH, R. and COCHRANE, W. P. (1972). *J. Chromatog.*, **70**, 37.

GRIGOR'EVA, L. V., DUBOVOI, V. P., MAKEEN, A.M. and CHKANIKOV, D. I. (1975). *Khim. Sel'sk. Khoz.,* **13**, 617.

GROVER, R. (1976). *Weed Sci.*, **24**, 26.

GRUEBNER, P. (1971). *Arch. Pflschutz,***7**, 309.

GRYZLOVA, G. K. (1976). *Khim. Sel'sk. Khoz.*, **14**, 62.

GUPTA, R. K. and DUTTA, T. R. (1974). *Ind. J. Weed Sci.*, **6**, 48.

GYSIN, H. and KNÜSLI, E. (1960). In *Advances in Pest Control Research* (Ed. R. L. Metcalf), Interscience Publishers, New York, p. 289.

GZIK, A. and GRASER, H. (1972). *Biochem. Physiol. Pflanzen.*, **163**, 483.

HAGIN, R. D. and LINSCOTT, D. L. (1972). *Abstr. Mtg. WSSA St. Louis*, 65.

HALL, M. and MALLEN, D. N. B. (1976). *J. Chromat. Sci.*, **14**, 451.

HAMBLETON, L. G. (1973). *J. Assoc. Offic. Anal. Chem.*, **56**, 567.

HAMILTON, R. H. (1964). *J. Agric. Fd. Chem.*, **12**, 14.

HAMILTON, R. H. and MORELAND, D. E. (1962). *Science*, **135**, 373.

HAMMOND, H. (1973). *J. Assoc. Offic. Anal. Chem.*, **56**, 596.

HARGER, T. R., RIECK, C. E. and STANDIFORD, E. C. (1975). *Proc. 28th Ann. Mtg. S. Weed Sci. Soc.*, 295.

HARGREAVES, P. A. and RAPKINS, S. H. (1976). *Pestic. Sci.*, **7**, 515.

HATJULA, M. L. and RAISANEN, S. (1976). *Bull. Environ. Contam. Toxicol.*, **16**, 355.

HAWTON, D. and STOBBE, E. H. (1971a). *Weed Sci.*, **19**, 42.

HAWTON, D. and STOBBE, E. H. (1971b). *Weed Sci.*, **19**, 555.

HAZEMOTO, N., KAMO, N. and KOBATAKE, Y. (1976). *J. Assoc. Offic. Anal. Chem.*, **59**, 1097.

HEEDE, M. VAN DEN and HEYNDRICKS, A. (1976). Meded. Fakult. Landbouw. Gent, **41**, 1457.

HELLING, C. S. (1976). *Suppl. Proc. N.E. Weed Sci. Soc., Boston*, 44.

HERHOLDT, J. A. (1969). *Diss. Abstr. Int. B.*, **30**, 1978-B.

HOAGLAND, R. E., GRAF, G. and HANDEL, E. D. (1974). *Weed Res.*, **14**, 371.

HODGSON, R. H. (1971). *Weed Sci.*, **19**, 501.

HODGSON, R. H., DUSBADEK, K. E. and HOFFER, B.(1974). *Weed Sci.*, **22**, 205.

HODGSON, R. H., FREAR, D. S., SWANSON, H. R. and REGAN, L. A. (1973). *Weed Sci.*, **21**, 542.

HODGSON, R. H. and HOFFER, B. L. (1974). *Abstr. 1974 Mtg. Weed Sci. Soc. Amer.*, 17.

HOFBERG, A. H., HEINRICKS, L. C. and GENTRY, G. A. (1976). *J. Assoc. Offic. Anal. Chem.*, **59**, 758.

HOFBERG, A. H. and MURPHY, R. J. (1973). *J. Assoc. Offic. Anal. Chem.*, **56**, 586.

HOFFER, B. L. and HODGSON, R. H. (1978). *Abstr. 1978 Mtg. Weed Sci. Soc. Amer.*, 71.

HOLLY, K. (1968). *Proc. 9th Brit. Weed Control Conf.*, **9**, 1285.

HORMANN, W. D. and EBERLE, D. O. (1972). *Weed Res.,* **12**, 199.

HULL, R. J. and WEISENBERG, M. R. (1967). *Pl. Physiol., Lancaster,* **42**, 49 (Suppl.).

ISHIZUKA, K. and MITSUI, S. (1966). *Abstr. a. Mtg. Agric. Chem. Soc., Japan*, 62.

JACOBSOHN, R. and ANDERSON, R. N. (1972). *Weed Sci.*, **20**, 74.

JAMES, C. S. and PRENDEVILLE, G. N. (1969). *J. Agric. Fd. Chem.*, **17**, 1257.

JAWORSKI, E. G. (1964). *J. Agric. Fd. Chem.*, **12**, 33.

JEFFCOAT, B. and HARRIS, W. N. (1973). *Pestic. Sci.*, **4**, 891.

JEFFCOAT, B., HARRIS, W. N. and THOMAS, D. B. (1977). *Pestic. Sci.*, **8**, 1.

JENSEN, K. I. N. and STEPHENSON, G. R. (1974). In *Minutes, Canada Weed Committee Eastern Section, 30 Oct.–1st Nov. 1973*, Sainte-Foy, Quebec, Canada, Canada Weed Committee, 40.

JOHNSON, L. D., GERHARDT, K. O. and AUE, W. A. (1972). *Sci. Total Environ.*, **1**, 108.

JORDAN, L. S. and JOLLIFFE, V. A. (1973). *Pestic. Sci.*, **4**, 467.

JORDAN, L. S., ZURQUIYAH, A. A., LERX, W. A. and LEASCH, J. G. (1975). *Arch. Environ. Contam. Toxicol.*, **3**, 268.

JOSEPOVITS, G., DOBROVOLSZKY, A. and JENEY, K. (1976). In *The Development of a Pesticide as a Complex Scientific Task* (Ed. L. Barki), Medicina, Budapest, Hungary, p. 61.

KASSEBEER, VON, H. (1971). *Z. Pflkrankh. Pflpath. Pflschutz.*, **78**, 158.

KAUFMAN, D. D. (1976). In *Herbicides: Chemistry, Degradation and Mode of Action* (Eds. P. C. Kearney and D. D. Kaufman), Marcel Dekker, Inc., New York, p. 665.

KEARNEY, P. C. and KAUFMAN, D. D. (1976). *Herbicides: Chemistry, Degradation and Mode of Action*, Vols. I and II, 2nd edn., Marcel Dekker, Inc., New York, 1036 pp.

KEELEY, P. E. and THULLEN, R. J. (1971). *Weed Sci.*, **19**, 297.

KLAMBT, H. D. (1961). *Planta*, **57**, 339.

KLOTZ, F. K. and DUYSEN, M. E. (1972). *Pl. Physiol., Lancaster, Suppl.*, **49**, 63.

KOONS, J. R. and DAY, E. W. Jr. (1972). *J. Chromat. Sci.*, **10**, 176.

KRUMZDOROV, A. M. (1974). *Agrokhimiya*, **11**, 128.

KRUMZDOROV, A. M. (1975a). *Agrokhimiya*, **12**, 130.

KRUMZDOROV, A. M. (1975b). *Agrokhimiya*, **12**, 132.

KRUMZDOROV, A. M. (1976). *Sel'-Khoz. Biol.*, **11**, 383.

KUDELOVA, A. and BERGMANNOVA, E. (1975). *Ochrana. Rostlin*, **11**, 111.

KURATLE, H., RAHN, E. M. and WOODMANSEE, C. W. (1969). *Weed Sci.*, **17**, 216.

LAMOUREAUX, G. L., SHIMABUKURO, R. H., SWANSON, H. R. and FREAR, D. S. (1970). *J. Agric. Fd. Chem.*, **18**, 81.

LAMOUREAUX, G. L., STAFFORD, L. E. and SHIMABUKURO, R. H. (1972). *J. Agric. Fd. Chem.*, **20**, 1004.

LAMOUREAUX, G. L., STAFFORD, L. E., SHIMABUKURO, R. H. and ZAYLSKIE, R. G. (1973). *J. Agric. Fd. Chem.*, **21**, 1020.

LAMOUREAUX, G. L., STAFFORD, L. E. and TANAKA, F. S. (1971). *J. Agric. Fd. Chem.*, **19**, 346.

LAWRENCE, J. F. (1976). *J. Chromat. Sci.*, **14**, 557.

LEACH, R. W. A., BIDDINGTON, N. L., VERLOOP, A. and NIMMO, W. B. (1971). *Ann. Appl. Biol.*, **67**, 137.

LEAFE, E. L. (1962). *Nature (London)*, **193**, 485.

LEE, I. N. and ISHIZUKA, K. (1976). *Arch. Environ. Contam. Toxicol.*, **4**, 161.

LEE, S. S. and FANG, S. C. (1973). *Weed Res.*, **13**, 59.

LEMIN, A. J. (1966). *J. Agric. Fd. Chem.*, **14**, 109.

LINSCOTT, D. L., HAGIN, R. D. and DAWSON, J. E. (1968). *J. Agric. Fd. Chem.*, **16**, 844.

LIU, L. C., SHIMABUKURO, R. H. and NALEWAJA, J. D. (1978). *Weed Sci.*, **26**, 642.

LOCKE, R. K. and BARON, R. L. (1972). *J. Agric. Fd. Chem.*, **20**, 861.

LONG, J. W., THOMPSON, L. JR. and RIECK, C. E. (1974). *Weed Sci.*, **22**, 91.

LOOS, M. A. (1976). In *Herbicides: Chemistry, Degradation and Mode of Action* (Eds. P. C. Kearney and D. D. Kaufman), Vol. 1, Marcel Dekker, Inc., New York, p. 1.

LUCKWILL, L. C. and LLOYD-JONES, C. P. (1960a). *Ann. Appl. Biol.*, **48**, 613.

LUCKWILL, L. C. and LLOYD-JONES, C. P. (1960b). *Ann. Appl. Biol.*, **48**, 626.

LUND-HØIE, K. (1970). *Weed Res.*, **10**, 367.

MCLEOD, H. A. and WALES, P. J. (1972). *J. Agric. Fd. Chem.*, **20**, 624.

MCRAE, D. H., YIH, R. Y. and WILSON, H. F. (1964). *Weed Sci. Soc. Amer. Abstr.*, 87.

MAGALHAES, A. C., ASHTON, F. M. and FOY, C. L. (1968). *Weed Sci.*, **16**, 240.

MATSUNAKA, S. (1969). *Residue Rev.*, **25**, 45.

MATSUNAKA, S. (1976). In *Herbicides: Chemistry, Degradation and Mode of Action* (Eds. P. C. Kearney and D. D. Kaufman), Vol. 2, Marcel Dekker, Inc., New York, p. 710.

MINE, A. (1974). *3rd Int. Cong. Pestic. Chem. IUPAC., Helsinki*, 172.

MOSINSKA, K. and KOTARSKI, A. (1972). *Chemia Analityezna*, **17**, 327.

MUMMA, R. O. and HAMILTON, R. H. (1975). In *Bound and Conjugated Pesticide Residues* (Eds. D. D. Kaufman, G. G. Still, G. D. Paulson and S. K. Bandal), Amer. Chem. Soc. Symp. Ser., No. 29, 68.

MUNRO, H. E. (1972). *Pestic. Sci.*, **3**, 371.

NALEWAJA, J. D., BEHRENS, R. and SCHMID, A. R. (1964). *Weeds*, **12**, 269.

NASHED, R. B. and ILNICKI, R. D. (1970). *Weed Sci.*, **18**, 25.

NASHED, R. B., KATZ, S. E. and ILNICKI, R. D. (1970). *Weed Sci.*, **18**, 122.

NAYLOR, A. W. (1976). In *Herbicides: Physiology, Biochemistry, Ecology* (Ed. L. J. Audus), Academic Press, London and New York, p. 397.

NAZAROVA, T. A., PAVOLVA, N. N., MAKEEV, A. M. and CHKANIKOV, D. I. (1978). *Khim. Sel'sk. Khoz.*, **16**, 34.

NIKI, Y., KUWATSUKA, S. and YOKOMICHI, I. (1976). *Agric. Biol. Chem.*, **40**, 683.

ONLEY, J. H., YIP, G. and ALDRIDGE, M. H. (1968). *J. Agric. Fd. Chem.*, **16**, 426.

PARKE, D. V. and WILLIAMS, R. T. (1969). *Brit. Med. Bull.*, **25**, 256.

PILLMOOR, J. B. and GAUNT, J. K. (1981). In *Progress in Pesticide Biochemistry* (Eds. D. H. Hutson and T. R. Roberts), **1**, 147.

PONT. V., JARCZYK, H. J., COLLET, G. F. and THOMAS, R. (1974). *Phytochemistry*, **13**, 785.

PRENDEVILLE, G. N., ESHEL, Y., JAMES, C. S., WARREN, G. T. and SCHREIBER, M. M. (1968). *Weed Sci.*, **16**, 432.

PROBST, G. W., GOLAB, T., HERBERG, R. J., HOLZER, F. J., PARKA, S. J., VAN DER SCHANS, C. and TEPE, J. B. (1967). *J. Agric. Fd. Chem.*, **15**, 592.

PROBST, G. W., GOLAB, T. and WRIGHT, W. L. (1976). In *Herbicides: Chemistry, Degradation and Mode of Action* (Eds. P. C. Kearney and D. D. Kaufman), Vol. II, Marcel Dekker, Inc., New York, p. 453.

PROBST, G. W. and TEPE, J. B. (1969). In *Degradation of Herbicides* (Eds. P. C. Kearney and D. D. Kaufman), Marcel Dekker, Inc., New York, pp. 255–82.

PURKAYASTHA, R. and COCHRANE, W. P. (1973). *J. Agric. Fd. Chem.*, **21**, 93.

PUTNAM, A. R. (1976). *Weed Sci.*, **24**, 425.

QUE HEE, S. S. and SUTHERLAND, R. G. (1975). *Weed Sci.*, **23**, 119.

QUIMBY, P. C. and NALEWAJA, J. D. (1961). *Weed Sci.*, **19**, 598.

RACUSEN, D. (1958). *Arch. Biochem. Biophys.*, **74**, 106.

RAKITIN, Y. V., ZEMSKAYA, V. A., VORONINA, E. I. and CHERNIKOVA, L. M. (1966). *Sov. Pl. Physiol.*, **13**, 30.

RAY, B. R. and WILCOX, M. (1967). *154th Mtg. Amer. Chem. Soc. Chicago, Abstr.*, A-28.

RAY, B. R. and WILCOX, M. (1969). *Physiol. Pl.*, **22**, 503.

RAY, T. B. and STILL, C. C. (1975). *Pestic. Biochem. Physiol.*, **5**, 171.

RICE, R. P. and PUTNAM, A. R. (1980). *Weed Sci.*, **28**, 176.

RIDEN, J. R. and HOPKINS, T. R. (1961). *J. Agric. Fd. Chem.*, **9**, 47.

RIDEN, J. R. and HOPKINS, T. R. (1962). *J. Agric. Fd. Chem.*, **10**, 455.

RIECK, C. E., MAKSYMOWICZ, W., GOSSETTI, B. J. and CAMPER, N. D. (1971). *Proc. 24th Ann. Mtg. S. Weed Sci. Soc.*, 363.

ROBINSON, D. E. (1974). *Diss. Abstr. Int. B*, **34**, 3590.

ROBOCKER, W. C. and ZAMORA, B. A. (1976). *Weed Sci.*, **24**, 435.

RODER, C. H. and LAASS, H. (1976). *Nach. Deuts. PflSchutz.*, **28**, 170.

ROGERS, R. L. (1969). *Abstr. Weed Sci. Soc. Amer.*, No. 171.

ROGERS, R. L. (1971). *J. Agric. Fd. Chem.*, **19**, 32.

ROGERS, R. L. and FUNDERBURK, H. H. (1967). *J. Agric. Fd. Chem.*, **15**, 577.

ROTH, W. (1957). *Compt. Rend.*, **245**, 942.

ROTH, W. and KNÜSLI, E. (1961). *Experientia*, **17**, 312.

RUEPPEL, M. L., BRIGHTWELL, B. B., SCHAEFER, J. and MARVEL, J. T. (1977). *J. Agric. Fd. Chem.*, **25**, 517.

RUEPPEL, M. L., SUBA, L. A. and MARVEL, J. T. (1976). *Biomed. Mass. Spectr.*, **3**, 28.

RUSNESS, D. G. and STILL, G. G. (1975). *Abstr. 170th Nat. Mtg. Amer. Chem. Soc. PEST*, 21.

SACHS, R. M. and MICHAEL, J. L. (1971). *Weed Sci.*, **19**, 558.

SANDBERG, C. L., MEGGETT, W. F. and PENNER, D. (1980). *Weed Res.*, **20**, 195.

SCHAFER, D. E. and CHILCOTE, D. O. (1970). *Weed Sci.*, **18**, 725.

SCHLAGBAUER, B. G. L. and SCHLAGBAUER, A. W. J. (1972a). *Residue Rev.*, **42**, 1.

SCHLAGBAUER, B. G. L. and SCHLAGBAUER, A. W. J. (1972b). *Residue Rev.*, **42**, 85.

SCHROEDER, R. S., PATEL, N. R., HEDRICH, L. W., DOYLE, W. C., RIDEN, J. R. and PHILLIPS, L. V. (1972). *J. Agric. Fd. Chem.*, **20**, 1286.

SCHULTZ, D. P. and TWEEDY, B. G. (1971). *Weed Sci.*, **19**, 133.

SCHUPHAN, I. and EBING, W. (1977). *Weed Res.*, **17**, 181.

SCKERL, M. M. and FRANS, R. E. (1969). *Weed Sci.*, **17**, 421.

SHIMABUKURO, R. H. (1967). *J. Agric. Fd. Chem.*, **15**, 557.

SHIMABUKURO, R. H. (1975). *Environ. Quality and Safety*, **4**, 140.

SHIMABUKURO, R. H. (1976). *Proc. 5th Asian-Pacific Weed Sci. Soc. Conf. Tokyo, Japan, 1975*, 183.

SHIMABUKURO, R. H., FREAR, D. S., SWANSON, H. R. and WALSH, W. C. (1971). *Pl. Physiol., Lancaster*, **47**, 10.

SHIMABUKURO, R. H., KADUNCE, R. E. and FREAR, D. S. (1966). *J. Agric. Fd. Chem.*, **14**, 392.

SHIMABUKURO, R. H. and SWANSON, H. R. (1969). *J. Agric. Fd. Chem.*, **17**, 199.

SHIMABUKURO, R. H., SWANSON, H. R. and WALSH, W.C. (1970). *Pl. Physiol., Lancaster*, **46**, 103.

SHIMABUKURO, R. H., WALSH, W. C., LAMOUREAUX, G. L. and STAFFORD, L. E. (1973). *J. Agric. Fd. Chem.*, **21**, 1031.

SHRIVER, J. W. (1972). *Diss. Abstr. Int. B*, **33**, 1338.

SHRIVER, J. W. and BINGHAM, S. W. (1973). *Weed Sci.*, **21**, 212.

SILBER, J., and PREVITALI, SILBERA, N. (1976). *J. Agric. Fd. Chem.*, **24**, 679.

SLIFE, F. W., KEY, J. L., YAMAGUCHI, S. and CRAFTS, A. S. (1962). *Weeds*, **10**, 29.

SMITH, A. E. and OSWALD, T. H. (1978). *Abstr. 1978 Mtg. WSSA*, 67.

SMITH, F. W. and SHEETS, T. J. (1967). *J. Agric. Fd. Chem.*, **15**, 577.

SMITH, G. N. and DYER, D. L. (1961). *J. Agric. Fd. Chem.*, **9**, 155.

SMITH, L. W., BAYER, D. E. and FOY, C. L. (1969). *Weed Sci.*, **16**, 523.

SMITH, L. W. and CHANG, F. W. (1973). *Weed Res.*, **13**, 339.

SPARACINO, C. M. and HINES, J. W. (1976). *J. Chromat. Sci.*, **14**, 549.

STILL, G. G. (1968a). *Pl. Physiol., Lancaster*, **43**, 543.

STILL, G. G. (1968b). *Science*, **159**, 992.

STILL, G. G. and HERRETT, R. A. (1976). In *Herbicides: Chemistry, Degradation and Mode of Action* (Eds. P. C. Kearney and D. D. Kaufman). Marcel Dekker, Inc., New York, p. 541.

STILL, G. G. and KUZIRIAN, O. (1967). *Nature (London)*, **216**, 799.

STILL, G. G. and MANSAGER, E. R. (1971). *J. Agric. Fd. Chem.*, **19**, 879.

STILL, G. G. and MANSAGER, E. R. (1972a). *Phytochemistry*, **11**, 515.

STILL, G. G. and MANSAGER, E. R. (1972b). *J. Agric. Fd. Chem.*, **20**, 402.

STILL, G. G. and MANSAGER, E. R. (1973a). *J. Agric. Fd. Chem.*, **21**, 787.

STILL, G. G. and MANSAGER, E. R. (1973b). *Phytochemistry*, **11**, 575.

STILL, G. G. and MANSAGER, E. R. (1974). *Abstr. 168th Nat. Mtg. Amer. Chem. Soc. PEST*, 1.

STILL, G. G. and MANSAGER, E. R. (1975). *Pestic. Biochem. Physiol.*, **5**, 515.

STOLLER, E. W. (1968). *Weed Sci.*, **16**, 384.

STOLLER, E. W. (1969). *Pl. Physiol., Lancaster*, **44**, 854.

STOLLER, E. W. and WAX, L. M. (1968). *Weed Sci.*, **16**, 283.

SUPIN, G. S., VAINTRAUB, F. P. and MAKAROVA, C. V. (1971). *Gigiena i Sanitatiya*, **5**, 61.

SUZUKI, K., NAGAYOSHI, H. and KASHIWA, T. (1976). *Agric. Biol. Chem.*, **40**, 845.

SWANSON, C. R., HODGSON, R. H., KADUNCE, R. E. and SWANSON, H. R. (1966). *Weeds*, **14**, 323.

SWANSON, C. R. and SWANSON, H. R. (1968). *Weed Sci.*, **16**, 137.

SYNERHOLM, M. E. and ZIMMERMAN, P. W. (1947). *Contrib. Boyce Thompson Inst.*, **14**, 369.

THOMAS, E. W., LOUGHMAN, B. C. and POWELL, R. G. (1963). *Nature (London)*, **199**, 73.

THOMAS, E. W., LOUGHMAN, B. C. and POWELL, R. G. (1964). *Nature (London)*, **204**, 286.

THOMPSON, J., SLIFE, F. W. and BUTLER, H. S. (1970). *Weed Sci.*, **18**, 509.

THOMPSON, L., JR. (1972a). *Weed Sci.*, **20**, 153.

THOMPSON, L., JR. (1972b). *Weed Sci.*, **20**, 584.

TSAI, W. F. (1974). *Diss. Abstr. Int. B*, **35**, 1478.

TUINSTRA, L. G. M. T., ROOS, A. H. and BRONSGEEST, M. (1976). *Meded. Fakult. Landb. Wet. Gent*, **41**, 433.

UGULAVA, N. A. and UGREKHELIDZE, D. SH. (1973). *Soobs, Akad. Nauk Gruz. SSR*, **70**, 205.

VERLOOP, A. (1972). *Residue Rev.*, **43**, 55.

VERLOOP, A. and NIMMO, W. B. (1969). *Weed Res.*, **9**, 357.

VERLOOP, A. and NIMMO, W. B. (1970). *Weed Res.*, **10**, 59.

VOSS, G. and GEISSBÜHLER, H. (1971). *2nd Int. IUPAC Congr. Pestic. Chem.*, **4**, 525.

VOSS, G. and GEISSBÜHLER, H. (1971). *2nd Int. IUPAC Congr. Pestic. Chem.*, **4**, 525.

WAHLROOS, OE. and VIRTANEN, A. I. (1959). *Acta Chem. Scand.*, **13**, 1906.

WAIN, R. L. and SMITH, M. S. (1976). In *Herbicides: Physiology, Biochemistry, Ecology* (Ed. L. J. Audus), Vol. II, Academic Press, London and New York, p. 279.

WATHANA, S. and CORBIN, F. T. (1972). *J. Agric. Fd. Chem.*, **20**, 23.

WEINTRAUB, R. L., REINHART, J. H. and SCHERFF, R. A. (1956). In *A Conference on Radioactive Isotopes in Agriculture*, A.E.C. Ref. No. TID-7512, p. 203.

WILCOX, M., MORELAND, D. E. and KLINGMAN, G. C. (1963). *Physiol. Pl.*, **16**, 565.

WILLARD, J. L. and PENNER, D. (1976). *Residue Rev.*, **64**, 67.

WILLIAMS, J. H. B. (1976). *Studies on the Metabolism of MCPA in Plants*, Ph.D. Thesis, Univ. Wales.

WOOLSON, E. A. (1976). In *Herbicides: Chemistry, Degradation and Mode of Action* (Eds. P. C. Kearney and D. D. Kaufman), Vol. 1, Marcel Dekker, Inc., New York, p. 741.

WRIGHT, T. H., RIECK, C. E. and HARGER, T. R. (1974). *Proc. 27th Ann. Mtg. Sci. Soc.*, 347.

WYRILL, J. B. and BURNSIDE, O. C. (1976). *Weed Sci.*, **24**, 557.

YIH, R. Y., MCRAE, D. H. and WILSON, H. F. (1968a). *Science*, **161**, 376.

YIH, R. Y., MCRAE, D. H. and WILSON, H. F. (1968b). *Pl. Physiol., Lancaster*, **43**, 1291.

YIP, G. (1972). *J. Assoc. Offic. Anal. Chem.*, **55**, 287.

ZAKI, M. A., TAYLOR, H. F. and WAIN, R. L. (1967). *Ann. Appl. Biol.*, **59**, 481.

CHAPTER SEVEN

Herbicides and the Environment

Legislation and the use of herbicides

Conditions governing the registration of pesticides (including herbicides) vary from country to country but most highly industrialised countries lay down stringent conditions that must be met before a particular pesticide can be marketed. In this chapter it is possible to outline the legislation in only a few countries. It should not be assumed that conditions governing the registration and use of pesticides are any less stringent in countries which have not been mentioned. Indeed they may be even more stringent in some cases. The examples that are given here are merely to illustrate how large a part concern for the environment plays in the considerations of those who produce, those who legislate for and those who use pesticides. Too often a picture is painted of pesticides being sprayed indiscriminately around the countryside with little or no control being exercised. This is a travesty of the truth. Readers requiring information on the registration and control of use of pesticides in a particular country should apply to the appropriate Government Department of the country concerned.

Hahn (1972), Makepeace (1972), Wright (1973) and Anon (1973), have reviewed the regulatory schemes in the countries of Western Europe noting that there are widely divergent schemes in the various countries. Although it might be desirable that one scheme should cover all of the countries there are political, agricultural and historical reasons that make this goal a fairly distant one. How a country operates its safety scheme is not however entirely its own affair. International trade means that food grown in one country will have to meet the requirements of the countries to which it exports its goods. It is not surprising then that although the mechanism of approval may differ, the fundamental requirements that have to be met do not differ radically from one country to another. In all countries the burden of proof of a pesticide's safety (and efficacy) lies with the manufacturer who accepts that it is the government's duty to regulate its handling and use.

EUROPE

All European countries, with the exception of the U.K., *prohibit by law* the sale of any pesticide unless it has been registered for the specific uses described on the manufacturer's instruction label. The U.K. operates a voluntary scheme which will be dealt with in some detail later in this chapter.

In the Federal Republic of Germany the applicant must answer seventy detailed questions; in the Netherlands fifteen questions; but both cover the same broad headings of composition, analytical methods, toxicity of the active agent and its formulation and residue data. The questions are searching and each may require an enormous amount of investigative work, e.g. behaviour in soil is one question but it might need 100 pages of data to provide the answer. The responsibility for registration varies from country to country, e.g. in Sweden authority rests with the National Poisons and Pesticides Board but the registration of the more poisonous Class I pesticides is issued by the National Board of Health and Social Welfare. In Italy the Ministry of Health bears the responsibility, while in France and Germany two committees deal with toxicology and biology. France is unique in allowing certain industrial representatives to sit on these two committees. Special rules govern information to be shown on labels – the formula must be shown together with recommendations to doctors in case of accident and in the Benelux countries, in addition, poisoning symptoms and first aid measures must also be listed (Wright, 1973).

Labels must also state the minimum interval in days or weeks which must elapse between the last spraying and harvest. Field residue studies are therefore necessary in each country to support registration as the majority of Western European countries have fixed statutory tolerances.

Environmental pollution plays an ever-increasing role in assessing safety. In Denmark and the Netherlands the law imposes strict control over contamination by dumping, disposal of packages and cleaning spray equipment in water catchment areas. In France the legal minimum distance between a spray operation and a source of water supply is specified.

UNITED KINGDOM

In the U.K. the Pesticides Safety Precautions Scheme – non statutory – (formerly entitled the Notification of Pesticides Scheme) was drawn up in negotiation between the Ministry of Agriculture, Fisheries and Food, the then Ministry of Health, the corresponding Scottish Departments and the Industrial Associations concerned. It came into effect in 1957. In 1965 it was revised and in 1970 it was extended to cover Northern Ireland and its scope widened to include the use of pesticides and related chemicals in forestry as well as in agriculture, horticulture, home-gardening and food storage. In 1972 a new category of use, animal husbandry, was added to the Scheme covering certain general animal hygiene products and ectoparasiticides not for direct administration and therefore not covered by the Medicine Act of 1968. In 1975 the Scheme was further extended to encompass certain 'non-agricultural' uses of pesticides not previously covered. The Scheme is described in considerable detail by Jones *et al.* (1977). Its purpose is to safeguard human beings (whether they are users of the chemicals, consumers of treated produce, users of treated products or other members of the public), livestock, domestic animals and wildlife (including bees, fish, mammals and birds) against risks arising from the use of pesticides. Under the Scheme, distributors (i.e. manufacturers, importers, formulators, servicing companies, etc.) proposing to introduce new pesticides and new uses for pesticides in the U.K. are required:

1. to notify such new pesticides to Departments
2. to ascertain and disclose such information as may be required by the Departments to enable them to advise on precautionary measures
3. not to introduce such products until agreement has been reached on appropriate precautionary measures
4. to include the agreed precautions and the British Standards Institute common name (or the chemical name) of the active ingredient on the label
5. to notify any substantial change in the text or layout of the label
6. to withdraw a product if recommended to do so by the Departments

The Scheme applies to all active ingredients formulated as pesticides, i.e. insecticides, fungicides, herbicides, rodenticides and similar substances. Full details of the Scheme may be found in a pamphlet drawn up by the Ministry of Agriculture, Fisheries and Food (1975). Included also is a Toxicity Data Guide offered as a guide to notifiers who wish to present adequate data on a product's toxicity. The properties of a chemical that may need to be investigated include:

1. The physicochemical properties which would influence risks in the field or persistence as a residue.

2. Acute toxicity; LD_{50} values by single doses and apparent mode of toxic action.
3. Skin penetration and absorption; percutaneous toxicity; irritancy of liquid or vapour to body surface.
4. Cumulative effects of known functions of LD_{50} values over short periods representative of user exposures.
5. Effects arising from prolonged exposure, chronic toxicity
6. Delayed effects, arising usually after a silent development period.
7. Metabolic studies.
8. Potentiation of, or by, other toxic chemicals under special circumstances.
9. Diagnostic and therapeutic possibilities.

The single dose toxicity is established with an approximate LD_{50} for common laboratory species, e.g. rats and mice, and tests are also made on other species including at least one non-rodent species. The time course and characteristics of the poisoning with behavioural changes and pathological changes at post mortem and the mode of action, e.g. effect on nervous system, liver damage, etc. are noted. Oral and percutaneous routes are always used comparing the purest ingredients with the formulated products and where the pesticide is volatile inhalation tests are carried out. Toxicity is established by subcutaneous, intraperitoneal or intravenous injections. Regarding cumulative toxicity, doses are given daily, 5 days a week for 2–4 weeks. Chronic toxicity tests are normally carried on for at least 6 months including checks on carcinogenicity. Metabolic studies include rate of breakdown in the tissues and the nature of degradation products. As well as laboratory findings additional information is required such as close attention to any effects on users during field trials; also on the likely scale of use, the crops to be treated and residues at the time of harvest.

Although this Scheme is a voluntary one there are a number of laws which can be invoked should the occasion arise. It is not proposed to discuss any of these in detail but adequate summaries may be found in Jones *et al.* (1977). They include the Health and Safety at Work, etc. Act 1974 and the Health and Safety (Agriculture) (Poisonous Substances) Regulations, the Pharmacy and Poisons Act 1933, the Rivers (Prevention of Pollution) Acts 1951 and 1961, the Deposit of Poisonous Waste Act 1972 and the Control of Pollution Act 1974.

The Agricultural Chemicals Approval Scheme is also a voluntary Scheme under which proprietary brands of agricultural chemicals can be officially approved. The purpose of this Scheme is to enable users to select and advisers to recommend, efficient and appropriate crop protection chemicals (including herbicides) and to discourage the use of unsatisfactory products. The Scheme is operated on behalf of the Government Agricultural Departments of the United Kingdom by the Agricultural Chemicals Approval Organisation. Approval is granted for specific uses under U.K. conditions when the Organisation is satisfied that the product fulfils the claims made on the label. Approval cannot be given to a product containing a new chemical or to a new use for an existing chemical until it has first been considered and cleared under the Pesticides Safety Precautions Scheme. There are proposals to merge the two Schemes (Royal Commission on Environmental Pollution Report September 1979).

The Ministry of Agriculture, Fisheries and Food publish annually (generally in February) a booklet *List of Approved Products and Their Uses for Farmers and Growers* which includes herbicides and growth regulators as well as insecticides, fungicides and molluscicides. The British Agrochemicals Association (1975) have published a very informative leaflet which summarises the stages in the discovery and development of a new pesticide at a typical manufacturer's Research Station. In 1975 the cost was around £3m spent over a period of 7 years (it is now nearer £10m). The leaflet also describes in some detail the work of the ecologists (employed by the manufacturing company) whose job it is to concern themselves with the safety to wildlife and the environment. The safety work alone necessary to satisfy the manufacturers and to obtain government clearance and registration can cost £500 000 or more. An example of the work on direct effects on beneficial animals is toxicity testing on honey bees. In the laboratory, batches of bees are fed on honey with different strengths of pesticide to see if they are affected. They are also sprayed with the pesticide and any effects noted.

In order to measure possible accumulation in food chains a model ecosystem is created with plants, caterpillars, insects and small animals, and fish, and radiolabelled pesticides are used to measure possible accumulation. Field studies are carried out on six metre square plots covering different soil types, in different parts of the country, to determine the effect of high and low concentrations of the pesticide on the soil microflora, soil microarthropods and earthworms using a variety of techniques. These tests may be continued over a number of years.

Field tests are also carried out on wild birds using census methods.

UNITED STATES

In the United States the Environmental Protection Agency (E.P.A.)

was established in December 1972. Turin (1976) has described its history. Federal involvement with pesticides began with the Insecticide Act of 1910 which prevented fraudulent efficacy claims and authorised the seizure and banning of compounds dangerous to human health. The Federal Insecticide, Fungicide and Rodenticide Act of 1947 (F.I.F.R.A.) broadened the scope of products regulated and established a product registration procedure which prohibited the distribution and sale of chemicals in inter-state commerce, unless shown to be 'safe' when used as directed and effective for the purpose shown on the label. Further amendments were made in 1959, 1961 and 1964. Growing awareness of residues and environmental problems led to the creation of E.P.A. and the transferring to it the authority to administer F.I.F.R.A. and to establish tolerance levels for pesticide residues in food and food products. F.I.F.R.A. was thus changed from a labelling law into a comprehensive regulatory Act to control more fully the manufacture, distribution and use of pesticides. The law was further amended in 1975 to include the Secretary of the Department of Agriculture in the review process of the major pesticide decisions affecting agriculture.

The E.P.A.'s Office of Pesticide Programs (*Pesticide Review*, 1976) has a broad range of responsibilities including development of strategic plans for controlling the adverse effects of pesticides; registering all pesticide products under the Federal Insecticide, Fungicide and Rodenticide Act (F.I.F.R.A.); establishing tolerance levels for pesticide residues occurring in food and feed; monitoring pesticide residue levels in humans, food and non-target fish and wildlife and their environments. E.P.A. is also responsible for reviewing environmental impact statements concerning pesticide use, and developing policies and regulations which will lead to a more judicious and environmentally acceptable use of pesticides. The E.P.A. is a very active body and reports of its various activities appear from time to time in the *Pesticide Review*.

The E.P.A. has the authority to proceed against persons who engage in misusing pesticides by applying them in a manner inconsistent with labelling and as described by Evrard (1974), it classifies pesticides into two categories: (a) General – can be used by the general public and (b) Restricted Use – can be used by certified applicators only, due to possible environmental and human hazards. The certification of individuals for the application of pesticides is done by the State but only after the certification plan has been submitted to and approved by the E.P.A.

The guidelines for the registration of pesticides in the U.S.A. are constantly undergoing revision to meet ever-changing situations. The latest issued by the E.P.A. (1978) is a considerable document which it is not possible to summarise in a book of this kind. It should however be required reading for all environmentalists who believe that the registration and permission to use pesticides is easily attained. The requirements are stringent and the scientific tests demanding.

CANADA

In Canada registration of a pesticide is governed by the Pest Control Products Act of 1969. The Act has four major clauses (Kobylnyk, 1974).

1. No person shall manufacture, store, display, distribute or use any control product under unsafe conditions related to human health or environmental quality.
2. No person may package, label or advertise any control product which could be deceptive in character, value, quantity, composition, merit or safety.
3. No person shall sell or import any control product unless it is registered.
4. No person shall export or convey from province to province any control product which is not made in a registered manufacturing establishment.

Data relating to a wide range of tests covering chemical and physical properties, efficacy, toxicology, residues, environmental effects and analytical methods have to be submitted to the appropriate government agencies for evaluation. Three marketing and use classes have been developed in order to direct products to persons capable of using them without undue occupational, contamination or environmental risk. The classes are:

1. *Restricted Class* – this is a small class of control products which require specialised knowledge and equipment. Included are products used over wide areas which if not adequately managed could endanger public health or other biological resources.
2. *Commercial Class* – this is the largest class of control products. Included are products for agricultural, industrial or other commercial uses.
3. *Domestic Class* – this class includes control products destined for non-commercial uses in and around the home. All control products are labelled on the basis of four types of risk which could be encountered during handling and use: (a) toxicity, (b) flammability, (c) explosiveness, (d) corrosiveness.

Safety of pesticides

In view of the strict conditions governing the registration of pesticides it is perhaps not surprising that few casualties result from their use. According to figures issued by the Health and Safety Executive and quoted in the Annual Report of the British Agrochemicals Association for 1979/80 there were 101 fatal accidents and diseases in agriculture in 1975, 108 in 1976, 105 in 1977, 73 in 1978 and 94 in 1979. Over the same period the corresponding figures for fatal accidents due to pesticides were zero in every case. For non-fatal accidents and diseases in agriculture the totals from all causes were 5273 in 1975, 5272 in 1976, 4835 in 1977, 4606 in 1978 and 4085 in 1979. The corresponding figures for pesticides over the same period were 18, 4, 12, 15, 7.

The safety record of pesticides used in forestry is even better. There were no fatalities and no non-fatal accidents recorded in 1979. In the garden according to the Home Accident Surveillance System for 1978 there were a total of 60 534 accidents investigated. Of these thirteen involved herbicides (compared with 209 involving the lawn mower) and these could have been avoided if the advice given on the labels had been followed.

PARAQUAT

Paraquat is potentially a very dangerous chemical but only at high doses. According to Calderbank (1968) application of the bipyridylium herbicides (paraquat and diquat) from normal spraying equipment can be accomplished with a minimum of hazard and without elaborate precautions. Consumer hazards are also negligible because residues in food crops are non-existent or very small, except in some desiccation outlets. Furthermore, when given orally, the chemicals are rapidly and completely excreted and there is no accumulation in the animal body. Other safety factors operating are the low levels required for herbicidal effect and their immediate inactivation in the soil.

There have however been a number of fatalities, numbered in hundreds, through people drinking paraquat accidentally or intentionally. According to Calderbank (*loc. cit.*) oral dosing of animals with high levels of paraquat first results in inflammation of the mouth and throat, difficulty in swallowing and reluctance to eat. This is followed by vomiting and diarrhoea some days later, breathing becomes laboured and cyanosis develops. Death may ultimately result from asphyxia. The lungs are affected showing pulmonary congestion and oedema, followed by proliferation of alveolar and bronchiolar lining cells and accumulation of macrophages. This is followed by increasing pulmonary fibrosis causing interference with gas exchange (Clark *et al.*, 1966). According to Bullivant (1966) the pathological features in humans are similar to those described above for laboratory animals. There is no specific cure for paraquat poisoning. Paraquat may also affect the nails which in severe cases may crack at the base and come off. Concentrated paraquat may cause splash injury to the eye leading to inflammation and irritation of the conjunctiva, leading to the shedding of epithelial cells from the conjunctiva and cornea, but the injury is superficial and if treated recovery should be complete (Calderbank, 1968).

HERBICIDES AND CANCER

The question of herbicides causing cancer or having long-term adverse effects is always a very difficult one to answer. How do you determine accurately the cause of an illness whose symptoms are revealed years after the alleged exposure? Presumably only by building up enough case evidence and by feeding experiments on test animals and in the case of herbicides it must be said that this evidence is not so far forthcoming. Cancers may be caused in test animals but only by using massive doses that bear no relationship to the quantities used in the field. Lewert (1976) in a closely argued defence of pesticides quotes the *Sixth Annual Report of the Council on Environmental Quality 1975*. Both mortality data and surveys on incidence conducted by the National Cancer Institute, U.S.A. indicate three trends within the overall increase – a rise in cancer for white and black males, an approximately steady rate for black females and a decline for white females. Cancer incidence has remained relatively uniform for some body sites since 1930 but has changed markedly for others. The most dramatic changes are the nearly twentyfold increase in lung cancer for males and a two-thirds decrease of stomach cancer in both males and females. The increase in lung cancer is primarily attributed to heavier smoking, coupled with impacts from other environmental pollutants; the stomach cancer decline is understood less, but it may be linked to diet. Lewert makes the point that diet which consists largely of pesticide-protected foods may be linked to a *decrease* in cancer over the past forty-odd years. Furthermore, he states, if pesticides were a cancer threat we should expect to see the greatest incidence of the disease down on the farm where pesticide users are regularly exposed and where run-off from fields may enter the waterways. This is not the case. Lastly, remarking that pesticides have had their greatest increase within the past 25 years he notes that during the same period the life expectancy of Americans has increased enormously from 63 in 1940 to 72 in 1974.

2,4,5-T, DIOXIN AND CANCER

2,4,5-T (2,4,5-trichlorophenoxyacetic acid) is widely used as a brushwood killer (see chapter 1). Many allegations have been made that 2,4,5-T, or rather an impurity – TCDD (2,3,7,8-tetrachlorodibenzo-*p*-dioxin) or dioxin (as it is more commonly called), can cause cancer, increase birth defects and induce nervous disorders. 2,4,5-T itself has a very low order of toxicity. According to Dr B. E. Day of the University of California it has less potential for causing birth defects than such everyday chemicals as aspirin, vitamin A, vitamin C and common salt. According to R. G. Harvey of the University of Wisconsin there is an 8000-fold safety factor in the normal use of 2,4,5-T over the amount necessary to cause embryonic effects in rats. If a 130 lb woman ate 3.3 lb of food every day that contained 0.2 p.p.m. of 2,4,5-T her total consumption would be 0.3 mg/day which is more than 8000 times less than the levels shown to cause birth defects in six animal species including rats, mice, sheep and monkeys. A survey of food samples collected over a period of 13 years showed only three samples that contained the herbicide and in those three the levels were 0.001, 0.008 and 0.19 p.p.m. (Lewert, 1976). According to Goring (1979) the acute toxicity of 2,4,5-T to mammals is about 100 times greater than sugar, ten times greater than table salt and half as great as caffeine. On the average (in the U.S.A.) people eat about $\frac{1}{20}$ of a lethal dose of sugar and about $\frac{1}{50}$ of a lethal dose of salt, and heavy coffee drinkers enjoy a substantially smaller margin for caffeine. These low safety margins are typical of many other natural foods. Goring states that 2,4,5-T is not found in either food or drinking water but assuming that it were at a level equal to the present analytical limit of detection then people would consume about 20 μg/day which is about one two and a half millionth of a lethal dose.

TCCD or dioxin is, on the other hand, a very toxic chemical. It is a weak carcinogen at toxic levels. The oral LD_{50} is 0.0021–3.0 mg/kg body weight depending on species, the guinea pig being the most susceptible and the dog the least (Shadoff *et al.*, 1977). It is cumulative. Rats fed for 13 weeks on food containing 0.1 μg/kg body weight per day lost weight (Kociba *et al.*, 1976). When Rhesus monkeys were fed 0.5 μg/kg in the diet for 6–9 months there was a marked effect on reproduction and some monkeys died (Allen, 1977). It is teratogenic for experimental animals and long-term studies with rats and mice are in progress under the auspices of the International Agency for Research on Cancer (1977).

2,4,5-T was used by the United States forces as a component of 'Agent Orange' for the defoliation of forests in the Vietnamese War. It contained up to 45 mg/kg TCDD. Following the use of 'Agent Orange' there were allegations that it was causing liver cancers among the Vietnamese and that Vietnamese mothers were bearing children with birth deformities such as cleft palate and spina bifida. Furthermore it was alleged that U.S. soldiers were developing liver complaints and cancers on their return from Vietnam. The International Agency for Research on Cancer (1977) reported that people were harmed by 'Agent Orange'. Effects included skin rashes, fever and deaths of children, but the U.S. National Academy of Science (1974) and later the World Health Organisation (1977) considered that the reports of increased birth defects in South Vietnam were inconclusive due to inadequate data.

The British Advisory Committee on Pesticides publishing their *Review of the Safety for Use in the U.K. of the Herbicide 2,4,5-T* in March 1979 stressed that 'the dose makes the poison'. The maximum allowable level of TCDD in the trichlorophenol used for the manufacture of 2,4,5-T was set up by the Advisory Committee in 1970 at 0.1 mg/kg. This limit has now also been set by the F.A.O. and is adopted in many countries. The Committee point out that this means that the application of the formulated product at the highest overall spray rate used in forestry would result in less than 0.5 mg of TCDD being distributed per hectare of soil and vegetation and that this is equivalent to one grain of sugar over a football pitch, although a higher concentration is used for localized spot treatment of tree bases and stumps. Furthermore they state that the Laboratory of the Government Chemist has recently used several procedures for determining the TCDD content of samples of six of the commercial formulations currently on sale in the U.K. The results indicate that TCDD could not be detected in three of the products. Another two products contained materials which greatly interfered with the analysis and out of three samples of the final product, two showed no evidence of the presence of TCDD and the third had it present at about the limits of determination. The Committee is satisfied that there is no evidence that amounts to which individuals might be exposed as a result of the use of 2,4,5-T would result in any harmful effects and was reassured by the results of surveys in the U.K. of operators using 2,4,5-T. These surveys took place in 1970, 1975 and 1976 and no operator showed any evidence of chloracne, a form of dermatitis, which is the hallmark of TCDD poisoning in humans.

The Committee looking at other aspects came to the conclusion that foliage burnt after treatment with 2,4,5-T does not release dangerous quantities of TCDD; that there is no evidence of bioaccumulation of 2,4,5-T in the soil; that the effects occurring in Vietnam, even if they could be substantiated (which is not necessarily accepted), have little relevance today since the herbicide used contained an extremely high level of TCDD impurity (up to 450 times the permitted level in the U.K.); that the industrial mishap at a pharmaceutical firm's factory in Seveso, Italy, in 1976 where dioxin escaped into the atmosphere was not relevant to the 2,4,5-T discussion since 2,4,5-T was neither present nor being produced; and that there is no reason for change in the official recommendations for use of 2,4,5-T products which contain 0.1 mg/kg TCDD or less in the trichlorophenol from which it is manufactured, although, like all other compounds under the Pesticides Safety Precautions Scheme, it should be reviewed in the light of any new data or future studies.

THE NULL HYPOTHESIS

White-Stevens (1971) states

'Over the past 25 years when pesticide application to crops, livestock, industrial plants, homes, forests, swamps and indeed the total environment have increased over 1000-fold, there has not been one single medically annotated cause of sickness, let alone cancer or death from the use of any (U.S.) registered pesticide when it has been applied strictly in accordance with label recommendations. There have been unfortunate accidents and deaths due to exposure of workers, children drinking, careless use, etc. These are not indictments of pesticides but the irresponsibility of the user.'

Dealing with the impossibility of proving that a chemical never does any harm he states further

'It is recognised in all scientific research that the proof of the negative or null hypothesis is impossible, namely that there is no difference between two or among more than two treatments whatever they may be. If there are no significant differences shown still the null hypothesis cannot be accepted. The public assume that science can prove the absence of a deleterious quality of a chemical as readily as it can prove its beneficial properties. Such a demand is a complete impossibility to meet.'

Herbicides and micro-organisms

The fertility of soils depends to a great extent on the micro-organisms therein. Bacteria, actinomycetes and fungi convert residue organic materials into a form more readily assimilable by plants, releasing minerals or small organic molecules to be taken up by the roots. In addition the nitrogen fixing bacteria, both free-living and symbiotic add substantial nitrogen fractions, thereby enriching the soil. Micro-algae carrying out photosynthesis add to the organic matter and in water their role in food chains is well known. It is important that any substance added to the soil or to water should not adversely affect the activity of these micro-organisms (Fletcher, 1966b).

A proportion of applied herbicides reaches the soil either through direct application, through run-off from plants or through the accumulation of debris from sprayed plants. Fletcher (1962) has discussed why it is important to determine what effect these herbicides may have on the soil microflora. He makes the following points:

1. There is the possibility that they may interfere with the activities of beneficial and pathological soil micro-organisms.
2. The activity of a number of herbicides against higher plants is partly dependent on their persistence in the soil.
3. The persistence of herbicides in the soil determines the 'lag period' before the following crop can be planted.
4. Partial decomposition may be essential to the activity of a herbicide in the soil.
5. Studies of the effect of herbicides on soil micro-organisms and of the effect of soil micro-organisms on herbicides may prove useful in determining the mode of action of herbicides on higher plants.

EFFECT OF HERBICIDES ON SOIL MICRO-ORGANISMS

Major reviews on the effect of herbicides on soil micro-organisms have been written by Fletcher (1960, 1966a); Domsch (1963); Fletcher *et al.* (1968) and Anderson (1978a, 1978b). Fletcher (1960) has pointed out that in their investigations authors differ in the expression of quantities used during their experiments. Thus some have used molar quantities, some parts per million, some percentages, some kilograms per hectare and some pounds per acre. This makes comparisons extremely difficult but in his review he converted molar solutions and percentages to parts per million, whilst noting that in some papers it was difficult to determine whether the stated amounts related to acids or salts so that the calculated parts per million were approximations. For the sake of simplicity kilograms per hectare and pounds per acre could be considered as equivalent measures (1 kg per hectare ≡ 2.2 lb per 2.5 acres). The important (and most difficult part) was to relate parts per million to pounds per acre. He suggested that as a working scheme most readers might accept an earlier calculation (Fletcher, 1956). One acre of top-soil, 8–9 inches in depth, contains approximately 3 million lb of moist soil. This refers to 'mineral' soil with about

3–12% organic matter. Therefore a convenient figure for the weight of top-soil of an arable field to the deliberately vague specification of 'plough depth' would be 2¼ million lb or 1000 tons. If the amount of herbicide reaching the soil is 1 lb the concentration in the top soil would be 0.5 p.p.m. Assuming a 20% soil water content, no adsorption and complete solution then the concentration becomes 2.0–2.5 p.p.m.≡ 1 lb/acre. Although this calculation is recognised as being no more than an approximation and it should be recognised that there will probably be higher (and lower) 'pockets' of concentration, it is however a fairly useful one since by using it one can have a rough idea of the maximum concentration in the soil after a known application. It can also be used to help relate concentrations used *in vitro* to the 'real' situation in the field.

Soil is such a variable medium, however, that such relationships are speculative at best and it should be borne in mind that herbicides in general are much more active against micro-organisms in culture than they are in the field.

Walker *et al.* (1977) whilst noting that data in the literature indicate that most, if not all, the herbicides in current use in Britain can affect the soil microflora to a certain extent state that where adverse effects have been noted they have almost invariably been associated with concentrations higher than those encountered in agricultural practice and have usually been transient in nature.

Anderson (1978) in a major review of the effect of pesticides on non-target soil micro-organisms finds that in very general terms few herbicides have any great or prolonged adverse effect on the total bacterial component of soils and that whilst individual bacterial species are frequently adversely affected to some degree the effects are seldom permanent. The usual pattern is one of an initial decrease in total numbers followed by a return to normal or even an increase in numbers, although the effects can depend upon the nutrient status of the soil. He notes that nitrification is one of the most extensively studied soil processes and the accumulation of data with respect to pesticide effects makes it difficult to generalise. The most inhibitory herbicides have been found to be sodium chlorate and calcium cyanamide; the former can inhibit nitrifiers for up to a year while the latter has been reported to almost eliminate nitrification (Audus, 1970).

Denitrification is an important soil activity yet Anderson states that it is one of the least investigated. He notes that no particular group of herbicides is predominantly inhibitory or stimulatory.

Some herbicides can harm rhizobia and the nodulation of legumes and he recommends that herbicide applications to legumes or to soil designated for legume growth should be carefully evaluated prior to recommending their use. Under soil conditions many herbicides at normal field rates have little or no effect on free-living microbial nitrogen fixation.

When submitting a herbicide for registration the Company has to provide data on its effect on soil micro-organisms. There is therefore a wealth of data available but it is too extensive to quote in a chapter of this kind. Suffice it to say that a herbicide would not be registered if it showed permanent adverse effects.

EFFECTS ON MICRO-ALGAE

As one would expect the effects of herbicides on soil algae can be quite severe and in general they are very inhibitory towards them although comparatively few studies have been undertaken.

Evidence of food-chain biological magnification problems with herbicides is lacking and in general these compounds probably exert only a temporary harmful effect on natural populations of micro-algae in aquatic environments (Mullison, 1970).

On the other hand herbicides such as 2,4,-D, dalapon, dichlobenil, diquat, maleic hydrazide, acrolein, TCA, aminotriazole, some substituted ureas and triazines are used for aquatic weed control and algal control in drainage and irrigation waters (Hawkins, 1972). Monuron has been used in field tests against a wide range of algae. Maloney (1958) obtained good results against filamentous algae though there was the danger of seepage to surrounding vegetation. Diuron has also been used. Jordan *et al.* (1962) and Frank (1970) found it to be the most effective of a number of herbicides tested. Its use however may be restricted by long persistence and possible accumulation in fish (Hawkins, 1972). Yamagishi and Hashizume (1974) found that simetryn and prometryne effectively controlled green algae in rice paddy fields and Robson *et al.* (1976) found that the methylthio-1,3,5-triazines, terbutryne and cyanatryn (4-[1-cyano-1-methylethylamino]-6-ethylamino-2-methylthio-1,3,5-triazine) were the most active of some twelve herbicides that they tested against troublesome fresh-water algae, killing them at levels of 0.05 p.p.m. Wright (1978) notes that other chemicals used for algal control include paints containing tin compounds; fentin (triphenyltin acetate) and dichlone (2,3-dichloro-1,4-naphthoquinone) are used in rice paddies, sodium hypochlorite, formalin or dichlorophen (4,4′-dichloro-2,2′-methylenediphenol) on exposed concrete surfaces and copper sulphate in the concrete mix of underwater structures. Numerous commercial algicides are used in swimming pools many of which contain quaternary ammonium compounds or organic halides as active ingredients; and quaternary compounds, amines, phenates and chlorine are used in the treatment of cooling waters.

Fitzgerald (1975) considers that most algicides if used at minimal effective concentrations do not pose a threat to aquatic environments due to their biodegradability at such levels.

AQUATIC AND SOIL MICRO-FAUNA

Edwards (1978) notes that although there is a considerable literature on the influence of pesticides on soil micro-fauna there is much less information on interactions of pesticides with the micro-fauna in the aquatic environment. He states further that our knowledge of the roles of the micro-fauna in terrestrial and aquatic systems is still very poor so that functional interpretation of the effects of pesticides on these organisms is extremely difficult. Bearing these observations in mind however after reviewing the literature he has concluded that herbicides tend to have little direct effect on the soil micro-fauna; there are few reports on their effects on aquatic micro-fauna but a few herbicides – simazine and diquat – have been shown to be toxic to plankton, while fenoprop and sodium arsenite were not. There can however be indirect effects in that the number of micro-arthropods are often changed drastically after the use of herbicides because their available food supply is altered (Edwards, 1970). Similarly, although there is little evidence, it would seem likely that aquatic herbicides must affect some of the micro-fauna by removing some of their food supply. McEwen and Stephenson (1979) have listed the effects of a number of herbicides on some aquatic organisms (table 7.1).

EFFECTS ON SOIL INVERTEBRATES

Some herbicides are directly toxic to earthworms while others have no effect. Martin and Wiggans (1959) found in immersion experiments with 100 p.p.m. concentrations that monuron was fatal while 2,4-D was harmless. When TCA, atrazine and monuron were applied to grassland there was a reduction in the earthworm population but this was attributed to a general reduction in vegetative cover except in the case of TCA where there might have been a direct toxicity (Fox, 1964), but Ilijin (1969) found that applying TCA, dalapon, or simazine at normal agricultural rates had no effect on the earthworm (or ant) populations. Propham, chlorpropham and DNOC have been reported to cause a reduction in numbers of *Lumbricus* (van der Drift, 1963). Paraquat on the other hand appears to induce increased populations

Table 7.1 Acute toxicity (p.p.b.) to aquatic organisms.

Herbicide	Effect on phytoplankton (% decrease in CO_2 fixation after 4 h in presence of 1 p.p.m.)	Daphnia EC_{50} (48 h)	Gammarus LC_{50}	Pteronarcys LC_{50}	Crangon LC_{50}	Bufo LC_{50} (tadpoles)
Aminotriazole		23				
Atrazine		3 600				
Dalapon	0	11 000		>100 000		
Dicamba			1 000			
Dichlobenil		3 700	1 500	4 400		
Diquat	−45				>10 000	
Endothal		46 000	2 000			
MCPA	0	100 000				
Monuron	−94	106 000				
Paraquat	−53	3 700	18 000	>100 000		54 000
Picloram	0	>380 000	48 000	120 000		
Simazine			21 000	50 000		
Trifluralin		240	8 800	130 000		
2,4-D		320 000	1 800 000			
2,4,5-T (acid)		>1 500				

(After McEwen and Stephenson, 1979)

of earthworms compared with those found in ploughed fields (Edwards, 1970).

Walker *et al.* (1977) whilst noting that a relatively small amount of work has been published on the effect of herbicides on soil invertebrates suggest that any effects that there are are indirect rather than the result of direct toxicity to the organisms and that such changes that may occur in this way are of little more than academic interest. Few herbicides affect worms, which make up the major weight of living invertebrate tissue in soil and Walker *et al.* (*loc. cit.*) state that whilst TCA and DNOC have been reported to decrease their numbers they have done so only when applied at doses far in excess of those recommended for agricultural practices. The use of paraquat has been reported to lead to an increase in both worm and slug numbers. Decreased numbers of nematodes and predating and saprophagous mites (*Acarina*) have been reported after treatment of plots with simazine, and populations of springtails (*Collembola*) have decreased after soil treatment with DNOC (Walker *et al.*, 1977). They conclude that with the possible exception of DNOC and some of the triazines, few of the herbicides that have been examined cause serious changes in the population of soil invertebrates.

Brown (1978) has reviewed the effect of herbicides on soil nematodes. He notes that applications of 2,4-D to vegetable seed beds controlled the rootknot nematode *Meloidogyne* (Burgis and Beckenbach, 1948) but that treatment of red clover and oat plants with 2,4-D made them more susceptible to *Ditylenchus dipsaci* and made callus tissues of red clover and alfalfa more suitable for the reproduction of *Aphelenchoides ritzemabosi* (Webster, 1967). Application of amitrol (aminotriazole) and dalapon killed a high percentage of *Anguia agrostis* which forms galls in the flower heads of colonial bentgrass (Courtney, Peabody and Austenson, 1962) and a mixture of amitrol and simazine caused a reduction of 75% in the nematode population of orchard soils (Wilcke, 1968).

MICROBIAL DEGRADATION OF HERBICIDES

Fletcher *et al.* (1968) and Walker *et al.* (1977) have described how herbicides may be lost from the soil either by physical removal of the unchanged molecule or by degradation (fig. 7.1). *Physical removal* may occur (a) by volatilisation; the thiocarbamates are the most volatile of the herbicides, losses of 20% in 30 months having been reported for EPTC under tropical conditions, (b) by leaching, which is influenced by soil drainage; the solubility of the herbicide and lack of adsorption are important factors in the loss of the phenoxyalkanoic

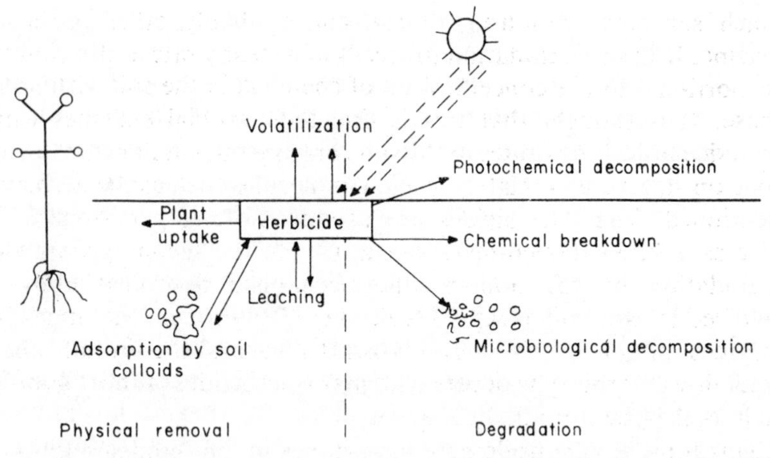

Fig. 7.1 Processes of herbicide inactivation (after Fletcher *et al.*, 1968).

acids and the chlorinated aliphatic acids from soils and (c) by uptake by plants. Adsorption on soil colloids, such as occurs with paraquat, is included under the heading of physical removal for although the herbicide is not actually lost from the soil, the process of adsorption causes a loss in concentration. *Degradation* or breakdown of the herbicide may be either non-biological or by biological means. Dealing first with non-biological breakdown a number of dinitro-anilines have been shown to decompose when exposed to ultraviolet light. A number of triazines contain chemical groupings which may be susceptible to hydrolysis although evidence suggests that the breakdown is very slow except when organic surfaces catalyse the reaction.

Conclusive evidence that the breakdown of herbicides (and other pesticides) in soil is mainly microbiological has been obtained by adding small amounts of metabolic poisons such as the cytochrome oxidase inhibitor, sodium fluoride, to soils and observing the subsequent prevention of breakdown of the herbicide.

Further evidence is got by the isolation of organisms from the soil that are capable of breaking down herbicides. A great many organisms capable of doing so have now been isolated. They include bacteria, fungi and actinomycetes. All herbicides seem to be susceptible to microbial attack though the rate of breakdown varies with the herbicide, with the soil type and with variations in soil factors such as pH, temperatures and moisture content.

Walker *et al.* (1977) point out that microbial breakdown of herbicides falls into one of two categories: (1). In the first group to which simazine, monuron, diuron and probably other ureas and triazines belong, degradation proceeds at a steady rate approximately proportional to the concentration of chemical in the soil with no lag phase. It is thought that this is due to microbial enzymes which degrade natural substrates in the soil carrying out similar transformations on structurally related 'foreign' molecules such as the herbicides mentioned. These herbicides are usually of long persistence. The process is known as co-metabolism. (2). In the second group, slow degradation by the non-specific mechanism described above is followed by a period of rapid breakdown. It is thought that microbial enzymes adapt to the new substrate during the initial phase. Breakdown of this type occurs with many herbicides of short duration such as the phenoxyalkanoic acids.

This topic is of considerable importance in that biodegradation of herbicides is vital to their lack of accumulation in the environment. The subject has also given considerable insight into the relationship between chemical structure and biodegradability. It is well established that the number of chlorines in the molecule affects decomposition of the halogenated aliphatic acids (table 7.2). Both Jensen (1957) and Kaufman (1966) have found that monochloroacetic acid is degraded more readily than dichloroacetic acid which in turn is degraded more readily than trichloroacetic acid. They have also found the same to be true of the corresponding propionic acids. The positioning of the

Table 7.2 Effect of number of chlorines on the microbial decomposition of chlorinated acetic acids.

	% decomposition on day				
	2	4	8	16	32
Monochloroacetate	0	4	98	100	100
Dichloroacetate	0	3	63	100	100
Trichloroacetate	0	0	1	39	97

(After Kaufman, 1966)

	Number of days to achieve 80% breakdown in soil
Monochloroacetic acid	2–4
Dichloroacetic acid	8–14
Trichloroacetic acid	14–28

(After Jensen, 1957)

chlorine molecule on the other hand seems to be of more importance in the phenoxyacetic group (table 7.3). Using the soil perfusion technique Audus (1960) in an investigation of the phenoxyacetic group of herbicides and their homologues found that whereas it took some 35 days for phenoxyacetic acid (POA) to be 80% reduced in soil, the same percentage breakdown was effected in 12 days by a *para*-chloro substitution (4-CPA) whereas the *ortho*-chloro substitution (2-CPA)

Table 7.3 Effect of positioning of the chlorine on the rate of breakdown of phenoxyacetic herbicides and their homologues (after Audus, 1960).

Compound	Days to reach 80% detoxication (100 p.p.m.)
Phenoxyacetic acid	35
4-Chlorophenoxyacetic acid (*para*)	12
2-Chlorophenoxyacetic acid (*ortho*)	>360

increased the breakdown period to beyond 360 days, and indeed Audus was not successful in enriching his soil for 2-CPA breakdown. Working with the phenylcarbamate herbicides Kaufman (1966) (table 7.4) has also found that both the number and the position of the chlorine substituents affect the rate of breakdown of chlorophenylcarbamate compounds. Thus IPC (isopropyl-*N*-phenylcarbamate) which has no chlorine was decomposed more rapidly followed by

Table 7.4 Effect of number and positioning of chlorine on microbial decomposition of chlorophenylcarbamate compounds in soil (after Kaufman, 1966).

Compound	% decomposition on day			
	4	8	12	16
Isopropylphenylcarbamate	5	64	100	100
Isopropyl-3-chlorophenylcarbamate (*meta*)	1	25	71	100
Isopropyl-4-chlorophenylcarbamate (*para*)	0	2	9	27
Isopropyl-2-chlorophenylcarbamate (*ortho*)	0	1	2	5

CIPC (isopropyl-*N*-3-chlorophenylcarbamate). In this latter group the *meta*-chloro-substituted compound was more rapidly decomposed than either the *ortho*- or *para*-chloro-substituted compounds. With regard to halogen substitution it seems that the type of halogen substituent also affects the rate of breakdown. Working with the halogenated aliphatic acids Hirsch and Alexander (1960) found that bromine-substituted compounds were more easily broken down than corresponding chlorine-substituted compounds, whereas Kaufman (1966) found the opposite to be true in the substituted phenylcarbamate compounds. It is obvious that generalisations with regard to pesticides are rather difficult to make, but such data are extremely valuable in environmental terms in providing guidelines towards the production of herbicides that will be readily degraded in the soil.

It is not surprising that there is an immense literature on the subject of soil microbes/herbicides interactions involving studies from soil systems and with organisms isolated from soil in pure culture, with mixed populations and in some cases with microbial enzymes. The subject has recently been extensively reviewed by Cripps and Roberts (1978) to which the reader is referred for more detailed descriptions.

The persistence of herbicides in soils

The rate of breakdown of herbicides in the soil is dependent on a great many factors including soil type, nutrient status, pH, moisture content and temperature. It is difficult to generalise and this should be borne in mind when assessing findings. Many sophisticated techniques have been developed and used for the determination of pesticide residues in soil, water and plant and animal tissues and it is sometimes said that the finding of residues at parts per million, billion and trillion is more a measure of the skill of the chemist than a measure of the persistence of the herbicide.

The majority of herbicides have a relatively short life in the soil (see table 7.5). Under warm moist conditions 2,4-D persists for only 2–3 weeks and MCPA for 6–8 weeks although 2,4,5-T is more resistant to breakdown and may persist in the soil for 2 months to one year. In all cases microbial breakdown takes place by degradation of the side chain to the phenol, ring hydroxylation and cleavage, and subsequent degradation to succinate and other naturally occurring organic acids (Loos, 1975). Among other herbicides which break down readily in the soil are CDAA (*N*,*N*-diallyl-2-chloroacetamide) and propachlor (3–6 weeks), alachlor (6–10 weeks), endothal (a few weeks), glyphosate (a few weeks), aminotriazole (half-life of only 2–3 weeks).

Table 7.5 Relative persistence in soil of some common herbicides when applied at doses recommended for selective weed control.

Persistence period		
Less than 3 months	3–6 months	Greater than 6 months
Aminotriazole	Chlorbromuron	Atrazine
Aziprotryn	Chlortoluron	Bromacil
Carbetamide	Cycloate	Chlorthiamid
Chlorpropham	Di-allate	Dichlobenil
Cyanazine	Dinitramine	Diuron
Dalapon	EPTC	Lenacil
Metoxuron	Isoproturon	Methazole
Prometryne	Linuron	Metribuzin
Propachlor	Methabenzthiazuron	Simazine
Propham	Metobromuron	Terbacil
Terbutryne	Monolinuron	TCA
	Propyzamide	Trifluralin
	Pyrazon	
	Tri-allate	
	Trietazine	

(After Walker *et al.*, 1977)

A number of herbicides volatilise readily from soil surfaces. Among them are the nitroanilines (e.g. trifluralin, benefin (benfluralin), butralin, dinitramine, isopropalin, oryzalin) which are also susceptible to photolysis and degradation by soil micro-organisms. Reduction of the nitro groups and *N*-dealkylation reactions are important in the early stages (Probst *et al.*, 1975). The carbamates (including thio and dithio) are also quite volatile and readily degraded in soil persisting for only a few weeks.

At the other end of the scale a number of herbicides are quite persistent. Fifty per cent of a 16 lb/acre application of TBA was found in a Kansas soil 11 years afterwards but this was exceptional. Picloram is one of the most persistent herbicides, its half-life varying from only a few months in warm moist soils to four years in cold semi-arid regions (McEwen and Stephenson, 1979). Paraquat and diquat may persist in soils for a number of years, breakdown being hindered by their adsorption on to soil colloids. Some of the 'total' herbicides are fairly resistant to breakdown. Simazine or atrazine applied at less than 2 lb/acre are normally degraded by the end of the growing season. Rates higher than 2 lb/acre may carry over into the next year and high rates (10–20 lb/acre) may leave residues for up to three years. According to McEwen and Stephenson (1979) they are adsorbed on to soil, clay or other organic matter. There may be some photodecomposition but microbial breakdown for most triazines is slow with no lag phase. It involves conversion to 2-hydroxy and *N*-dealkylated derivatives, with eventual ring cleavage and liberation of carbon and nitrogen.

Many of the ureas are moderately persistent in soil varying from a few months to more than a year. Some photodecomposition takes place and a number of micro-organisms have been isolated which are capable of effecting breakdown by using the herbicides as a carbon source. The uracils are also moderately persistent in soil with an average half-life of several months.

Hance (1979) in reviewing the processes by which herbicides are lost from the soil environment concludes that any long-term effects attributed to the repeated use of herbicides arise from the consequences of their short-lived biological activity rather than from their persistence in the environment. In other words, the inherent toxicity of the herbicide in bringing about long-term effects is more important than its longevity. Fryer (1977) who, as Director of the A.R.C. Weed Research Organisation at Oxford, has long and wide experience of all aspects of herbicide usage, states that of the thirty-four new herbicides introduced into British agriculture during 1970–79 only three have an LD_{50} (oral) to rats of less than 1000 mg/kg, i.e. they are less toxic than aspirin or caffein. (Reference to table 7.11 will show that this standard is well maintained for succeeding years.) Fryer states further that phytotoxic residues of herbicides used correctly in agriculture do not build up in the soil and there is no evidence that repeated application is liable to affect soil fertility adversely. Whilst at high concentrations herbicides under laboratory conditions can affect both species composition of the soil microflora and their biochemical behaviour there is no reliable evidence that herbicides at rates used in normal practice have or are likely to have any adverse effects of practical significance. The techniques used and the difficult problems of interpretation of data have been reviewed by Grossbard (1973) and by Greaves *et al.* (1976).

EFFECT OF HERBICIDES ON INSECTS

2,4-D, 2,4,5-T, fenoprop, 2,4-DB, dicamba, chloramben, picloram, ethephon, EPTC and dalapon are non-toxic to bees even when fed to them at concentrations of up to 1000 p.p.m. w (Morton and Moffett, 1971, 1972). On the other hand paraquat, MAA and cacodylic acid are moderately toxic at 10 p.p.m. w, MSMA and DSMA at 100 p.p.m., and bromoxynil and endothal only at 1000 p.p.m.

These authors conclude that herbicides are more likely to injure colonies by depriving them of the plants on which they forage for nectar and pollen than through direct effects as poisons.

There is some evidence that many species of butterflies have disappeared from large areas of Britain as a result of grassland improvement. Some insects feed almost exclusively on certain plants and if these plants are destroyed by herbicides then the source of food goes too. For example the larvae of five species of butterflies – the Peacock (*Inachis io*), the Red Admiral (*Vanessa atalanta*), the Map butterfly (*Araschnia levana*), the Comma butterfly (*Polygonia c-album*) and the Small Tortoiseshell (*Aglais urticae*) have the nettle (*Urtica* spp.) as their food source in the U.K. A plea is therefore made to farmers not to attempt to eliminate all their nettles for the sake of tidiness. A few clumps left growing will provide for these beautiful butterflies.

Similarly the Painted Lady (*Vanessa cardui*) feeds almost exclusively on thistles (*Cirsium* spp.), the Glanville fritillary (*Melitaea cinxia*) on certain species of plantain (*Plantago* spp.), the Heath fritillary (*Mellicta athalia*) on cow wheat (*Melampyrum pratense*) and the Marsh fritillary (*Euphydryas aurinia*) on devils-bit scabious (*Succisa pratensis*). Many weeds are equally valuable sources of nectar for bees, e.g. dandelion (*Taraxacum officinale*), charlock (*Brassica sinapis*), poppy (*Papaver* spp.) and rosebay willow herb (*Chamaenerion angustifolium*), while plants such as coltsfoot (*Tussilago farfara*), poppy and others are visited for their pollen (Fletcher, 1974).

A number of entomophagous insects have shown varying degrees of susceptibility to herbicides in the laboratory which would be likely to be hazardous to them in the field. Thus testing alachlor, propachlor, and chlorfenprop-methyl against them, Franz and Tanke (1979) found that the egg parasite *Trichogramma cacoeciae* was particularly sensitive. The Hovering fly (*Epistrophe balteata*) was also sensitive but the fly *Chrysopa carnea* appeared to be completely resistant.

Stevenson (1978) has listed the toxicity of some sixty pesticides to bees. The list includes some herbicides and their contact and LD_{50}s are given.

Effect of herbicides on fish

A number of factors affect the toxicity of a herbicide to fish. Among them are:

1. The temperature of the water. In general the higher the temperature the greater the toxicity, although the effects are not dramatic.

2. The duration of exposure. This is important; in general the longer fish are exposed to a herbicide (if it is toxic) then the greater is the mortality.
3. The age of the fish. The effect of this is variable.
4. The salinity of the water.
5. Water hardness.
6. Species of fish.

Herbicides have been detected in fish but only in low concentrations. Schultz and Whitney (1974) monitored fish from a canal in Florida where 7000 acres had been sprayed with 2,4-D at 4.48 kg (acid equivalent)/ ha. Of sixty samples of fish and analysed, three had greater than 0.01 p.p.m. (0.162 was the highest) and eight had residues of 0.01–0.10 p.p.m. Forty-nine of the samples had residues either not detectable or less than 0.010 p.p.m. Similarly, low residues were found in 307 fish analysed from nine ponds treated with 2,4-D at 2.24, 4.48 or 8.96 kg/ ha (Schultz and Harman, 1974).

Cope *et al.* (1970) exposed bluegills (*Lepomis macrochirus*) to 2,4-D in six ponds and the fish and the pond environment were studied for 5 months to measure persistence and chronic effects. No 2,4-D was detected in the fish after 4 days. Small quantities were found in the pond water for 6 weeks after application. In other experiments ⅕ of the fish treated with 10 p.p.m. 2,4-D died within 8 days. Mortality was negligible at 5 p.p.m. or less.

According to Frank (1971) monitoring studies in the U.S.A. indicate that contamination of natural waters occurs infrequently and at very low levels. Residues in water from the use of herbicides for the control of aquatic weeds are relatively high initially but decline quite rapidly and are often not detectable after a few days or weeks. With few exceptions aquatic herbicides do not accumulate.

Frank (1972) and McEwen and Stephenson (1979) have listed the disappearance of aquatic herbicides from treated water (see tables 7.6 and 7.7).

Frank *et al.* (1970) applied dalapon (6.7–20 lb/acre), TCA (3.8–5.9 lb/acre) and 2,4-D (1.9–3 lb/acre) to a canal and determined the concentrations 0–4.25 miles downstream 0–600 min after treatment. The concentrations found were 23–365 p.p.b., 31–128 p.p.b and 25–61 p.p.b. respectively. These concentrations reduced rapidly by dilution following passage of water so that negligible concentrations remained after the water had travelled a distance of 20–25 miles. Very low levels of dalapon persisted for several hours but TCA and 2,4-D were reduced to traces or non-detectable amounts within 30–60 min. The low concentrations were not considered to be hazardous to animals.

Table 7.6 Disappearance of aquatic herbicides from treated water; days after application in parentheses (Frank, 1972).

Compound	p.p.m. applied	Highest p.p.m. found	Final p.p.m. remaining	Usual application rate
2,4-D dimethylamine	1.5	0.14 (1)	0.004(41)	1.5–2.0 p.p.m.
Copper (sulphate)	0.5	0.42(1)	0.19(3)	0.1–2.0 p.p.m.
Diquat	0.62	0.49	0.001(8)	0.25–1.5 p.p.m.
Endothal	1.2	0.79(4)	0.001(36)	0.5–4.0 p.p.m.
'Silvex'* (butoxy ethanol ester)	2.9	1.6(7)	0.02(182)	1.5–2.0 p.p.m.
Fenac (granular)	1.0	0.71(8)	0.07(160)	15–20 lb/acre
Dichlobenil (granular)	0.58	0.32(36)	0.004(160)	10–15 lb/acre
2,4-D ester (granular)	1.33	0.067(18)	0.001(36)	20–40 lb/acre†

Fate of Organic Pesticides in the Aquatic Environment

* Fenoprop or 2,4,5-TP
† Equivalent to 1.8–3.6 p.p.m.

Table 7.7 Relative persistence in natural waters.

Non-persistent (half-life less than 2 weeks)	Slightly persistent (half-life 2–6 weeks)	Moderately persistent (half-life 6 weeks–6 months)
2,4-D	Aminotriazole	Ametryn
DNOC	CDAA	Atrazine
Diquat	CDEC	Bromacil
Endothal	Chloramben	Dichlobenil
IPC	Chlorpropham	Diphenamid
	CIPC	Linuron
	Dalapon	Prometon
	Dicamba	Propazine
	DNBP (dinoseb)	Simazine
	EPTC	Terbacil
	Fenuron	Trifluralin
	MCPA	
	Monuron	
	Propham	
	TCA	

(After McEwen and Stephenson, 1979)

Suzuki *et al.* (1977) detected thiobencarb residues of 0.11–10 μg/l in water samples immediately after the herbicide had been aplied to rice paddies but these residues were not detected for long after completion of treatment. Schulze *et al.* (1973) monitored twenty stations, one on each of twenty rivers in the western United States, monthly from October 1968 to September 1971. Herbicides were detected in all streams, the highest concentration being 0.99 p.p.b. 2,4-D. The range of 2,4,5-T was 0.01–0.05 p.p.b.

Richard *et al.* (1975) found that a 3″ rainfall immediately after the corn planting season in Iowa was reflected in 12 p.p.b. of herbicide in an adjacent river the following day but levels declined to 3.9 p.p.b. 2 days later and to 0.4 p.p.b. in 1 week. McEwen and Stephenson (1979) after reviewing the literature conclude that, for most pesticides, persistence in water is brief, many being hydrolysed and microbially degraded. They note that Goring *et al.* (1975) have identified forty-eight genera of microbes capable of degrading one or more pesticides. They note also that Hiltibran (1967) tested twenty-two herbicides on eggs and small fry of bluegill, green-sunfish, small-mouthed bass, lake chubsucker and stoneroller and found that fertilised eggs hatched even in high concentrations (several p.p.m.) and that small fry were not killed at dosages likely to be encountered in normal application procedures. They conclude (table 7.8) that the acute toxicity of herbicides to fish is low and that problems of direct fish kill should not be encountered. In contrast to the comparatively low concentrations of insecticides used for mosquito or black fly control, herbicides are recommended at high rates (diquat 0.5–1.0 p.p.m.; 2,4-D and 2,4,5-T 2.0 p.p.m.; endothal 1–3 p.p.m.; dichlobenil 10 lb/acre; fenac 20lb/acre). In spite of this the margin of safety seems adequate and fish kills are not observed. Their findings are in line with those of the *U.S. Fish Pesticide Research Lab. Report* (1975) which states that the acute toxicity 96 h LC_{50} values for most herbicides for trout is $>$1000μg/litre but 2,4-D esters, dinitramine and endothal amine show values between 100–1000 μg/litre. The fish toxicity studies of herbicides commonly used in water have been summarised by Tooby (1976) and in this paper he discusses the possible risks to fisheries from the herbicides available in the U.K.

INDIRECT EFFECTS

As has been aptly pointed out by Robson and Barret (1977) even if a herbicide could be developed which was specific to one species of weed, with no toxic effect on other weeds or animals, it would still produce an effect on the ecosystem. The chief factors are that plants

Table 7.8 Acute toxicity to some freshwater fish. LC_{50} in p.p.b. (values for 48 h unless indicated otherwise).

Herbicide	Fathead minnow	Rainbow trout	Bluegill	Goldfish	Channel catfish	Carp
Aminotriazole			>50 000			
Atrazine		12600				
Dalapon	440 000	>500 000	11 500			
Dicamba		35 000	130 000			
Dichlobenil		20 000	20 000			
Diquat	140 000†	20 000	145 000	3 500†		
Endothal	480 000		280 000	175 000		175 000
MCPA			10 000†			
Monuron					75 900*	
Paraquat			400 000*			
Picloram	64 000*	34 000*	26 500*	32 500		
Simazine		56 000	118 000			
Trifluralin		152	210	100		
2,4-D			960	1 300		
2,4,5-T (acid)		1 300	500			

(After McEwen and Stephenson, 1979)
* 24 h exposure
† 96 h exposure

die *in situ* (in contrast to manual weed removal) and regrowth is eliminated or very much reduced.

Weeds killed by herbicides in or near water may lead to decomposition and result in reduced oxygen levels in the water. Much depends upon the rate of movement of the water and stagnant pools set up anaerobic conditions much more readily than do fast-flowing rivers. Although McCraren *et al.* (1969) found that ponds treated with diuron for weed control had reduced oxygen levels for some 30 days after treatment leading to some mortality of bluegills, most reports on the use of herbicides in water indicate remarkably little mortality in a wide range of species. McEwen and Stephenson (1979) consider that when macrophyte growth is removed phytoplankton becomes abundant and since their photosynthetic rate is high, they may play an important role in restoring oxygen levels.

Effect of herbicides on birds and their eggs

EFFECT ON BIRDS

There is no evidence that herbicides affect birds directly. Most birds appear to have a high tolerance level (see tables 7.9 and 7.11) Paraquat,

Table 7.9 Acute toxicity to birds.

Herbicide	LD_{50} (mg/kg) (orally)		LC_{50} (p.p.m.) (fed to 2-week-old chicks for 5 days)			
	mallard	pheasant	quail	coturnix	mallard	pheasant
Aminotriazole	2000					
Atrazine	2000		750		5000	5000
Dalapon				5000	5000	5000
Dicamba		740				
Dichlobenil	2000	1189		5000		1750
Diquat	564			1500	5000	3750
Monuron				5000	5000	4500
Picloram	2000	2000			5000	5000
Simazine				5000	5000	5000
Trifluralin	2000	2000				
2,4-D	1000	472	5000	5000	5000	5000
2,4,5-T (acid)				5000	5000	1775

(After McEwen and Stephenson, 1979)

diquat and ethalfluralin appear to be the most toxic but even they have LD_{50}s of 260 mg/kg (hens), 200–400 mg/kg (hens) and >200 mg/kg (mallard ducks), respectively.

Palmer (1972) found that chickens could tolerate up to ten daily doses of 25 mg/kg barban. Dobson (1954) placing chickens on grass runs which had been sprayed with MCPA, 2,4-D or 2,4,5-T found no subsequent evidence of toxicity. Roberts and Rogers (1957) found that turkeys were not affected when fed on alfalfa sprayed with 2,4,5-T or added to their feed at the rate of 250 p.p.m. Horn (1974) carried out surveys of game birds in several areas of Braunschweig (Germany) during 1956–73 and found that extensive use of 2,4-D and other growth regulators in cereals did not adversely affect development and growth of pheasants and indeed there was an increase in the pheasant population during this period.

Safety studies have shown that most herbicides do not involve hazards for food and wildlife. Birds show good tolerance of relatively high rates of herbicides (Mathys, 1975). The use of herbicides may have indirect effects on birds by affecting their food sources and nesting sites due to the destruction of vegetation. There appears to have been little work done on this aspect of herbicide usage.

EFFECT ON EGGS AND EMBRYOS

The effect of herbicides on egg hatching and the subsequent development of the chicks has been investigated by a number of workers. Two techniques have been used (1) spraying the eggs with, or dipping them in, herbicide (2) injecting the herbicide into the air sac or directly into the yolk.

Somers *et al.* (1973) sprayed aqueous solutions of (1) 2,4-D amine at recommended field rates (2.8 kg/ha) and at ten times the recommended field rate and (2) a mixture of 2,4-D tri-isopropanolamine and picloram (formulated as 'Tordon' 101) in a 4:1 ratio at ten times the recommended field rate on to fertile hens' eggs. No treatment had any effect on hatching or early chick performance.

The same workers (Somers *et al.*, 1978) sprayed weak-shelled hens' eggs with 2,4,5-T preparations contaminated with excessive amounts of TCDD, at ten times the normal recommended field rate and found that there was no effect on the development and growth of the chicks.

Similarly applying 2,4-D, 2,4,5-T and picloram at ten times the rate to three stages of embryonic development had no adverse effect.

Aqueous solutions of 2,4-D and picloram at the recommended field rate were sprayed on fertile pheasants' eggs preceding incubation. No treatment affected hatching success, incidence of malformed embryos or chick mortality, thus confirming earlier studies with domestic chicken eggs (Somers *et al.*, 1974). These findings are also in line with those of Grolleau *et al.* (1974) who dusted eggs of quail, partridge and red-legged partridge with 2,4-D at rates equivalent to 1.2, 2.4 and 6 kg/ha. There was no effect on the hatching or growth of the young birds and no increase in the rate of deformities.

Pheasant and quail eggs were dipped in solution containing proprietary products of 2,4-D, 2,4,5-T and MCPA and hatched 24 hours later in an incubator. Hatching was unaffected by concentrations twelve times higher than normal field applications and only concentrations thirty times higher than normal caused a decrease in hatching. There was no evidence of abnormalities (Spittler, 1976). Kopischke (1972) found that application of 2,4-D at concentrations normally required for weed control did not adversely affect hatchability of eggs and did not cause deformities or death of chicks of either pheasant or bantam fowl. Whitehead and Pettigrew (1972) found no teratogenic effects on chicks which had hatched from eggs from which hens had been fed up to 150 p.p.m. 2,4-D from the 28th to the 48th week. Gyrd-Hansen and Dalgaard-Mikkelsen (1974) injected 2,4-D, 2,4,5-T, MCPA, mecoprop and dichlorprop into the yolk of hens' eggs. All of the herbicides at about 2 mg per egg decreased the percentage hatch and viability of the chick. Immersion in 1% solution had no effect.

The findings of Lutz-Ostertag and Lutz (1970) appear to be at variance with the findings of most other workers since they found that 2,4-D sprayed over pheasants' eggs in nests was extremely toxic to the embryos. Furthermore there were considerable abnormalities among the chicks which hatched. De Witt *et al.* (1963) also found that feeding aminotriazole, dalapon and MCPA to quail, pheasant and mallard ducks had an adverse effect on the hatchability of their eggs.

Dunachie and Fletcher (1967, 1970) using the egg injection technique examined the effect of twenty-five herbicides at various concentrations on the hatching rate (and teratogenicity) of domestic hens' eggs. They used concentrations up to 500 p.p.m. (per egg weight minus shell). Of the compounds tested seventeen had no effect at 50 p.p.m. and of these eight had none at 100 p.p.m. Except for feather blanching caused by some of the substituted phenoxy acids, no teratogenic effects were found. The most toxic herbicide to hen embryos was paraquat. At 0.3 p.p.m. no embryos developed and even at the 0.15 p.p.m. level only one-third hatched. Morfamquat and diquat also showed toxicity although of a lower order (10 p.p.m.). Their findings are summarised in table 7.10. Lutz-Ostertag and Henon (1975) have also reported high embryonic mortality in chicken and quail eggs

Table 7.10 Effect of herbicides on the hatching of hens' eggs.

Percentage hatch	Herbicide
1. Dose given: 200 p.p.m.=10 mg/egg (selected compounds only).	
100	Dalapon
90	Atrazine. HCl, chlorpropham
80	Mecoprop, dicamba, prometryne
50	2,4-D, TBA
30	Monolinuron, linuron
20	MCPA, MCPB
0	Desmetryne, dichlorprop
2. Dose given: 100 p.p.m.=5 mg/egg (all compounds).	
100	Atrazine.HCl, chlorphencarb, dalapon, MABSC*, monolinuron, prometryne, simazine.HCl
90	Dicamba, TBA
80	Mecoprop, aminotriazole.HCl, chlorpropham
70	2,4-D, 2,4-DB, dichlorprop
60	Linuron
50	MCPA, MCPB, desmetryne
10	Diquat, prefix
0	Bromoxynil, DNOC, ioxynil, paraquat, pentachlorophenol
3. Dose given: 10 p.p.m.=0.5 mg/egg (all compounds except prometryne and desmetryne).	
100	Aminotriazole.HCl, atrazine.HCl, chlorphencarb, chlorpropham, dalapon, ioxynil (10% at 25 p.p.m.), linuron, MABSC*, MCPA, MCPB, monolinuron, prefix (50% at 50 p.p.m.), simazine.HCl
90	2,4-D, dicamba, dichlorprop, mecoprop, TBA
80	2,4-D, pentachlorophenol (50% at 50 p.p.m.)
70	Bromoxynil
30	DNOC (0% at 20 p.p.m.; 80% at 5 p.p.m.)
10	Diquat (60% at 5 p.p.m.)
0	Paraquat (0% at 0.25 p.p.m.; 40% at 0.15 p.p.m.; 100% at 0.1 p.p.m.)

(After Dunachie and Fletcher, 1967)
* Methyl-4-aminobenzene sulphonylcarbamate

treated with paraquat. In a follow-up investigation of the work by Dunachie and Fletcher (*loc cit.*) Fletcher (K.) (1967) set up experiments to test the passage of paraquat into the egg through the hen and its effect when thus present on hatchability. His findings showed that concentrations of paraquat of about 0.1 p.p.m. caused a statistically significant diminution of hatch. In order to get 0.1 p.p.m. into an egg by feeding a fowl, rather than directly with a syringe he determined that it must be present in the drinking water of the fowl at a concentration of 40 p.p.m., four times the concentration that the manufacturers recommend for standard spraying.

Few herbicide residues have been found in eggs. Foster *et al.* (1972) fed atrazine and linuron at 0.5 p.p.m. for eight weeks to poultry and subsequently could not detect any residues in eggs, fat and other tissues. 2,4-D residues have been found in eggs of quail, pheasants and mallard ducks after feeding at 1250 p.p.m. (De Witt, 1965) and Fletcher (1967) found residues of paraquat (around 0.1 p.p.m.) after feeding hens at 40 p.p.m. for 14 days.

Erne (1966) found some residues in the liver, kidney, lung, spleen and eggs of chickens which had been fed 2,4-D and 2,4,5-T but in general there was a rapid elimination via the kidneys. Egg production may be affected by herbicides. Chickens in grass runs which had been sprayed with MCPA, 2,4-D or 2,4,5-T showed a reduction in egg production (but not in fertility or hatchability) (Dobson, 1954) but Whitehead and Pettigrew (1972) found that feeding hens up to 150 p.p.m. 2,4-D for 20 weeks had no effect on egg production nor on egg-shell thickness.

In a major review of the physiological and biological effects of pesticide (including herbicide) residues in poultry (including chickens, turkeys, ducks, geese, Japanese quail, pheasants and mallard ducks) Foster (1974) has provided a comprehensive compendium for the research worker. Whilst noting that no large-scale losses have been attributable to pesticides he makes a plea for more uniformity of approach so that data can be more easily compared. He points out that very little research has been carried out on the effect of metabolites of pesticides. He would also like to see more reasonable exposure levels based on realistic residue studies and the choice of more appropriate means of exposure, e.g. in feed or by dusting, spraying, immersion, or vaporisation. There is need, he states, for more research on the effects of chronic ingestion which in the poultry industry means 15 years.

Effect of herbicides on mammals

There is no evidence of acute toxicity resulting in death in mammals due to the application of herbicides at approved rates. Some deaths may occur through their misuse but this is also true of many chemicals other than herbicides.

Most herbicides have a very low toxicity to mammals (table 7.11) and since rats and other mammals are included in the test systems it is very unlikely that any herbicide would be accepted for registration if it did not show a high margin of safety. Only some twenty to twenty-five out of those listed have LD_{50}s of less than 1000 mg/kg and of these only five have LD_{50}s of less than 100 mg/kg – acrolein (46), allyl alcohol (64), dinoseb (58), DNOC (25–40) and endothal (51).

Most herbicides are excreted very rapidly, e.g. excretion of ^{14}C molinate fed to rats (72 mg/kg) was 95% complete 48 h after dosing (Baun *et al.*, 1978); the dose of diuron fed to rats was entirely eliminated in the faeces (66%) and urine (34%) within 48 h (Kato *et al.*, 1978); Crayford *et al.* (1976) fed benzoylprop-ethyl, flamprop-isopropyl and flamprop-methyl to lactating cows and to pigs and hens at doses ranging from 0.3 to 3 mg/kg in the diet, i.e. approximately 10–300 times the total residue found in treated cereals in the field. Residues in milk in most cases were well below 0.001 mg/kg, in muscle >0.003 mg/kg and in eggs 0.008 mg/kg, decreasing by 50% in approximately 3 days to 0.001 mg/kg 4 days after the termination of treatment. The elimination of the herbicides from the animal was rapid in every case their metabolites being ideally suited for excretion via the kidneys and bile into the urine and faeces but unsuitable for transport into milk and eggs. For the results of a very comprehensive investigation of the toxicity of herbicides to cattle, sheep and chickens the reader is referred to Palmer and Radeleff (1969). The following is their summary and conclusions. Results of studies of the toxicity of twenty-nine organic herbicides to cattle, sheep and chickens have been presented. A total of 126 yearling cattle, 190 one- to two-year-old sheep, and 700 six-week-old chickens were studied. Repeated doses of the herbicides were administered in gelatine capsules or as water-diluted solutions by drench or in pipettes. The usual period of study was 10 days or until toxicological effects appeared. However in a number of instances, longer- and shorter-term studies were made.

The signs of poisoning by most of the herbicides included anorexia and reduced weight gains. In many instances digestion in the rumen seemed to have been impaired. Macroscopic lesions were somewhat variable and non-specific. The liver and kidneys were most frequently involved. An arbitrary yield of forage and rate of consumption were selected to evaluate the hazard for cattle, sheep and chickens likely to exist under the most severe conditions of use. The most common rates of aplication of many of these herbicide formulations are not a hazard to cattle, sheep or chickens. The maximum rates of application of some herbicides approach or surpass a hazardous level in one or more of the test species.

Paulson (1975) in a major review article has dealt with the metabolic fates of herbicides (covering all the major groups) in animals and dealing with the rates and routes of elimination, their isolation and identification in milk, tissues, eggs, urine and faeces. He notes that a large percentage of most of the herbicides is eliminated in the urine and faeces as the parent compound, but more commonly as metabolites, within a few days of treatment.

INDIRECT EFFECTS

Some weeds not normally eaten by animals may become more attractive when treated with 2,4-D or MCPA. Among them are pigweed (*Amaranthus retroflexus*), ragweed (*Senecio jacobea*) and jimson weed (*Datura stramonium*). The attractiveness has been variously attributed to greater succulence and to higher concentrations of sugar after treatment (Willard, 1950). Some of these weeds, e.g. ragweed, are highly poisonous and death of the animal eating the weed may result. Others which are not in themselves very poisonous may prove to be so after treatment. It is now known that the toxicity is not due to the 2,4-D but rather to high nitrate levels which subsequently develop in the treated plants (Fertig, 1952; Berg and McElroy, 1953). Nitrate levels in sugar-beet leaves which had been accidentally sprayed with 2,4-D showed high potassium nitrate levels some 4.5% in excess of the 1.5% maximum which had been established as the safe feeding level in forage crops (Stahler and Whitehead, 1950). As pointed out by McEwen and Stephenson (1979) nitrates are not in themselves toxic but become so when reduced to nitrites in the rumen. The nitrites cause the formation of methaemoglobin in the blood resulting in an interference with oxygen transport. As a result, anorexia follows and when severe brings about the death of the animal.

MUTAGENICITY

There is no evidence that any of the herbicides used at the correct levels are carcinogenic, teratogenic or mutagenic in humans or animals although some have been shown to be so when administered in large doses to laboratory animals. Epidemiological studies have not substantiated any claims.

The evaluation of mutagenicity is by its nature a difficult exercise and is generally carried out using microorganisms. One hundred and ten herbicides were evaluated for point mutations in a variety of microbial test systems in comparison with known mutagens such as 5-bromouracil and 2-aminopurine. Except for inconclusive evidence relating to four herbicides within one test system, where the rates were slightly in excess of controls (but not borne out in other test systems), the mutagenic rates did not differ from normal rates (Anderson, Leighty and Takahashi, 1972).

Table 7.11 Toxicity data.

(Note: figures are expressed as LD_{50} mg/kg (oral) for mammals and birds and LC_{50} mg/l for fish (96 h) unless otherwise stated. The abbreviations d=day, w=week, y=year are used.)

Herbicide	Rats	Rabbits	Mice	Guinea pigs	Dogs	Birds	Fish	Bees	Sensitivity	Feeding trials No effect – time – mg/kg food
Inorganic salts										
Ammonium sulphamate	3900									Rats – 90d – 400
Borax	4500–6000									Lethal dose to human infants 5–6 g
Disodium octaborate				5300						
Sodium chlorate	1200								Skin irritant	
Sodium metaborate	2330									
Sulphuric acid									Eye and skin irritant – very corrosive	

Haloalkanoic acids								
Chlorfenprop-methyl	1072–1321 Dermal >2000	500–1000	1000	500–1000	>500 Cat >1000	Chickens ca. 1500 Canary 1250–2500	Harlequin 2 (24 h) 1.3 (48 h) Goldfish 1–10	Rats – 90d – 1000
Chloroacetic acid	650		165				Rainbow trout 2000 (48 h)	Rats – several months –700
Dalapon	7570–9330					LC_{50} (mg/l, 2w) Coturnix 5000 Mallard 5000 Pheasant 5000	Trout >500 (48 h) Bluegill 11.5 (48 h) Minnow 440 (24 h)	Rats – 2y – 15
TCA	3200–5000 (sodium salt) 400 (acid)		5640 (sodium salt)					
Phenoxyalkanoic acids								
Phenoxyacetics								
2,4-D	375					Mallard 1000 Pheasant 472	Bluegill 0.96 (48 h) Goldfish 1.3 (48h)	

Table 7.11 Toxicity data (*contd*)

Herbicide	Rats	Rabbits	Mice	Guinea pigs	Dogs	Birds	Fish	Bees	Sensitivity	Feeding trials No effect – time – mg/kg food
2,4-D (*contd*)						LC_{50} (mg/l, 5d) Quail 5000 Coturnix 5000 Pheasant 5000				
2,4-DES sodium	730						Toxic			Rats – 2y – 2000
MCPA	700		550				Bluegill 10		Slight kidney enlargement	Rats – 2.10d – 100 Cows – 21d – 30
2,4,5-T	300				100	LC_{50} (mg/l, 5d) Coturnix 5000 Mallard 5000 Pheasant 1775	Trout 1.3 Bluegill 0.5			Rats – 90d – 10 Dogs – 90d – 10
Phenoxybutyrics										
2,4-DB	700		ca. 400							

2,4-DB sodium	1500							
MCPB	680		700					
Phenoxypropionics								
Clofop-isobutyl	1308							Rats, dogs, 90d – 32
Dichlorprop	800		400 Dermal 1400		Bluegills 1–165 (depending on salt or ester)			Rats – 98d – 12.5
Diclofop-methyl	563–580 Dermal 5000							Rats – 2y – 6.3 Dogs – 90d–80
Fenoprop (2,4,5-TP)	650 ca. 2940 (as triethanol-amine salt)	Dermal >3200 (as triethanol-amine salt)		Mallard Quail >12 800 (as triethanol-amine salt)				
Fluazifop-butyl	3000 Dermal >5000	Dermal >5000	1500	Mallard duck >17 000	Rainbow trout 1.6	No effect orally 240 mg	Slight irritant to rat and rabbit skin,	

Table 7.11 Toxicity data (*contd*)

Herbicide	Rats	Rabbits	Mice	Guinea pigs	Dogs	Birds	Fish	Bees	Sensitivity	Feeding trials No effect – time – mg/kg food
Fluazifop-butyl (*contd*)								No effect contact 120 mg	weak sensitiser guinea pig skin	
Mecoprop	700–1500		600–650							Rats – 3w – 65
Napropamide	>5000	Dermal >2000					LC_{50} (96 h) 10 p.p.m.			
Aromatic acids										
Chloramben	5620 Dermal >3160									Rats – 2y – 10 000
Chlorfenac	576–1780	Dermal 1440–2160								Rats – 2y – 2000
Dicamba	2900±800					Pheasant 740	Rainbow trout 28 Bluegill 23			Rats – 2y – 500 Dogs – 2y – 50

Amides									
Butam	6210	>2000		2025				An eye but not a skin irritant to rabbits	
Diphenamid	1373 Dermal >225		1717						Rats and dogs – 2y – 2000
Naptalam	8200 1770 (sodium salt)								Rats – 90d – 1000 Dogs – 90d – 1000
Propyzamide	Male 8350 Female 5620	Dermal >3160							
Nitriles									
Bromofenoxim	1217 Dermal >3000				Slightly toxic	Toxicity varies with species			Rats and dogs – 90d – 300
Bromoxynil	190–365		110		Pheasant 50 Hen 120–240 Chickens 80–100	Harlequin (48 h) 5 p.p.m.	No toxicity to bees sprayed 0.22%		Rats – 90d – 15.6–16.5

Table 7.11 Toxicity data (*contd*)

Herbicide	Rats	Rabbits	Mice	Guinea pigs	Dogs	Birds	Fish	Bees	Sensitivity	Feeding trials No effect – time – mg/kg food
Bromoxynil octanoate	250	325	245		>50	Pheasant 150	(36%) Rainbow trout 0.15 Goldfish 0.46 Catfish 0.063			Rats – 90d – 312 Dogs – 90d – 5
Chlorthiamid	757 Dermal >1000	≃300	500		>1000	Chickens 500	(24 h) Harlequin 41			Rats – 90d – 100
Dichlobenil	3160	Dermal 1350	2126	501		LD_{50} Mallard 2000 Pheasant 1189 LC_{50} (mg/kg, 2w) Coturnix 5000 Pheasant 1750	Trout 20 (48 h) Guppies >18 Pumpkin-seed 10–20 Bluegill 20 (48 h) Bass 10–20			Rats – 3m – 50
Ioxynil	110–190		190–240			Pheasants 75	(48 h) Harlequin 3.3			Rats – 30d – 111 (growth rate depressed at 333)

				Hens 200 Chickens 125	(sodium salt)	
Ioxynil octanoate	390		2300	Mallard >1200 Pheasant 1000 Pigeon 125	Harlequin 4.0	Rats – 90d – 4 Dogs – 90d – 4.5
Anilides						
Group 1						
Monalide	>4000 Dermal >800 g	Dermal >800 g				Rats – 28d – 150
Pentanochlor	>10 000					Rats – 140d – 2000
Propanil	1285–1483	Dermal 7080			(48 h) Carp 0.42 Goldfish 0.35 Japanese killifish 0.55	

Table 7.11 Toxicity data (*contd*)

Herbicide	Rats	Rabbits	Mice	Guinea pigs	Dogs	Birds	Fish	Bees	Sensitivity	Feeding trials No effect – time – mg/kg food
Group 2										
Alachlor	1200 Dermal >2000									
Butachlor	1740	4000 (perc)					(4d) Trout 0.52 Bluegills 0.74 Carp 0.76		Mild skin and eye irritant	
Diethatyl-ethyl	Albino 2300–3700	Dermal 4000								
Dimethachlor	1600 Dermal >3170									
Metolachlor	2780 Dermal >3170					Insignificant toxicity	Slightly toxic	Insignificant toxicity		
Propachlor	780	Dermal 380								

Group 3								
Benzoylprop-ethyl	1555 Dermal >1000	>1000	716	1500–2000	>2000	Mallard >2000 Domestic fowl >1000	Harlequin (100 h) 5	Rats – 90d – 1000 Dogs – 90d – 300
Flamprop-isopropyl (racemate)	>3000 (perc) >600		2554		2000		Low toxicity	Rats – 90d – 500
Flamprop-isopropyl (R)=(–) – enantiomorph	>4000		>4000			Domestic hen >2000	Trout 3.3	Rats – 90d – 50 Dogs – 90d – 30
Flamprop-methyl	1210 (perc) >294		720			Fowl >1000	Rainbow trout 4.7	Rats – 2y – 2.5 Dogs – 1y – 15
Nitrophenols								
Dinoseb	58	Dermal 80–200						Rats – 6m – 100
Dinoseb acetate	60–65							Rats and dogs – 90d – 50
Dinoterb			ca. 25	150				
DNOC	25–40							

Table 7.11 Toxicity data (*contd*)

Herbicide	Rats	Rabbits	Mice	Guinea pigs	Dogs	Birds	Fish	Bees	Sensitivity	Feeding trials No effect – time – mg/kg food
Nitrophenyl ethers										
Bifenox	>6400	Dermal >20 000								High dosages no effect on rats and dogs
Fluorodifen	9000 Dermal >3000						Toxic response varies with species			
Nitrofen	630–755	300–510								
Oxyfluorfen	Oral 5000 Dermal 10 000				5000					
Nitroanilines										
Benfluralin	>10 000	>2000	>5000		>2000	Chicken >2000	Bluegills 0.37 No effect		No effect on skin or eyes 200 mg/kg	Rats – 90d – 1250 Dogs – 90d – 500
Butralin	126 000 (tech) 2500 (formu-lated)	10 200 (tech) 4600 (formu-lated)				LC_{50} (8d) Bobwhite quail Mallard duck	(48 h) Bluegills 4.2 Rainbow trout		It is classified as an extreme eye irritant	

Dinitramine	3000	Dermal >6800			Mallard duck 10 000 Bobtail quail 1200	Trout 6.6 Bluegills 11 Catfish 3.7		Rats and dogs – 90d – 2000
Ethalfluralin	>10 000	Dermal >2000	>10 000	>200	Mallard duck >200 Quail >200	Bluegills 1.012 Trout 0.0075 Goldfish 0.1	Slight eye irritant	Rats – 90d – 1100 Dogs – 90d – 27.5
Isopropalin	>5000	>2000	>5000		Mallard duck >2000 Quail >1000	Fathead minnow >0.1 Goldfish >0.15		Rats – 90d – 250 Dogs – 90d – 50
Nitralin	>2000	>2000 (perc)	>2000		Mallard duck >2000 Partridge >2000	Rainbow trout 27 Bluegill sunfish 31		Rats – 2y – 2000 Dogs – 2y – 2000
Oryzalin	>10 000	Dermal >2000		>1000	Chickens 1000	Goldfish >1.4		Rats – 90d – 750 Dogs – 90d – 750
Pendimethalin	1050– 1250	Dermal >5000	1340– 1620	>5000				

Table 7.11 Toxicity data (*contd*)

Herbicide	Rats	Rabbits	Mice	Guinea pigs	Dogs	Birds	Fish	Bees	Sensitivity	Feeding trials No effect – time – mg/kg food
Profluralin	2200	>10 250 Acute dermal				Non-toxic	Highly toxic	Toxic		
Trifluralin	>10 000	>2000	500		>2000	Chickens >2000 Mallard 2000 Pheasant 2000	Bluegills 0.058 Fathead minnows 0.094 Goldfish 0.59			Rats – 2y – 2000 Dogs – 2y – 1000
Carbamates										
Asulam	>5000 Dermal >1200	>4000	>5000			Chickens Quails >2000 Mallard Pheasant Pigeon >4000	Trout Bluegill Catfish Goldfish >5000			No effect on bees of contact or feeding 2% (w/v) Rats – 90d – 400
Barban	1300–1500 Dermal >1600	2500 Dermal >2300	1300–1500	850			Relatively susceptible			Rats – 2y – 5000 Adverse effects only at this the highest level
Carbetamide	11 000	Dermal 500 Non-toxic	1250		1000					Rats – 90d – 3200 Dogs – 90d – 12 800

Chlorbufam	2500									
Chlorpropham	5000–7500	5000								
Desmedipham	>10 250 a.i. Dermal 2025–10 250						Rainbow trout 3.8 Bluegill 13.4		Slight-to-medium skin irritant	Rats – 90d – 200 Dogs – 90d 200
Phenisopham	>4000	Dermal >1000	>5000			Mallard 4433				
Phenmedipham	>8000 Dermal >4000		>8000	4000	>4000	Chickens >3000	First toxic reaction Guppies 10 p.p.m. Trout 5 p.p.m. Carp 15 p.p.m.			Rats – 120 – 125, 250, 500 (reduced food intake)
Propham	5000									Rats – 1m – 10 000
Thiocarbamates										
Butylate	4000–4660	Dermal >2000								Rats, Dogs – 90d – 40
Cycloate	2000–3190	Dermal >4640			>240	Quail 56 g (7d)	Rainbow trout 4.5			Dogs – 90d – 240

Table 7.11 Toxicity data (*contd*)

Herbicide	Rats	Rabbits	Mice	Guinea pigs	Dogs	Birds	Fish	Bees	Sensitivity	Feeding trials No effect – time – mg/kg food
Di-allate	395	Dermal 2000–2500			510					Rats, dogs – daily – 125
EPTC	1652	Dermal ca. 10 000	3161			Bobwhite quail 20 000	Killifish >10 Bluegill 27 Trout 19			Rats – 21d – 326 (some loss of wt; excitability)
Molinate	720	Dermal >2000	795			LC_{50} Mallard (5d) >9300 Quail (9w) >1000				
Pebulate	1120	Dermal 4640								
Sulfallate	850								Eye and skin irritant – very corrosive	
Thiobencarb	1903		560			White leghorn 673	Carp (48 h) 3.6			Rats – 2y – 100

	Dermal 2900					Bobtail quail 7800 Mallard 10 000	Channel catfish (4d) 6.1 Bluegills (4d) 3.4		
Tiocarbazil	>10 000	Rabbits Hares >10 000	8000	>10 000		Chickens Quails Pheasant >10 000			Rats, dogs – 2y – 1000 (except for slight loss in weight in male dogs); Rats – 3 generations–300
Tri-allate	1675–2165	Dermal 2225–4050		310	20 000 No effect		Non-toxic to trout and bluegills	No irritation to skin	
Vernolate	1625–1710	Dermal 4640			>1000	LC_{50} (7d) Quail 12 000 p.p.m.	Trout 10.8 Bluegills 8.4 Sunfish 11.0		Rats – 14w – 32
Ureas									
Benzthiazuron	1280 Dermal 500								Rats – 60d – 130
Buthidazole	1483–1581						Bluegill 122 Rainbow trout 74.7	Not irritant to eyes of rats	

Table 7.11 Toxicity data (*contd*)

Herbicide	Rats	Rabbits	Mice	Guinea pigs	Dogs	Birds	Fish	Bees	Sensitivity	Feeding trials No effect – time – mg/kg food
Buthidazole (*contd*)							Channel catfish 239			
Buturon	3000									Rats – 120d – 500
Chlorbromuron	>5000 Dermal >2000						Slightly toxic			Rats, dogs – 90d – >316
Chloroxuron	>3000 Active dermal >3000					Slightly toxic	Hardly toxic to most fish but very toxic to catfish			
Chlortoluron	>10 000 Dermal >2000					Low toxicity	Low toxicity			Rats – 90d – 53 Dogs – 90d – 23
Cycluron	2600									
Difenoxuron	>7750 Dermal >2150		>1000				Moderately toxic			

Diuron	3400						May cause irritation to eyes	Rats – 2y – 250 Dogs – 2y – 125
Ethidimuron	>5000 Dermal >1000	>2500		Quail 300–400 Canaries 1000		Not harmful		Rats – 90d – >1000
Fenuron	6400							Rats – 90d – 500
Fenuron–TCA	4000–5700							
Fluometuron	6416– >8000 Dermal >2000	>10 000		Negligible toxicity				Rats – 1y – 10 Dogs – 1y – 15
Isocarbamid	>2500 Dermal >500		>500	Low toxicity	Low toxicity	Not harmful		Rats – 90d – 800 Dogs – 90d – >5000
Isoproturon	>4640 Dermal >3170	>3350			Trout 31 Guppy 90 Bluegill >100 p.p.m. Catfish 9			Rats – 90d – 400 Dogs – 90d – 50
Linuron	1500		ca. 500					Rats – 2y – 125 Dogs – 2y – 125

Table 7.11 Toxicity data (*contd*)

Herbicide	Rats	Rabbits	Mice	Guinea pigs	Dogs	Birds	Fish	Bees	Sensitivity	Feeding trials No effect – time – mg/kg food
Methabenz-thiazuron	>2500 Dermal >500		>1000		>1000 Cats >1000		Goldfish >20 (48 h)	Not harmful	No skin effect	Rats – 90d – 150
Metobromuron	2500 Dermal >3000					Non-toxic	Very slightly toxic			Rats – 2y – 250
Metoxuron	3200 Dermal >2000						Moder-ately toxic	Safe		Dogs – 90d – 2500 Chickens – 42d – 1250
Monolinuron	2250				>500		*Lebistea reticulatus* 25–30	Not toxic		Rats – 2y – 250
Monuron	3600					LC_{50} (mg/15d) Coturnix 5000 Mallard 5000 Pheasant 5000	Catfish (24 h) 75.9		30% paste No effect guinea pig	Rats – 250 to 500 Dogs – 250 to 500
Monuron–TCA	2300–3700								It is an irritant to skin and mucous membranes	

Neburon	>11 000							
Siduron	7500							Rats – 96d – 7500
Tebuthiuron	644	286 Dermal 200	579	>500 Cats 200	Bobwhite quail Mallard duck Chickens >500	(24 h) Solofish 160 Minnow 160 Trout 144 Bluegill 112		Rats – 2y – 1600 Mice – 2y – 1600 Dogs – 90d – 1000 Chickens – 30d – 1000
Thiazafluron	464 (80% a.i.) Dermal >3170		630		Practically non-toxic	Practically non-toxic	Non-poisonous	
Heterocyclic nitrogen compounds – triazines								
Ametryn	1100–1750				No detectable hazard to wild birds	No detectable hazard to fish, oysters or shrimps		Rats – 2y – 1000
Atrazine	1859–ca. 3080	Dermal 7500			Mallard 2000 LC_{50} (mg/l, 5d) Quail 750	Trout 12.6		

Table 7.11 Toxicity data (*contd*)

Herbicide	Rats	Rabbits	Mice	Guinea pigs	Dogs	Birds	Fish	Bees	Sensitivity	Feeding trials No effect – time – mg/kg food
Aziprotryn	3600–5833 Dermal >3000						Slightly toxic			Rats, dogs – 90d – >50
Cyanazine	182–334	141	380			Chickens 750 Quail 400–500	Fathead minnow 21 Harlequin (24 h) 16 (48 h) 10 Rainbow trout 9			Rats – 2y – 12 Oral LD_{50} for dogs not possible to obtain because of vomiting after administration of doses as low as 25
Desmetryne	ca. 1390		1750							Rats – 90d – 100 (slight decrease in body wt.)
Dimethametryn	3000						Slightly toxic			
Dipropetryn	>10 000 (dermal)					Slightly toxic	Moder-ately toxic	Non-toxic		Dogs – 160d – 400
DPX 4189	5545								No skin irritation. Mild	

									temporary eye effects. Low dermal and inhalation toxicity	
Eglinazine-ethyl	>10 000	3000	>10 000	3375						
Metamitron	1832–3343 Dermal >1000		1450–1463			LC_{50} Canaries >1000 Hen >500 Quail 1000–7500	Goldfish >100	Not harmful	Not an irritant	Rats – 90d – 460 Dogs – 90d – 500
Methoprop-tryne	>5000		2400				Low toxicity			Rats – 90d – 60
Metribuzin	2200–2345 Dermal >1000	>500	698–711	250	Cat >500	Hen >1000 Quail >500 Bobtail quail >715 Canary 500–1000	Goldfish Catfish Rainbow trout >10 p.p.m.			Rats – 2y – 100 Dogs – 2y – 100
Proglinazine-ethyl	>8000	>3000	>8000						Not dermal eye irritant to rabbits	

Table 7.11 Toxicity data (*contd*)

Herbicide	Rats	Rabbits	Mice	Guinea pigs	Dogs	Birds	Fish	Bees	Sensitivity	Feeding trials No effect – time – mg/kg food
Prometon	3000	Dermal >2000								Rats – 28d – 1000 (LD_{50})
Prometryne	3150–5233 dermal >3100		3750			Pheasant >2000	Slightly toxic			Rats – 2y – 1250 Cattle – 4w – 100 Sheep – 4w – 100
Propazine	7700 Dermal >3100					Slightly toxic	Slightly toxic			Rats – 0.5y – 250
Secbumeton	2680						(24 h) Guppies 5			Rats – 90d – 2400
Simazine	>5000					LC_{50} (mg/l, 5d) Coturnix 5000 Mallard 5000 Pheasant 5000	Trout (48 h) 56 Bluegills (48 h) 118			Rats – 2y – 100
Simetryn	1830									

Terbumeton	483–651 Dermal >3170					Slightly toxic			Rats – 90d – 9 Dogs – 90d – 22
Terbuthyl-azine	2000–2160					Non-toxic to moderate			Rats – 90d – 5 Dogs – 90d – 5
Terbutryne	2400–2980 Dermal >2000		5000		Low toxicity	Moderate toxicity	Low toxicity		Rats – 90d – 50 Dogs – 90d – 40
Trietazine	2830–>4000				Quail 800	(24 h) Guppies 5.5			Rats – 90d – 16
Heterocyclic nitrogen compounds – pyridines									
3,6-Dichloro-picolinic acid	>4300–5000	Dermal >2000	>5000		LC_{50} (8 d) Mallard Bobtail quail 4640	Rainbow trout 103 Bluegills 125.4	LD_{50} 100		Rats – 90d – 150
Diquat	231		125	100–200	LD_{50} Hens 200–400 Mallard 564 LC_{50} (mg/l, 5d) Coturnix 1500	Minnow 140 Trout (48 h) 20 Bluegills (48 h) 145 Goldfish 3.5		It is an eye irritant and can cause temporary damage to nails and nose bleeding if inhaled	Rats – 2y – 25

Table 7.11 Toxicity data (*contd*)

Herbicide	Rats	Rabbits	Mice	Guinea pigs	Dogs	Birds	Fish	Bees	Sensitivity	Feeding trials No effect – time – mg/kg food
Diquat (*contd*)						Mallard 5000 Pheasant 3750				
Fluridone	>10 000		>10 000		>500 Cats >250					Rats – 90d – 1400
Paraquat	150	Dermal ca. 236	104		25–50	Hens 262	Bluegills (24 h) 400		Irritant to eyes, can cause temporary damage to nails and if inhaled may cause nose bleeding	Rats – 2y – 34 Dogs – 2y – 170
Picloram	8200	ca. 2000 Dermal >4000	2000–4000	3000		LD_{50} Mallard 2000 Pheasant 2000 LC_{50} (mg/l, 5d) Mallard 5000 Pheasant 5000	Bass (24 h) 17.9 (48 h) 13.1 Salmon (24 h) 29.0 (96 h) 21.0 Minnow (24 h) 64 Trout (24 h) 34 Bluegills (24 h) 26.5		Mild skin irritant Moderate eye irritant	

					(96 h) 21.0 Goldfish (48 h) 32.5			
Triclopyr	773	550 (perc) >2000		LC_{50} (p.p.m.) Mallard duck >5000 Japanese quail 3279 Bobtail quail 2935	Rainbow trout 117 Bluegills 148		Mild eye irritant	Rats – 90d – 30
Uniroyal S734	Oral 5200	Dermal 2000					Eye irritation mild to moderate Skin irritation rabbits – mild	
Heterocyclic nitrogen compounds – pyridazines								
Chloridazon (Pyrazone)	3600				Trout 17.6 p.p.m. acute oral			Rats – 2y – 300
Norflurazon	>8000	Dermal >2000		Bobwhite quail Mallard duck >1250	Goldfish Catfish >200			Rats – 90d – 50 Dogs – 90d – 12.5
Pyridate	ca. 2000	Dermal >3400		Pheasants Ducks >10 000	Rainbow trout 81	Non- toxic	Moderate irritant to rabbit skin	

Table 7.11 Toxicity data (*contd*)

Herbicide	Rats	Rabbits	Mice	Guinea pigs	Dogs	Birds	Fish	Bees	Sensitivity	Feeding trials No effect – time – mg/kg food
Pyridate (*contd*)						Bobwhite quail 1500	Bluegills 100			
3-*o*-Tolyloxy-pyridazine	3090		569 Dermal >1000				(48 h) Carp 62			Rats – 90d – 16.5 Mice – 90d – 42
Heterocyclic nitrogen compounds – pyrimidines (uracils)										
Bromacil	5200					LC_{50} (96 h) Mallard ducklings Bobwhite quail 10 g	Bluegills Rainbow trout ca. 80 Biologically safe conc. 10.1			Rats, dogs – 2y – 250
Lenacil	>11 000	Dermal >5000				Low toxicity	Low toxicity		Mild to moderate irritation to skin of guinea pig (50% wettable powder)	
Terbacil	>5000	Dermal >5000					Slightly toxic		Non-irritant to skin	Rats – 2y – 250 Dogs – 2y – 250

Heterocyclic nitrogen compounds – unclassified										
Aminotriazole	1100– 24 600 Dermal >10 000					Mallard 2000	Bluegill 50			Rats – 476d – 50 (but male rats developed enlarged thyroid after 90d)
Benazolin	4800 (sodium) 5000 (potassium)		3200			LC_{50} Bobtail quail >10 000 Mallard >10 000	Harlequin (24 h) 360 (48 h) 325 (as potassium salt) Trout ca. 8 Bluegill 381 (as potassium salt) 2.8 (as ethyl ester)	LD_{50} contact 480 (K salt) No deaths after spraying with 0.5% a.i.	Mild irritation of eye and skin of rabbit – 30% soln.	Rats – 90d – 300 Dogs – 90d – 300
Bentazone	ca. 1100 Dermal >2500								No skin irritation to rabbit	
Difenzoquat	470 g	Dermal 3540					Bluegill 696 Rainbow trout 694			
Isomethiozin	>10 000 (male) >2500 (female)	>1000	>2500		>500	Quail >5000 Hen >5000	Golden orf 8–10	Not harmful		90d feeding trials no effect level rats 100, dogs 500

Table 7.11 Toxicity data (*contd*)

Herbicide	Rats	Rabbits	Mice	Guinea pigs	Dogs	Birds	Fish	Bees	Sensitivity	Feeding trials No effect – time – mg/kg food
Isomethiozin (*contd*)	Dermal toxicity >1000 (24 h)						Goldfish 10–12 Carp 10–20 Rudd 10			
Methazole	1350	Dermal >900					Goldfish Rainbow Trout 3		Mildly irritating to skin and moderately to eyes	Rats – 90d – 50 Dogs – 90d – 50
Oxadiazon	>8000		>8000			Bobwhite quail 6000 Mallard duck 1000	Trout >9 Bullhead >15.4 Crayfish >15.4			Rats – 25 Dogs – 25
Heterocyclic compounds – other heteroatoms										
Endothal	51						Minnow (48 h) 480 Bluegill (48 h) 280 Goldfish (48 h) 175 Carp (48 h) 175		Moderate skin irritation	Rats – 2y – 1000

Ethofumesate	>6400 Dermal >1440		>1600	>1200	Quail 1600	Moderate toxicity	Low toxicity		Rats – 2y – >1000
NC 20 484	Oral 2013–3536								
Organoarsenic compounds									
Methylarsonic acid	900–1800 (MSMA) 900–1800 (DSMA)				LC_{50} MSMA Mallard 5000 p.p.m. Bobtail quail 3300 p.p.m. LD_{50} DSMA Bobtail quail 3160			Mildly irritating to rat skin	Dogs – 90d – 30
Sodium arsenite			Small rodents 10–50					Skin irritant	
Organophosphorus compounds									
Bensulide	770	Dermal 3950			Mallard (5d) >28 000 p.p.m. Quail (21d) 1000 p.p.m.	Stickleback (72 h) ca. 2 Goldfish 1.5 Trout 0.72			Rats – 90d – 250 Dogs – 90d – 625

Table 7.11 Toxicity data (*contd*)

Herbicide	Rats	Rabbits	Mice	Guinea pigs	Dogs	Birds	Fish	Bees	Sensitivity	Feeding trials No effect – time – mg/kg food
Butamifos	630–790		400–430				Carp (48 h) >1		Neither skin nor eye irritant	
Fosamine-ammonium	2400	Dermal >4000		7380		Mallard >10 000 Quail >10 000	Bluegill 670 Trout >1000 Minnow >1000		Not a skin irritant, eye irritant or skin sensitiser	Rats – 90d – 1000
Glyphosate	4900	Dermal >7940							Isopropyl-amine salt – mild-to-severe skin and severe eye irritant. Glyphosate – not skin but mild eye irritant	Rats – 90d – 2000
Piperophos	324									
Unclassified										
Acrolein	46	7.1					Highly toxic		Lachrymatory. Skin contact causes burns	
Alloxydim-sodium	2260–2322		3000–4600							

	Dermal 1689–1700		Dermal 2000–2450				
Allyl alcohol	64	Dermal 89	85			Lachrymatory. Intense skin irritant	
Metham-sodium	820	Dermal 800	285			Eye and skin irritant	
4-Methoxy-3,3′-dimethyl-benzophenone	>4000		>4000		Carp (48 h) 3.2 Goldfish 10		Rats – 90d – 1500 Mice – 90 d – 1000
NP 55	1500–2500		6000–6500				
Perfluidone	633	Dermal >4000	920		Bluegill 318 Rainbow trout 312	Mildly irritating to skin and eyes of rabbits	
Quinonamid	11 700–15 000				Guppies 5		Rats – 90d – 2000

References

ALLEN, R. J. (1977). *Food Cosmet. Toxicol.*, **15**, 401.

ANDERSON, J. R. (1978a and 1978b). In *Pesticide Microbiology* (Ed. I. R. Hill and S. J. L. Wright), Academic Press.

ANDERSON, K. J., LEIGHTY, E. G. and TAKAHASHI, M. T. (1972). *J. Agric. Fd. Chem.*, **20**(3), 649.

ANON (1973). *Agric. Pestic.*, Council of Europe, Strasbourg.

AUDUS, L. J. (1952). *J. Sci. Fd. Agric.*, **3**, 268.

AUDUS, L. J. (1960). In *Herbicides and the Soil* (Eds. E. K. Wodford and G. R. Sagar), Blackwells, London.

AUDUS, L. J. (1970). *Meded Fac. Landb. Rijksuniv. Gent*, **35**, 465.

BAIN, J. R. *et al.* (1978). *Advances in Pesticide Science*, Symposium papers 4th Int. Cong. Pest. Chem., Pergamon Press.

BERG, R. T. and MCELROY, L. W. (1953). *Can. J. Agric. Sci.*, **33**, 354.

British Advisory Committee on Pesticides (1979). *Review of the Safety for Use in the U.K. of the Herbicide 2,4,5-T.*

British Agrochemicals Assoc. (1975). *The Safety of New Pesticides in the Environment.*

BROWN, A. W. A. (1978). *Ecology of Pesticides*, John Wiley and Sons.

BULLIVANT, C. M. (1966). *Brit. Med. J.*, **1**, 1272.

BURGIS, D. S. and BECKENBACH, J. R. (1948). *Proc. Amer. Soc. Hort. Sci.*, **52**, 461.

CALDERBANK, A. (1968). 'The bipyridylium herbicides' in *Advan. Pest Control Res.* (Ed. R. L. Metcalf), Vol. 8, Interscience Publishers, p. 127.

CLARK, D. G., MCELLIGOTT, T. F. and WESTON HURST, E. (1966). *Brit. J. Ind. Med.*, **23**, 126.

COPE, O. B., WOOD, E. M. and WALLEN, G. H. (1970). *Trans. Amer. Fish Soc.*, **99**(1), 1.

COURTNEY, W. D., PEABODY, D. V. and AUSTENSON, H. M. (1962). *Plant Dis. Rptr.*, **46**, 256.

CRAYFORD, J. V., HARTHOORN, P. A. and HUTSON, D. (1976). *Pestic. Sci.*, **7**(6), 559.

CRIPPS, R. E. and ROBERTS, T. R. (1978). In *Pesticide Microbiology* (Eds. I. R. Hill and S. J. L. Wright), Academic Press.

DE WITT, J. B. (1965). In *Research in Pesticides* (Ed. C. O. Chichester), Academic Press.

DE WITT, J. B., STICKEL, W. H. and SPRINGER, P. F. (1963). *Fish and Wildlife*, Service Circular 167, U.S. Dept. of Interior.

DOBSON, N. (1954). *Agriculture*, **61**, 415.

DOMSCH, K. (1963). *Mitteil. Biolog. Bundes. Land-und Forstwrit, Berlin-Dahlem*, **107**, 1052.

DRIFT, VAN DER, J. (1963). *Tijdschr. Pflantezeikt*, **69**, 188.

DUNACHIE, J. F. and FLETCHER, W. W. (1967). *Nature (London)*, **215**(5180), 1406.

DUNACHIE, J. F. and FLETCHER, W. W. (1970). *Ann. Appl. Biol.*, **66**, 515.

EDWARDS, C. A. (1970). *Proc. 10th Brit. Weed Control Conf.*, **3**, 1052.

EDWARDS, C. A. (1978). In *Pesticide Microbiology* (Eds. I. R. Hill and S. J. L. Wright), Academic Press.

Environmental Protection Agency (1978). *Regis. of Pesticides in the U.S. – Proposed Guidelines Fed. Reg.*, **43**(132), 29 696.

ERNE, K. (1966). *Acta Vet. Scand.*, 7, 240.

EVRARD, T. E. (1974). *Hyacinth Cont. J.*, **12**(1), 2.

FERTIG, S. N. (1952). *Proc. N.E. Weed Control Conf.*, **6**, 13.

FITZGERALD, G. P. (1975). *Water Sewage Wks.*, **122**, 82.

FLETCHER, K. (1967). *Nature (London)*, **215**, 1407.

FLETCHER, W. W. (1956). *Nature (London)*, **177**, 1244.

FLETCHER, W. W. (1960). In *Herbicides and the Soil* (Eds. E. K. Woodford and G. R. Sagar), Blackwell.

FLETCHER, W. W. (1962). *Proc. 1st Irish Crop Prot. Conf.*, An. Foras Taluntais.

FLETCHER, W. W. (1966a). *Proc. 8th Brit. Weed Control Conf.*, 896.

FLETCHER, W. W. (1966b). *Landb. Tijdsch.*, **78**, 8, 274.

FLETCHER, W. W. (1974). *The Pest War*, Blackwell and Mott.

FLETCHER, W. W. *et al.* (1968). In *Weed Control Handbook* (Eds. J. D. Fryer and S. A. Evans), 5th edn., Blackwell.

FOSTER, T. S. (1974). *Residue Rev.*, **51**.

FOSTER, T. S. *et al.* (1972). *J. Econ. Entomol.*, **65**, 982.

FOX, C. J. S. (1964). *Can. J. Plant Sci.*, **44**, 405.

FRANK, P. A. (1970). *F.A.O. Int. Conf. Weed Control WSSA.*

FRANK, P. A. (1971). *Abstr. 161st Nat. Mtg. Amer. Chem. Soc.*

FRANK, P. A. (1972). In *Fate of Organic Pesticides in the Aquatic Environment,* Advan. Chem. Ser. No. 111, Amer. Chem. Soc., Washington, D.C., p. 143.

FRANK, P. A., DENRINT, R. J. and COMES, R. D. (1970). *Weed Sci.*, **18**(6), 687.

FRANZ, J. M. and TANKE, W. (1979). In *Absch. z. Schwerpunktprogramme Vertsalten und Nebenwirkungen von Herbiziden in Boden und in Kurturpflanen* (Ed. H. Borner), Boppard, G.F.R.

FRYER, J. (1977). In *Ecological Effects of Pesticides* (Eds. F. H. Perring and K. Mellanby), Academic Press.

GORING, C. A. I. (1979). *Indust. Veg. Turf and Pest Management*, **11**(1), 7.

GORING, C. A. I. *et al.* (1975). In *Environmental Dynamics of Pesticides* (Eds. R. Hague and V. H. Freed), Plenum Press.

GREAVES *et al.* (1976). *Herbicides and Microorganisms*, C.R.C. Critical Reviews in Microbiology, Cleveland Chemical Rubber Co.

GROLLEAU, G., LAVAUR, E. DE, and SIOU, G. (1974). *Ann. Zoo Ecol. Animale*, **6**(2), 313.

GROSSBARD, E. (1973). *Bull. Ecol. Res. Commun. (Stockholm)*, **17**, 457.

GYRD-HANSEN, N. and DALGAARD-MIKKELSEN, S. (1974). *Acta. Pharmacol. Toxicol.*, **35**(4), 300.

HAHN, S. (1972). *Proc. 11th Brit. Weed Control Conf.*, 1028.

HANCE, R. J. (1979). *Ann. Appl. Biol.*, **91**(1), 137.

HAWKINS, A. F. (1972). *Outlook on Agric.*, **7**, 21.

HILTIBRAN, R. C. (1967). *Trans. Amer. Fish Soc.*, **96**, 414.

HIRSCH, P. and ALEXANDER, M. (1960). *Can. J. Microbiol.*, **6**, 241.

HORN, A. VON (1974). *Nach. Deuts. Pflanzen*, **26**(10), 154.

ILIJIN, A. M. (1969). *Zool. Zh.*, **48**, 141.

Intern Agency for Res. on Cancer (1977). *Monographs on the Evaluation of the*

Carcinogenic Risk of Chemicals to Man, Vol. 15, W. H. O. Lyon.

JENSEN, H. L. (1957). *Can. J. Microbiol.*, **3**, 151.

JONES, P. J., BATES, J. A. R. and WOODMAN, M. J. (1977). *Weed Control Handbook* (Eds. J. D. Fryer and R. J. Makepeace), Vol. 1, 6th edn., Blackwell.

JORDAN, L. S., DAY, B. E. and HENDERSON, R. T. (1962). *Hilgardia*, **32**, 433.

KATO, Y. *et al.* (1978). *J. Pestic. Sci.*, **3**(1), 27.

KAUFMAN, D. D. (1966). In *Pesticides and Their Effects on Soils and Water* (Ed. S. A. Breth), Soil Sci. Soc. Amer., p. 85.

KOBYLNYK, R. W. (1974). *Proc. 21st Ann. Mtg. Agric. Pestic. Soc.*, 11.

KOCIBA, R. J. *et al.* (1976). *Toxicol. Appl. Pharmacol.*, **35**, 553.

KOPISCHKE, E. D. (1972). *J. Wildlife Management*, **36**(4), 1352.

LEWERT, H. V. (1976). *A Close Look at Pesticides for Those Who Want to Face the Facts*, Dow Chem. Co.

LOOS, M. A. (1975). In *Herbicides: Chemistry, Degradation and Mode of Action* (Eds. P. C. Kearney and D. D. Kaufman), Vol. 1, 2nd edn., Dekker, Inc., New York.

LUTZ-OSTERTAG, Y. and HENON, C. (1975). *Compte Rend. Herb. Seances de L'Acad. des Sciences*, Series D, 281.

LUTZ-OSTERTAG, Y. and LUTZ (1970). *Compte Rend.*, Series D, 271, 2418.

MCCRAREN, J. P., COPE, O. B. and ELLER, L. (1969). *Weed Sci.*, **17**, 497.

MCEWEN, F. L. and STEPHENSON, G. R. (1979). *The Use and Significance of Pesticides in the Environment*, John Wiley & Sons.

MAKEPEACE, R. J. (1972). *Proc. 11th Brit. Weed Control Conf.*, **3**, 1041.

MALONEY, T. E. (1958). *J. Amer. Water Wks. Assoc.*, **50**, 417.

MARTIN, L. W. and WIGGANS, S. C. (1959). *Oklahoma State Univ. Exp. Sta. Proc.*, Ser. P-334.

MATHYS, G. (1975). *European Plant Protection Organization Bull.*

Ministry of Ag., Fisheries and Food (1975). *Pesticides Safety Precaution Scheme* (Revised 1971).

MORTON, H. L. and MOFFETT, J. O. (1971). *Proc. West Soc. Weed Sci.*, **25**, 15.

MORTON, H. L. and MOFFETT, J. O. (1972). *Environ Entomol.*, **1**, 102.

MULLISON, W. R. (1970). *Weed Sci.*, **18**, 738.

PALMER, J. S. (1972). *J. Amer. Vet. Med. Assoc.*, **160**, 338.

PALMER, J. S. and RADELEFF, R. D. (1969). 'The toxicity of some organic herbicides to cattle, sheep and chickens' in *U.S.D.A. Prodn. Res. Rept.*, **106**.

PAULSON, G. P. (1975). *Residue Rev.*, **58**, 1.

Pesticide Review (1976). *U.S. Dept. Agric. Agric. Stab and Conv. Serv.*, Washington D.C.

PROBST, G. W., GOLAB, T. and WRIGHT, W. L. (1975). In *Herbicides: Chemistry, Degradation and Mode of Action* (Eds. P. C. Kearney and D. D. Kaufman), Vol. 1, 2nd edn., Marcel Dekker, Inc., New York.

RICHARD, J. J. *et al.* (1975). *Pestic. Monit. J.*, **9**, 117.

ROBERTS, R. E. and ROGERS, B. J. (1957). *Poultry Sci.*, **36**, 703.

ROBSON, T. O. and BARRET, P. R. F. (1977). In *Ecological Effects of Pesticides* (Ed. F. H. Perring and K. Mellanby), Academic Press.

ROBSON, T. O., FOWLER, M. C. and BARRET, P. R. F. (1976). *Pestic. Sci.*, **7**, 391.

SCHULTZ, D. P. and HARMAN, P. D. (1974). *Pestic. Monit. J.*, **8**, 173.

SCHULTZ, D. P. and WHITNEY, E. W. (1974). *Pestic. Monit.*, **7**, 146.

SCHULZE, J. A., MANIGOLD, D. B. and ANDREWS, F. L. (1973). *Pestic. Monit. J.*, **7**, 73.

SHADOFF, L. A. *et al.* (1977). *Bull. Environ. Contam. Toxicol.*, **18(4)**, 478.

SOMERS, J. D., MORAN, E. T. and REINHART, B. S. (1973). *Down to Earth*, **29**(3), 15.

SOMERS, J., MORAN, E. T. and REINHART, B. S. (1974; 1978). *Bull. Environ. Contam. Toxicol.*, **11**(4), 339; **19**(6), 648; **20**(1), 111; **20**(3), 289.

SPITTLER, H. (1976). *Z. Jagdwissenschaft*, **22**(4), 197.

STAHLER, L. M. and WHITEHEAD, E. I. (1950). *Science (N.Y.)*, **112**, 749.

STEVENSON, J. H. (1978). *Plant Pathol.*, **27**(1), 38.

SUZUKI, M., YAMATO, Y. and AKIYAMA, T. (1977). *Weed Res.*, **11**(3), 275.

TOOBY, T. E. (1976). *Proc. Symposium on Aquatic Herbicides*, Oxford, British Crop Prot. Council Monograph, **16**, 62.

TURIN, J. (1976). *Proc. S. Weed Sci. Soc.*, **29**, 13.

U.S. Fish Pesticide Research Lab. Report (1975). U.S. Dept. of Interior, Fish and Wildlife Service.

U.S. Nat. Acad. Sci. (1974). Committee on the effects of herbicides in Vietnam Report, *The effects of herbicides in S. Vietnam*, Part A, Summary and Conclusions.

WALKER, A. *et al.* (1977). 'Herbicides in soil' in *Weed Control Handbook* (Eds. J. D. Fryer and R. J. Makepeace), Vol. 1, 6th edn., Blackwells.

WEBSTER, J. M. (1967). *Plant Pathol.*, **16**, 23.

WHITEHEAD, C. C. and PETTIGREW, R. J. (1972). *Brit. Poultry Sci.*, **13**, 191.

WHITE-STEVENS, R. (1971). *Pesticides in the Environment* (Ed. R. White-Stevens), Vol. 1, Pt. 1, Marcel Dekker, Inc.

W.H.O. (1977). EHE/EHE/77.5, *Environmental Health Criteria for Tetrachloro-dibenzo-dioxin.*

WILCKE, E. (1968). *Z. Pflanzenkr. Pflanzenpathol. Pflanenschutz Sonderheft*, **4**, 163.

WILLARD, C. J. (1950). *N. Central Weed Control Conf. Proc.*, **7**, 110.

WRIGHT, N. (1973). *Outlook on Agric.*, **7**(3), 91.

WRIGHT, S. J. L. (1978). In *Pesticide Microbiology* (Ed. I. R. Hill and S. J. L. Wright), Academic Press.

YAMAGISHI, H. and HASHIZUME, A. (1974). *Zassokenkyu*, **18**, 39.

Index